A TEXTBOOK ON
DYNAMICS

FOR

B.A. & B.Sc. STUDENTS OF ALL INDIAN UNIVERSITIES

M. RAY, D.Sc., F.N.A., F.N.A.Sc.
Former-Principal, Agra College, Agra

and

G.C. SHARMA, M.A., Ph.D.
Professor and Head, Department of Mathematics
Agra University, Agra

S. CHAND
PUBLISHING

S Chand And Company Limited

(ISO 9001 Certified Company)

RAM NAGAR, NEW DELHI - 110 055

S Chand And Company Limited

(ISO 9001 Certified Company)

Head Office: 7361, RAM NAGAR, QUTAB ROAD, NEW DELHI - 110 055
Phone: 23672080-81-82, 66672000 Fax: 91-11-23677446
www.**schandpublishing.com**; e-mail: **info@schandpublishing.com**

Branches:

Ahmedabad : Ph: 27541965, 27542369, ahmedabad@schandpublishing.com
Bengaluru : Ph: 22268048, 22354008, bangalore@schandpublishing.com
Bhopal : Ph: 4209587, bhopal@schandpublishing.com
Chandigarh : Ph: 2625356, 2625546, 4025418, chandigarh@schandpublishing.com
Chennai : Ph: 28410027, 28410058, chennai@schandpublishing.com
Coimbatore : Ph: 2323620, 4217136, coimbatore@schandpublishing.com (Marketing Office)
Cuttack : Ph: 2332580, 2332581, cuttack@schandpublishing.com
Dehradun : Ph: 2711101, 2710861, dehradun@schandpublishing.com
Guwahati : Ph: 2738811, 2735640, guwahati@schandpublishing.com
Hyderabad : Ph: 27550194, 27550195, hyderabad@schandpublishing.com
Jaipur : Ph: 2219175, 2219176, jaipur@schandpublishing.com
Jalandhar : Ph: 2401630, jalandhar@schandpublishing.com
Kochi : Ph: 2809208, 2808207, cochin@schandpublishing.com
Kolkata : Ph: 23353914, 23357458, kolkata@schandpublishing.com
Lucknow : Ph: 4065646, lucknow@schandpublishing.com
Mumbai : Ph: 22690881, 22610885, 22610886, mumbai@schandpublishing.com
Nagpur : Ph: 2720523, 2777666, nagpur@schandpublishing.com
Patna : Ph: 2300489, 2260011, patna@schandpublishing.com
Pune : Ph: 64017298, pune@schandpublishing.com
Raipur : Ph: 2443142, raipur@schandpublishing.com (Marketing Office)
Ranchi : Ph: 2361178, ranchi@schandpublishing.com
Sahibabad : Ph: 2771235, 2771238, delhibr-sahibabad@schandpublishing.com

First Edition 1951

Subsequent Edition and Reprints 1953, 56, 61, 63, 67, 71, 72, 77, 79, 81, 82, 84, 85, 86, 88, 89, 90, 91, 93, 94, 95, 97, 99, 2000, 2001, 2002, 2006 (Thrice), 2008, 2009, 2011, 2013, 2014, 2016, 2018 (Twice)

Reprint 2019

ISBN : 978-81-219-0342-4 **Code :** 1014A 048

PRINTED IN INDIA

By Vikas Publishing House Pvt. Ltd., Plot 20/4, Site-IV, Industrial Area Sahibabad, Ghaziabad-201010 and Published by S Chand And Company Limited, 7361, Ram Nagar, New Delhi-110 055.

PREFACE TO THE THIRTEENTH EDITION

The book has been again thoroughly revised. A few examples have been added as per need of the topic. The chapters on Central Forces, Moment of Inertia an D' Alembert's Principle, have been revised. Efforts have been made to eliminate printing errors.

G.C. SHARMA

PREFACE TO THE TENTH EDITION

The book has been again revised. In this edition the chapter on Constrained Motion has been named as Circular and Harmonic Motions and a new chapter on Constrained Motion has been added. Chapter on Central Forces has been thoroughly revised and substantial addition has been made. A new chapter on Resisting Medium has been also added. It is hoped that in the new form, the book will be more helpful to the students for whom this is meant.

M. RAY

CONTENTS

1

Kinematics and Kinetics

A. KINEMATICS : THE GEOMETRY OF MOTION

1·1. Displacement.

When a point changes its position relative to surrounding objects and occupies different positions at different times, it is said to be in **motion** relative to those objects. In this book we shall deal with motion of a *particle*, a particle being defined *as a geometrical point having a definite mass.* If a particle at any particular instant is at a point P and at any subsequent instant comes to a point Q, then the length PQ is the **displacement** of the particle during that interval. Thus the displacement of a moving particle is the *change of its position* and is measured by the distance moved by it during a particular interval. Displacement has both sense and magnitude.

1·2. Motion in a straight line : Velocity and Acceleration.

In order to determine the position of a moving particle or a point in a straight line, we must have a *fixed point* on the line. The position of the moving point is then determined with respect to that fixed point. Let O be the fixed point on the line and let P be the position of the moving point at any particular instant and Q its position at any subsequent instant. Then PQ is the displacement of the moving point during that interval. If the time taken in moving from P to Q be denoted by t, then PQ/t measures the **average velocity** of the particle during the interval t. Thus the average velocity during a given interval is measured by the whole displacement divided by the whole time.

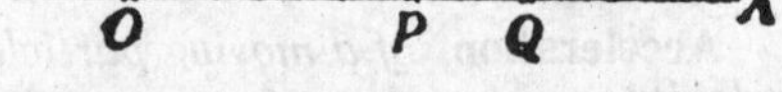

If a particle moves in a straight line and moves through equal distances in equal times, the particle is said to be moving with **uniform velocity**. If the velocity be uniform, then average velocity is the same as the uniform velocity. If, however, the velocity is variable, **it is defined as the rate of change of its displacement.**

Let a particle move along a straight line starting from a given point O on the line and let it come to the position P in time t where $OP=x$. Further suppose in a subsequent interval δt where δt is small, the particle moves through a distance $PQ(=\delta x)$ and comes to the position Q. Thus $\dfrac{\delta x}{\delta t}$ is the *average velocity* of the particle during the interval δt. As δt and consequently δx becomes smaller and smaller, the point Q approaches the point P and the quotient $\dfrac{\delta x}{\delta t}$ measures the rate of displacement of the particle. This quotient gives the velocity of the particle in the limit when $\delta t \to 0$. Thus if v be the velocity of the particle at time t, we have

$$v=\lim_{\delta t\to 0} \frac{\text{displacement in time } \delta t}{\delta t}$$

$$=\lim_{\delta t\to 0} \frac{\delta x}{\delta t}$$

$$=\frac{dx}{dt}$$

It is usual to denote differential co-efficients with regard to time by dots ; thus $\dot{x}$ means $\dfrac{dx}{dt}$ and $\ddot{x}$ means $\dfrac{d^2x}{dt^2}$.

Thus $$v=\frac{dx}{dt}=\dot{x}.$$

Velocity of a particle has got magnitude as well as direction. The magnitude is called the **speed.** If both speed and direction remain the same throughout a certain interval the velocity is uniform throughout that interval. If either of these changes, the velocity becomes variable.

Acceleration *of a moving particle is defined as the rate of change of velocity.*

If v be the velocity of the particle at time t when it is at the point P and $v+\delta v$ be its velocity at time $t+\delta t$ when it is at Q, then δv is the change of velocity in interval δt. Thus if f be the acceleration of the particle at time t, we have

$$f=\lim_{\delta t\to 0} \frac{\text{change of velocity in time } \delta t}{\delta t}$$

$$=\lim_{\delta t\to 0}\frac{\delta v}{\delta t}$$

$$=\frac{dv}{dt}.$$

But $$v=\frac{dx}{dt} \quad \therefore \quad f=\frac{d^2x}{dt^2}=\ddot{x}.$$

Also $$f=\lim\frac{\delta v}{\delta t} \qquad =\lim\frac{\delta v}{\delta x}\cdot\lim\frac{\delta x}{\delta t}$$

$$=\frac{dv}{dx}\cdot\frac{dx}{dt}=v\frac{dv}{dx}.$$

Thus any one of the three expressions $\frac{dv}{dt}$ or $\frac{d^2x}{dt^2}$ or $v\frac{dv}{dx}$ may be taken as the acceleration of the moving particle

Acceleration also has magnitude as well as direction. Negative acceleration is also known as **retardation.** Retardation implies decrease in the magnitude of velocity.

It should be very carefully noted that $\frac{dx}{dt}$ or $\dot{x}$ is the velocity of the particle *in the sense in which x increases,* similarly, $v\frac{dv}{dx}$ or $\ddot{x}$ is the acceleration of the particle *in the sense x increasing.*

Ex. 1. *Prove that, if a point moves with a velocity varying as any power not less than unity of its distance from a fixed point which it is approaching, it will take an infinite time to reach that point.*

Sol. Let O be the fixed point and P the position of the moving point at time t. If $OP=x$, the velocity is given to μx^n where μ is some constant. Since the point P is approaching O,

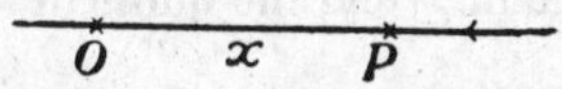

we must have $$\frac{dx}{dt}=-\mu x^n.$$

Therefore, $$\frac{dx}{x^n}=-\mu dt.$$

Integrating both sides, we get

$$-\frac{1}{(n-1)x^{n-1}}=-\mu t-C$$

where C is some constant, to be determined from the initial conditions of the problem.

Thus $$\frac{1}{(n-1)x^{n-1}}=\mu t+C.$$

If $n > 1$, then when $x \to 0$, $t \to \infty$
i.e., the moving point will take an infinite time before it reaches the point O.

When $n=1$, the equation is $\frac{dx}{dt} = -\mu x$

or $$\frac{dx}{x} = -\mu t$$

Integrating we get $$\log x = -\mu t - C$$

From this equation also, we see that when $x \to 0$ (as $\log x \to -\infty$ $x \to 0$) ; $t \to \infty$. Hence the same result.

1·3. Motion in a plane ; Velocity and Acceleration.

When a particle moves in a plane, its path is in general a curve. To fix up the position of the particle on its path at any instant, we must have a fixed point on this curve. Let A be a fixed point on the curve and let P be the position of the particle at time t ; and let arc $AP = s$.

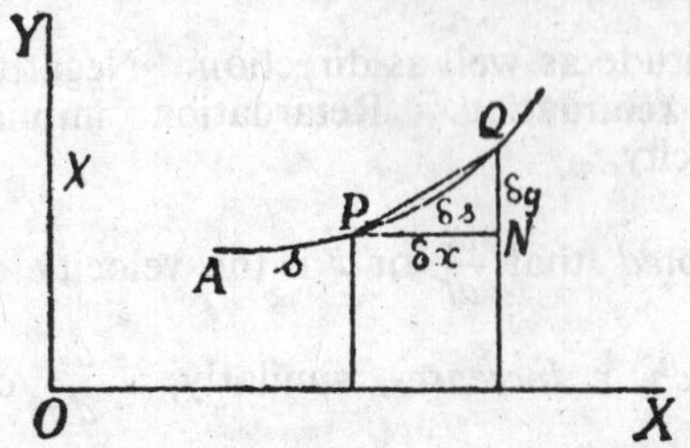

Let Q be the position of the particle at time $t + \delta t$, so that during the interval δt, the displacement of the particle is the chord PQ. Hence the average velocity of the particle during the interval δt

$$= \frac{\text{chord } PQ}{\delta t}.$$

As δt becomes smaller and smaller the point Q approaches the point P and the quotient $\frac{\text{chord } PQ}{\delta t}$ becomes the rate of displacement.

Hence the velocity of particle at time t

$$= \lim_{\delta t \to 0} \frac{\text{chord } PQ}{\delta t}$$

$$= \lim_{\delta t \to 0} \frac{\text{chord } PQ}{\delta s} \cdot \frac{\delta s}{\delta t}$$

$$= \lim_{\delta s \to 0} \frac{\text{chord } PQ}{\delta s} \times \lim_{\delta t \to 0} \frac{\delta s}{\delta t}$$

$$= 1 \times \frac{ds}{dt} = \frac{ds}{dt} = \dot{s}$$

Further as the point Q approaches the point P, the chord PQ becomes the tangent to the curve at P. Hence *the direction of the velocity at P is along the tangent to the curve at P and the velocity is* $\dot{s}$ *in the sense in which s increases.*

Now the velocity of a particle moving in a plane can be resolved into two components with respect to two axes fixed in that plane. Let OX and OY be two rectangular axes fixed in the plane with O as origin. Let the co-ordinates of the points P and Q be (x, y) and $(x+\delta x, y+\delta y)$, so that the displacement PQ has two components $PN=\delta x$, parallel to OX and $QN=\delta y$, parallel to OY.

The component of velocity parallel to OX

$$=\text{rate of displacement parallel to } OX$$

$$=\lim_{\delta t\to 0}\frac{PN}{\delta t}$$

$$=\lim_{\delta t\to 0}\frac{\delta x}{\delta t}=\frac{dx}{dt}=\dot{x}$$

The component of velocity parallel to OY

$$=\lim_{\delta t\to 0}\frac{QN}{\delta t}$$

$$=\lim_{\delta t\to 0}\frac{\delta y}{\delta t}=\frac{dy}{dt}=\dot{y}.$$

Thus if x, y be the co-ordinates of a particle moving in a plane, referred to two rectangular axes fixed in that plane, the components of velocity of the particle are $\dot{x}$ and y. Let these be represented by u and v so that $u=\dot{x}$, $v=\dot{y}$.

If V be the resultant velocity, then

$$V^2=\dot{x}^2+\dot{y}^2$$

and if ψ be the angle which the direction of motion makes with OX, then

$$\tan\psi=\frac{\dot{y}}{\dot{x}}=\frac{dy}{dx}$$

i.e., the direction of motion is the same as the direction of the tangent to the path.

We now find the components of acceleration of the particle parallel to OX and OY. Let (u, v) be the components of velocity of the particle at time t parallel to OX and OY and $(u+\delta u, v+\delta v)$ be those at time $t+\delta t$, so that changes in velocity during the interval δt parallel to OX, OY are δu, δv.

Hence the component of acceleration parallel to OX

$$=\lim_{\delta t\to 0}\frac{\text{change of velocity parallel to } OX \text{ in time } \delta t}{\delta t}$$

$$=\lim_{\delta t\to 0}\frac{\delta u}{\delta t}=\frac{du}{dt}=\frac{d}{dt}\left(\frac{dx}{dt}\right)=\frac{d^2x}{dt^2}=\ddot{x},$$

and the component of acceleration parallel to OY

$$=\lim_{\delta t\to 0}\frac{\delta v}{\delta t}=\frac{dv}{dt}=\frac{d^2y}{dt^2}=\ddot{y}.$$

The resultant acceleration is $\sqrt{\ddot{x}^2+\ddot{y}^2}$ and its direction makes an angle $\tan^{-1}\frac{\ddot{y}}{\ddot{x}}$ with OX.

Further we can write

$$\ddot{x}=\frac{d^2x}{dt^2}=\frac{du}{dt}=\frac{du}{dx}\cdot\frac{dx}{dt}=u\frac{du}{dx}$$

and
$$\ddot{y}=\frac{d^2y}{dt^2}=\frac{dv}{dt}=\frac{dv}{dy}\cdot\frac{dy}{dt}=v\frac{dv}{dy}.$$

Thus components of acceleration can also be written as $u\dfrac{du}{dx}$, $v\dfrac{dv}{dy}$ where u, v are the components of velocity.

Ex. 2. *The law of motion in a straight line being $s=\frac{1}{2}vt$, prove that the acceleration is constant*

$$s=\tfrac{1}{2}vt.$$

Sol. Differentiating with respect to t, we get

$$\dot{s}=\tfrac{1}{2}v+\tfrac{1}{2}\dot{v}t,\text{ but }\dot{s}=v,\ \dot{v}=\text{acceleration}=f,\text{ say}$$

then
$$v=\tfrac{1}{2}v+\tfrac{1}{2}ft$$

or
$$\tfrac{1}{2}v=\tfrac{1}{2}ft\quad\text{or}\quad v=ft$$

Differentiating again with respect to t, we get

$$\dot{v}=f+\dot{f}t$$

i.e.,
$$f=f+\dot{f}t$$

or
$$\dot{f}t=0\quad\therefore\quad\dot{f}=0\quad\therefore\quad f=\text{constant}.$$

Ex. 3. *If t be regarded as a function of the velocity v, prove that the rate of decrease of acceleration is given by $f^3\dfrac{d^2t}{dv^2}$, f being the acceleration.*

Rate of decrease of acceleration $=-\dfrac{df}{dt}$

Let $$t=F(v)$$

Differentiating with respect to t, we have

$$1=F'(v)\frac{dv}{dt}$$

$$\therefore \qquad f=\text{acceleration}=\frac{dv}{dt}=\frac{1}{F'(v)}.$$

Differentiating again

$$\frac{df}{dt}=-\frac{F''(v)}{\{F'(v)\}^2}\frac{dv}{dt}=-\frac{F''(v)}{\{F'(v)\}^3}=-f^3F''(v)$$

Again $$\frac{dt}{dv}=F'(v)$$

$$\therefore \qquad \frac{d^2t}{dv^2}=F''(v)$$

Hence $$\frac{df}{dt}=-f^3\frac{d^2t}{dv^2}$$

i.e. $$-\frac{df}{dt}=f^3\frac{d^2t}{dv^2}.$$

Ex. 4. Velocities of a moving point parallel to the axes of x and y are $u+ey$ and $v+ex$ respectively ; show that the path is a conic section.

Sol. Here $\dfrac{dx}{dt}=u+ey$ and $\dfrac{dy}{dt}=v+ex$

$\therefore$ dividing we get, $\dfrac{dy}{dx}=\dfrac{v+ex}{u+ey}$ or $(u+ey)\,dy=(v+ex)\,dx$.

Integrating, we get $(u+ey)^2=(v+ex)^2+C$ where C is some constant.

This represents a conic section.

EXAMPLES I (A)

1. The velocity of a particle moving in a straight line is given by the relation $v^2=ax^2+2bx+c$. Prove that the acceleration varies as the distance from a fixed point in the line.

2. The co-ordinates of a moving point at time t are given by $x=a(2t+\sin 2t)$, $y=a(1-\cos 2t)$; prove that its acceleratlon is constant.

3. If the time of a body's descent in a straight line towards a given point vary directly as the square of the distance fallen through, prove that the acceleration is inversely proportional to the cube of the distance fallen through.

1·4. Radial and Transverse Components of Velocity and Acceleration.

A particle is moving in a plane curve, to find components of velocity and acceleration at time t along and perpendicular to the radius vector drawn from a fixed point in the plane.

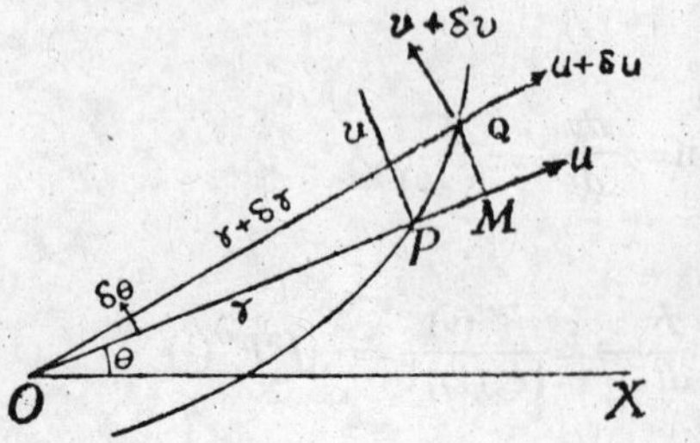

Take the fixed point O as the pole and line OX as the initial line. Let P be the position of the particle at time t, its co-ordinates be (r, θ) and Q be the position at time $t+\delta t$, its co-ordinates $(r+\delta r, \theta+\delta\theta)$, so that the chord PQ is the displacement in time δt. Draw QM perpendicular from Q to OP so that PM and QM are the components of the displacement PQ along and perpendicular to OP. Let u, v be the components of velocity along and perpendicular to OP.

Then

$$u = \lim_{\delta t \to 0} \frac{\text{displacement along } OP \text{ in time } \delta t}{\delta t}$$

$$= \lim_{\delta t \to 0} \frac{PM}{\delta t} = \lim_{\delta t \to 0} \frac{OM-OP}{\delta t}$$

$$= \lim_{\delta t \to 0} \frac{OQ \cos \delta\theta - OP}{\delta t}$$

$$= \lim_{\delta t \to 0} \frac{(r+\delta r) \cos \delta\theta - r}{\delta t}$$

$$= \lim_{\delta t \to 0} \frac{(r+\delta r)\, 1 - r}{\delta t},$$ small quantities of above the first order being neglected.

$$= \lim_{\delta t \to 0} \frac{\delta r}{\delta t} = \frac{dr}{dt} = \dot{r}$$

$$v = \lim_{\delta t \to 0} \frac{\text{displacement perp. to } OP \text{ in time } \delta t}{\delta t}$$

$$= \lim_{\delta t \to 0} \frac{QM}{\delta t} = \lim_{\delta t \to 0} \frac{OQ \sin \delta\theta}{\delta t}$$

$$= \lim_{\delta t \to 0} \frac{(r+\delta r) \sin \delta\theta}{\delta t} = \lim_{\delta t \to 0} \frac{(r+\delta r) \sin \delta\theta}{\delta\theta} \cdot \frac{\delta\theta}{\delta t}$$

$$= \lim_{\delta t \to 0} (r+\delta r)\frac{\delta\theta}{\delta t} \text{ because } \lim_{\delta\theta \to 0} \frac{\sin\delta\theta}{\delta\theta} = 1$$

$$= \lim_{\delta t \to 0} \frac{r\delta\theta}{\delta t}, \text{ neglecting the other term}$$

$$= r\frac{d\theta}{dt} = r\dot{\theta}.$$

Thus the components of velocity along and perpendicular to the radius vector are $\dot{r}$ and $r\dot{\theta}$, in the senses in which r and θ increase. These are called the radial and transverse or cross-radial components of velocity.

Now let the components of velocity along and perpendicular to OQ be $u+\delta u$, $v+\delta v$; (u, v) being those along and perpendicular to OP.

Thus the change of velocity along OP in time δt

$$= (u+\delta u)\cos\delta\theta - (v+\delta v)\sin\delta\theta - u$$

$$= (u+\delta u)\,.\,1 - (v+\delta v)\,.\,\delta\theta - u, \quad \text{neglecting higher powers of } \delta\theta$$

$$= \delta u - v\delta\theta, \quad \text{neglecting the other term.}$$

Similarly the change of velocity perpendicular to OP in time δt

$$= (u+\delta u)\sin\delta\theta + (v+\delta v)\cos\delta\theta - v$$

$$= (u+\delta u)\,\delta\theta + (v+\delta v)\,.\,1 - v, \text{ as before}$$

$$= u\delta\theta + \delta v, \quad \text{neglecting the other term.}$$

Therefore, radial acceleration

$$= \lim_{\delta t \to 0} \frac{\text{change of velocity along } OP \text{ in time } \delta t}{\delta t}$$

$$= \lim_{\delta t \to 0} \frac{\delta u - v\delta\theta}{\delta t}$$

$$= \frac{du}{dt} - v\frac{d\theta}{dt}$$

$$= \frac{d}{dt}\left(\frac{dr}{dt}\right) - r\frac{d\theta}{dt}\cdot\frac{d\theta}{dt}$$

$$= \frac{d^2r}{dt^2} - r\left(\frac{d\theta}{dt}\right)^2 = \ddot{r} - r\dot{\theta}^2.$$

Transverse acceleration

$$= \lim_{\delta t \to 0} \frac{\text{change of velocity perp. to } OP \text{ in time } \delta t}{\delta t}$$

$$= \lim_{\delta t \to 0} \frac{u\delta\theta + \delta v}{\delta t}$$

$$= u\frac{d\theta}{dt} + \frac{dv}{dt}$$

$$= \frac{dr}{dt} \cdot \frac{d\theta}{dt} + \frac{d}{dt}\left(r\frac{d\theta}{dt}\right)$$

$$= \frac{dr}{dt} \cdot \frac{d\theta}{dt} + \frac{dr}{dt} \cdot \frac{d\theta}{dt} + r\frac{d^2\theta}{dt^2}$$

$$= r\frac{d^2\theta}{dt^2} + 2\frac{dr}{dt} \cdot \frac{d\theta}{dt}$$

$$= r\ddot{\theta} + 2\dot{r}\dot{\theta}$$

also $$= \frac{1}{r}\frac{d}{dt}\left(r^2\frac{d\theta}{dt}\right)$$

$$= \frac{1}{r}\frac{d}{dt}(r^2\dot{\theta}).$$

Thus the components of acceleration are $\ddot{r} - r\dot{\theta}^2$ along OP in the sense r increasing and $\frac{1}{r}\frac{d}{dt}(r^2\dot{\theta})$ perp. to OP in the sense θ increasing.

Cor. If the particle describes a circle of radius a, then $r=a$ =constant, so that $u=\dot{r}=0$, $v=r\dot{\theta}=a\dot{\theta}$;

Radial acc. $=\ddot{r} - r\dot{\theta}^2 = 0 - a\dot{\theta}^2 = -a\dot{\theta}^2$, *i.e.*, $a\dot{\theta}^2$ towards the centre

and Transverse acc. $= r\ddot{\theta} + 2\dot{r}\dot{\theta} = a\ddot{\theta}$, *i.e.*, tangentially.

Alternative method. We have

$$x = r\cos\theta \;.\; y = r\sin\theta$$

$$\therefore \quad \left.\begin{aligned} \dot{x} &= \dot{r}\cos\theta - r\sin\theta\,\dot{\theta} \\ \dot{y} &= \dot{r}\sin\theta + r\cos\theta\,\dot{\theta} \end{aligned}\right\} \quad \ldots(1)$$

and $$\left.\begin{aligned} \ddot{x} &= \ddot{r}\cos\theta - 2\dot{r}\dot{\theta}\sin\theta - r\dot{\theta}^2\cos\theta - r\ddot{\theta}\sin\theta \\ \ddot{y} &= \ddot{r}\sin\theta + 2\dot{r}\dot{\theta}\cos\theta - r\dot{\theta}^2\sin\theta + r\ddot{\theta}\cos\theta \end{aligned}\right\} \quad \ldots(2)$$

Radial velocity $= \dot{x}\cos\theta + \dot{y}\sin\theta = \dot{r}$, from (1)

Transverse velocity $= \dot{y}\cos\theta - \dot{x}\sin\theta = r\dot{\theta}$, from (1)

Radial acceleration $= \ddot{x}\cos\theta + \ddot{y}\sin\theta = \ddot{r} - r\dot{\theta}^2$, from (2)

Transverse acceleration $= \ddot{y}\cos\theta - \ddot{x}\sin\theta = 2\dot{r}\dot{\theta} + r\ddot{\theta}$, from (2)

$$= \frac{1}{r} \cdot \frac{d}{dt}(r^2\dot{\theta}).$$

Ex. 4. *If the radial and transverse velocities of a particle are always proportional to each other, show that the path is an equiangular spiral.*

Sol Here $\frac{dr}{dt}=kr\frac{d\theta}{dt}$ where k is some constant

or $\frac{dr}{r}=kd\theta.$

Integrating we get, $\log r=k\theta+C$ where C is some constant

or $r=ae^{k\theta}$ where a is also a constant.

This is an equiangular spiral.

Ex. 5. *The velocities of a particle along and perpendicular to the radius from a fixed origin are λr and $\mu\theta$; find the path and show that the along and perpendicular to the radius vector, are*

$$\lambda^2 r-\frac{\mu^2\theta^2}{r} \text{ and } \mu\theta\left(\lambda+\frac{\mu}{r}\right).$$

Sol. Here $\frac{dr}{dt}=\lambda r$ and $r\frac{d\theta}{dt}=\mu\theta.$

Dividing we get, $\frac{rd\theta}{dr}=\frac{\theta\mu}{\lambda r}$ or $\frac{\mu}{\lambda}\frac{dr}{r^2}=\frac{d\theta}{\theta}.$

Integrating we get

$$-\frac{\mu}{\lambda}\cdot\frac{1}{r}=\log\theta+C, \text{ where } C \text{ is a constant.}$$

This gives the path.

Radial acceleration $=\ddot{r}-r\dot{\theta}^2.$

$$=\lambda\dot{r}-r.\frac{\mu^2\theta^2}{r^2}, \quad \because \quad \dot{r}=\lambda r \quad \therefore \quad \ddot{r}=\lambda\dot{r}=\lambda^2 r$$

$$=\lambda^2 r-\frac{\mu^2\theta^2}{r}, \qquad \text{and } \dot{\theta}=\frac{\mu\theta}{r}.$$

Transverse acceleration

$$=2\dot{r}\dot{\theta}+r\ddot{\theta}$$

$$=2\lambda r.\frac{\mu\theta}{r}+r\ddot{\theta}$$

By differentiating $r\dot{\theta}=\mu\theta$

we get $r\ddot{\theta}+\dot{r}\dot{\theta}=\mu\dot{\theta}$

i.e., $$r\ddot{\theta}=\mu.\frac{\mu\theta}{r}-\lambda r.\frac{\mu\theta}{r}=\frac{\mu\theta}{r}(\mu-\lambda r).$$

Hence transverse acc. $=2\lambda\mu\theta+\frac{\mu\theta}{r}(\mu-\lambda r)$

$$=\mu\theta\left(2\lambda+\frac{\mu}{r}-\lambda\right)=\mu\theta\left(\lambda+\frac{\mu}{r}\right).$$

Ex. 6. *A straight line of constant length moves with its ends on two fixed rectangular axes OX, OY and P is the foot of the perpendicular from O on the straight line. Show that the velocity of P perpendicular to OP is OP.$\dot{\theta}$ and along OP is 2 CP.$\dot{\theta}$ where C is the middle point of the line and θ is the angle COX.*

Sol. Since AOB is a right angled triangle, C is the middle point of the hypotenuse, $OC=CA=CB=a$ if $AB=2a$.

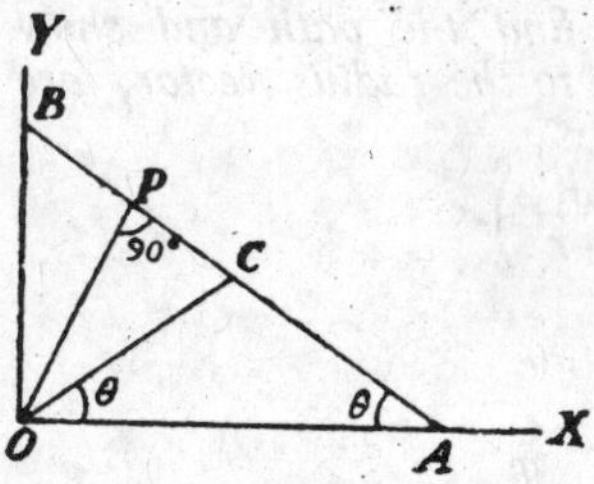

$$\angle COA=\theta=\angle CAO$$

$$\angle POX=90-\theta$$

and $\quad PO=OC\cos(90-2\theta)=a\sin 2\theta$

$\therefore$ Polar co-ordinates of P are

$$(a\sin 2\theta, 90-\theta)$$

Also $CP=a\cos 2\theta$

$\therefore$ Velocity of P along OP

$$=\frac{d}{dt}(a\sin 2\theta)=2a\cos 2\theta\dot{\theta}$$

$$=2.CP\dot{\theta}.$$

Velocity of P perp. to $OP=a\sin 2\theta\frac{d}{dt}(90-\theta)$

$$=-OP.\dot{\theta},$$

i.e., velocity along $PA=OP.\dot{\theta}$.

Ex. 7. *A boat which is rowed with constant velocity U starts from a point A on the bank of a river which flows with a constant velocity V; and it points always towards a point B on the other bank exactly opposite to A; find the equation of the path of the boat. If V=U, show that the path is a parabola whose focus is B.*

Sol. Let P be the position of the boat at time t. It has two velocities, U towards B and V downstream. Take the opposite bank as initial line and B as pole and P the point (r, θ), its components of velocity are $\dot{r}$ along BP produced and $r\dot{\theta}$ perpendicular to BP.

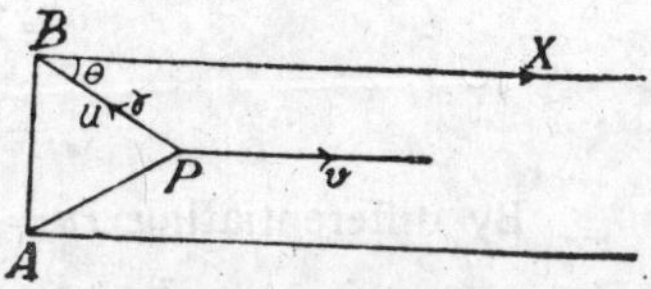

Hence resolving along and perp. to BP, we get

$$\dot{r}=V\cos\theta-U$$

and $$r\dot{\theta}=-V\sin\theta$$

Dividing we get $$\frac{rd\theta}{dr}=-\frac{V\sin\theta}{V\cos\theta-U}$$

or $$\frac{dr}{r}=\left(-\cot\theta+\frac{U}{V}\operatorname{cosec}\theta\right)d\theta.$$

Integrating we get $\log r=\log\operatorname{cosec}\theta+\frac{U}{V}\log\tan\theta/2+C_1$

where C_1 is constant.

This gives the path.

If $V=U$, it becomes $\log r=\log\operatorname{cosec}\theta+\log\tan\theta/2+C_1$

or $$\log r=\log\left(\frac{1}{2\cos^2\theta/2}\right)+C_1$$

or $$2r\cos^2\theta/2=A \text{ where } A \text{ is also a constant}$$

i.e., $\frac{A}{r}=2\cos^2\theta/2=1+\cos\theta$ which is a parabola, with B as focus.

Ex. 8. *A and B are two fixed points on the circumference of a circle and the distances from A and B of any other point P on the circumference are r and s respectively. If u and v are the components of P's velocity, as it moves round the circumference, along AP and BP respectively prove that α being the angle APB*

$$u\sin^2\alpha=\dot{r}-\dot{s}\cos\alpha,$$
$$v\sin^2\alpha=\dot{s}-\dot{r}\cos\alpha$$

and deduce that $ur+vs=0$

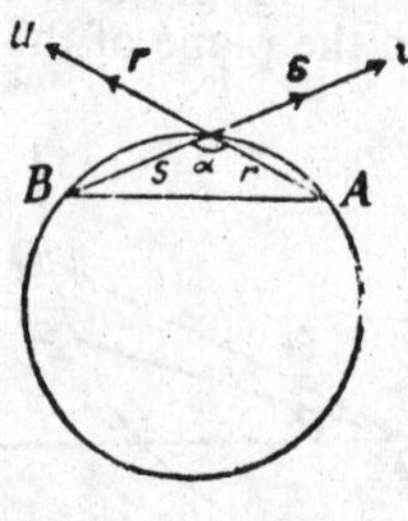

Sol. Resolving along AP,

$$\dot{r}=u-v\cos(180-\alpha)=u+v\cos\alpha$$

and resolving along BP

$$\dot{s}=v-u\cos(180-\alpha)=v+u\cos\alpha$$

$$\therefore\quad \dot{r}-\dot{s}\cos\alpha=u\sin^2\alpha$$

and $$\dot{s}-\dot{r}\cos\alpha=v\sin^2\alpha.$$

Also $(ur+vs)\sin^2\alpha=r\dot{r}-r\dot{s}\cos\alpha+s\dot{s}-s\dot{r}\cos\alpha$

$$=\tfrac{1}{2}\frac{d}{dt}(r^2+s^2)-\cos\alpha\frac{d}{dt}(rs)$$

$$=\tfrac{1}{2}\frac{d}{dt}(r^2+s^2-2rs\cos\alpha)$$

$$=\tfrac{1}{2}\frac{d}{dt}AB-0, \text{ hence}$$

EXAMPLES I (B)

1. If the radial and transverse velocities of a point are always proportional to each other and this holds for acceleration also, prove that its velocity will vary as some power of the radius vector.

2. The velocities of a particle along and perpendicular to a radius vector from a fixed origin are λr^2 and $\mu\theta^2$. Show that the equation to the path is

$$\frac{\lambda}{\theta}=\frac{\mu}{2r^2}+C,$$

and the components of accelerations are

$$2\lambda^2r^3-\mu^2\frac{\theta^4}{r} \text{ and } \lambda\mu r\theta^2+2\mu^2\frac{\theta^3}{r}.$$

3. Prove that the path of a point which possesses two constant velocities, one along a fixed direction and the other perpendicular to the radius vector drawn from a fixed point, is a conic section.

4. A particle moves along a circle $r=2a\cos\theta$ in such a way that its acceleration towards the origin is always zero. Prove that

$$\frac{d^2\theta}{dt^2}=-2\cot\theta\,.\,\dot{\theta}^2.$$

1·5. Angular Velocity and Acceleration.

Definition. The angular velocity of a point P about another point O is the rate of change of the angle which OP makes with some fixed direction

A particle is moving in a plane curve. Take a line OX fixed in the plane of the curve as initial line and O as pole. Let P be the position of the particle at time t, being given by the angle $XOP=\theta$.

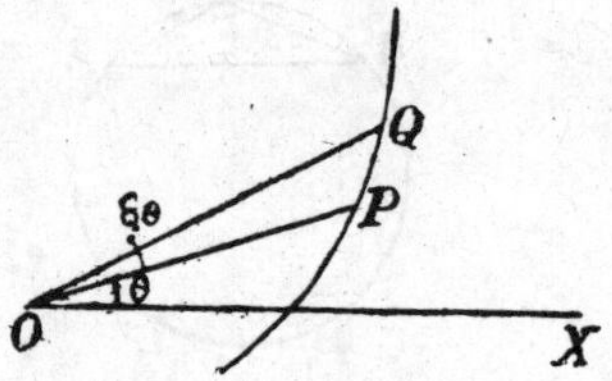

Let Q be the position at time $t+\delta t$, so that the angle described in time δt is $\delta\theta$.

Thus the average angular velocity of P about O is $\frac{\delta\theta}{\delta t}$.

As δt becomes smaller and smaller, Q approaches P and $\frac{\delta\theta}{\delta t}$ becomes the rate of change of θ. This is angular velocity.

Thus angular velocity of P about $O=\lim\limits_{\delta t\to 0}\frac{\delta\theta}{\delta t}=\frac{d\theta}{dt}=\dot{\theta}.$

Similarly angular acceleration is $\frac{d}{dt}\left(\frac{d\theta}{dt}\right)=\frac{d^2\theta}{dt^2}=\ddot{\theta}.$

Angular velocity and angular acceleration of direction of motion. Since the direction of motion (or the direction of velocity) is tangential, the angular velocity of direction of motion is the angular velocity of tangent. Tangent makes an angle ψ with the x-axis or the initial line, therefore, the angular velocity of motion is,

$$\frac{d\psi}{dt}=\psi'$$

and angular acceleration of direction of motion is,

$$\frac{d\psi'}{dt}=\psi''$$

1.5. (a) Relation between angular and linear velocities.

Let v be the velocity of the particle at P. It is along the tangent to the path at P.

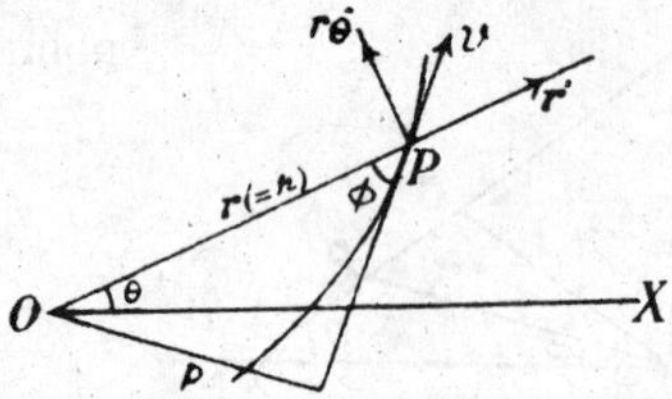

Components of velocity of P along and perp. to OP are $\dot{r}$ and $r\dot{\theta}$. Resolving perp. to OP, we get

$$r\dot{\theta}=v\sin\phi$$

where ϕ is the angle the tangent at P makes with OP.

Thus $\quad \dot{\theta}=\dfrac{v\sin\phi}{r}$

i.e., angular velocity of P about $O=\dfrac{\text{velocity of P resolved perp. to } OP}{OP}$

Also $\quad \dot{\theta}=\dfrac{v\sin\phi}{r}$ and $\sin\phi=\dfrac{p}{r}$ where p is perp. from O to the tangent at P.

$$\therefore \quad \dot{\theta}=\frac{vp}{r^2}$$

Ex. 9. *A particle describes an equiangular spiral $r=ae^{\theta}$ in such a manner that its acceleration has no radial component. Prove that its angular velocity is constant and that the magnitude of the velocity and acceleration is each proportional to r.*

Sol. Here $\quad \ddot{r}-r.\dot{\theta}^2=0$

From $\quad r=ae^{\theta}, \quad \dot{r}=ae^{\theta}\dot{\theta}=r\dot{\theta}$

$\quad \ddot{r}=\dot{r}\dot{\theta}+r\ddot{\theta}=r\dot{\theta}^2+r\ddot{\theta}$

$\therefore \quad r\ddot{\theta}=\ddot{r}-r\dot{\theta}^2=0$ as given above.

Hence $\quad \ddot{\theta}=0$

$\therefore \quad \dot{\theta}=\text{const.}=K$ say

Then $\quad \dot{r}=Kr$ so that $v^2=\dot{r}^2+r^2\dot{\theta}^2=2K^2r^2$

$\therefore \quad v$ varies as r

Since radial acc. is zero, only acc. is transverse

$$=2\dot{r}\dot{\theta}+r\ddot{\theta}=2\dot{r}\dot{\theta}=2K^2r$$

$\therefore$ acc. varies as r.

Ex. 10. *A rod moves with its ends on rectangular axes OX, OY. If x, y be a point P on the rod and if the angular velocity ω of the rod is constant, show that components of acceleration of P along the axes are $-x\omega^2$ and $-y\omega^2$ and the resultant acceleration is $OP.\omega^2$ towards O.*

Sol. If C be the middle point then $OC=CA=CB=a$, so that $\angle AOC=\theta=\angle OAC$.

It is given that $\dot{\theta}=\text{const.}=\omega$

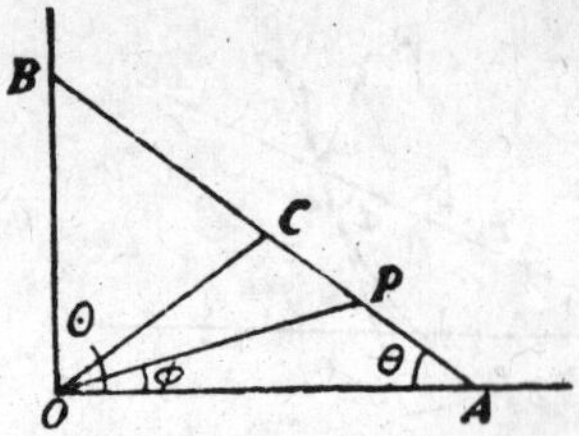

Let $CP=\text{fixed}=c$ for any particular point P chosen.

Then $x=a\cos\theta+c\cos\theta$

$=(a+c)\cos\theta$

$y=(a-c)\sin\theta$

$\dot{x}=-(a+c)\sin\theta\dot{\theta}$

$=-(a+c)\,\omega\sin\theta$

$\dot{y}=(a-c)\,\omega\cos\theta$

$$\ddot{x}=-(a+c)\,\omega^2\cos\theta=-x\omega^2,$$

$$\ddot{y}=-(a-c)\,\omega^2\sin\theta=-y\omega^2.$$

Resultant acc. makes angle with x-axis as

$$\tan^{-1}\frac{\ddot{y}}{\ddot{x}}=\tan^{-1}\frac{y}{x}=\phi$$

where ϕ is the angle OP makes with x-axis.

Hence the acc. is along PO as both $\ddot{x}$ and $\ddot{y}$ are negative, and is equal to $\sqrt{\ddot{x}^2+\ddot{y}^2}=\omega^2\sqrt{x^2+y^2}=\omega^2.OP$.

1·6. Tangential and Normal Components of Acceleration.

A particle is moving in a plane curve ; to find components of its acceleration along the tangent and the normal to the curve at any instant.

Let A be a fixed point on the curve, and P be the position of the particle at time t where $AP=s$. Let v be the velocity of the particle at P, it being entirely along the tangent at P ; so that $v=\dot{s}$.

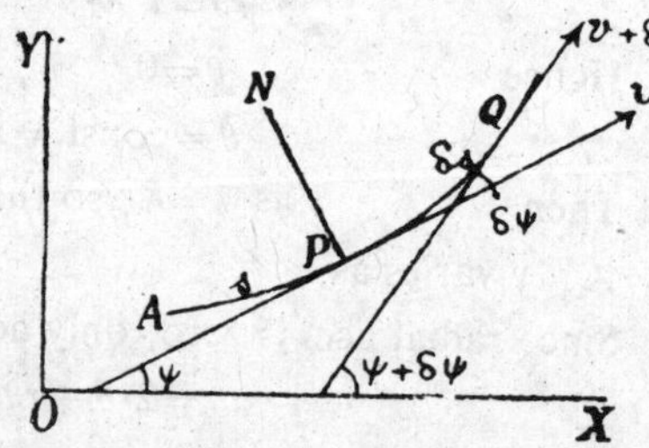

Let Q be the position of the particle at time $t+\delta t$ and $v+\delta v$ be the velocity there *i.e.*, along the tangent at Q.

Let the tangents at P and Q make angles ψ and

$\psi+\delta\psi$ with a fixed line OX so that $\delta\psi$ is the angle between the tangents.

Thus the change of velocity along the tangent at P in times δt

$$=(v+\delta v)\cos\delta\psi-v$$

$$=(v+\delta v).1-v, \text{ neglecting terms of second order}$$

$$=\delta v,$$

and the change of velocity along the normal at P in time δt

$$=(v+\delta v)\sin\delta\psi-0$$

$$=(v+\delta v)\,\delta\psi, \qquad \text{neglecting second order terms.}$$

$$=v\delta\psi, \qquad \text{neglecting the other term.}$$

Thus tangential acceleration

$$=\lim_{\delta t\to 0}\frac{\text{change of velocity along the tangent in time }\delta t}{\delta t}$$

$$=\lim_{\delta t\to 0}\frac{\delta v}{\delta t}=\frac{dv}{dt}=\frac{d^2s}{dt^2}$$

or $$=\frac{dv}{ds}\cdot\frac{ds}{dt}=v\frac{dv}{ds}.$$

Normal acceleration

$$=\lim_{\delta t\to 0}\frac{\text{change of velocity along the normal in time }\delta t}{\delta t}$$

$$=\lim_{\delta t\to 0}\frac{v\delta\psi}{\delta t}=\lim\frac{v\delta\psi}{\delta s}\quad\frac{\delta s}{\delta t}$$

$$=\lim_{\delta s\to 0}\frac{v\delta\psi}{\delta s}\cdot\lim_{\delta t\to 0}\frac{\delta s}{\delta t}=\frac{v}{\rho}\cdot v=\frac{v^2}{\rho}$$

where ρ is the radius of curvature at P.

Thus for a particle moving in a plane curve, component of acceleration along the tangent is $\dfrac{dv}{dt}$ or $\dfrac{d^2s}{dt^2}$ or $v\dfrac{dv}{ds}$, in the sense in which s increases ; and the component of acceleration along the normal is $\dfrac{v^2}{\rho}$ in the inward sense.

Cor. For a particle moving in a circle of radius a, $s=a\theta$, so that tangential acc. $=\ddot{s}=a\ddot{\theta}$, in the sense θ increasing and normal acc. $=\dfrac{v^2}{\rho}=\dfrac{\dot{s}^2}{a}=\dfrac{a^2\dot{\theta}^2}{a}=a\dot{\theta}^2$, towards the centre.

Ex. 11. *Prove that the acceleration of a point moving in a curve with uniform speed is $\rho\dot{\psi}^2$.*

A point describes a cycloid $s=4\ a \sin \psi$ with uniform speed v. Find its acceleration at any point s.

If v=const.,

components of acc. are $\dfrac{dv}{dt}=0$ and $\dfrac{v^2}{\rho}$.

Now $\quad \dot{\psi}=\dfrac{d\psi}{dt}=\dfrac{d\psi}{ds}\dfrac{ds}{dt}=\dfrac{v}{\rho}$

$\therefore \quad v=\rho\dot{\psi}$

Hence acc.$=\dfrac{\rho^2\dot{\psi}^2}{\rho}=\rho\dot{\psi}^2$.

For $\quad s=4a \sin \psi,\ \rho=\dfrac{ds}{d\psi}=4a \cos \psi$

$$\therefore \quad \text{acc.}=\frac{v^2}{\rho}=\frac{v^2}{4a \cos \psi}=\frac{v^2}{4a\sqrt{1-\sin^2 \psi}}$$

$$=\frac{v^2}{4a\sqrt{1-\dfrac{s^2}{16a^2}}}=\frac{v^2}{\sqrt{16a^2-s^2}}.$$

Ex. 12. *A point P describes, with a constant angular velocity of OP, an equiangular spiral of which O is the pole. Find its acceleration and show that its direction makes the same angle with the tangent at P as the radius vector OP makes with the tangent.*

Sol. Equation to equiangular spiral is $r=ae^{\theta \cot \alpha}$

Here $\quad \dot{\theta}=\text{constant}=c$, say.

$$\therefore \quad \dot{r}=a \cot \alpha e^{\theta \cot \alpha}\times\dot{\theta}=cr \cot \alpha$$

$$\ddot{r}=c\dot{r} \cot \alpha=c^2 \cot^2 \alpha .\ r$$

$$\therefore \quad \ddot{r}-r\dot{\theta}^2=c^2 \cot^2 \alpha\ r-c^2r=(\cot^2 \alpha-1)\ c^2r$$

Also $\quad 2\dot{r}\dot{\theta}+r\ddot{\theta}=2c^2r \cot \alpha+0=2c^2r \cot \alpha \quad \because \ddot{\theta}=0.$

These are the components of acceleration in polar co-ordinates hence the resultant acc.$=\sqrt{(\cot^2 \alpha-1)^2 c^4r^2+4c^4r^2 \cot^2 a}=c^2r \operatorname{cosec}^2\alpha$. If θ be the angle the direction of the resultant acc. makes with OP then $\tan \theta=\dfrac{2c^2r \cot \alpha}{(\cot^2 \alpha-1)\ c^2r}=\dfrac{2 \cot \alpha}{\cot^2 \alpha-1}=\tan 2\alpha \quad \therefore \quad \theta=2\alpha$

but the tangent makes an $\angle \alpha$ with OP $\therefore$ direction of the acceleration makes the same angle α with the tangent.

Ex. 13. *A particle is moving in a parabola with uniform angular velocity about the focus ; prove that its normal acceleration at any point is proportional to the radius of curvature of its path at that point.*

Sol Equation to parabola with the focus as pole is $p^2=ar$.

Here $\dot{\theta}=\text{const.}=c$, say but $\dot{\theta}=\dfrac{vp}{r^2}$ $\therefore$ $c=\dfrac{vp}{r^2}$ $\therefore$ $v=\dfrac{cr^2}{p}$.

Also $\rho=\dfrac{rdr}{dp}$.

Differentiating $p^2=ar$ with respect to p, we get

$$2p=a\frac{dr}{dp} \quad \therefore \quad \frac{dr}{dp}=\frac{2p}{a}$$

$$\therefore \qquad \rho=r.\frac{2p}{a}=\frac{2pr}{a}=\frac{2r.ar}{ap}=\frac{2r^2}{p}.$$

$$\text{Now normal acc.}=\frac{v^2}{\rho}=\frac{\dfrac{c^2r^4}{p^2}}{\dfrac{2pr}{a}}=\frac{ac^2r^3}{2p^3}=\frac{ac^2r^3}{2ar.p}=\frac{c^2r^2}{2p}=\frac{c^2}{4}.\rho$$

i.e., varies as ρ.

Ex. 14. *A particle describes a parabola with uniform speed ; show that its angular velocity about the focus S, at any point P, varies inversely as $SP^{3/2}$.*

Sol Here $v=\text{const.}=c$, say.

Equation to parabola with S as pole is $p^2=ar$.

$$\text{Now} \qquad \dot{\theta}=\frac{vp}{r^2}=\frac{cp}{r^2}=\frac{c\sqrt{ar}}{r^2}=\frac{c\sqrt{a}}{r^{3/2}}\propto\frac{1}{r^{3/2}}.$$

Ex. 15. *Prove that the angular acceleration of the direction of motion of a point moving in a plane is* $\dfrac{v}{\rho}\cdot\dfrac{dv}{ds}-\dfrac{v^2}{\rho^2}\cdot\dfrac{d\rho}{ds}$.

Sol. If the tangent at any point of the path makes angle with the fixed line OX, then ψ gives the direction of motion.

To find $\ddot{\psi}$.

$$\text{Now } \dot{\psi}=\frac{d\psi}{ds}\cdot\frac{ds}{dt}=v\frac{d\psi}{ds}=\frac{v}{\rho}$$

$$\therefore \quad \ddot{\psi}=\frac{d}{dt}\left(\frac{v}{\rho}\right)=\frac{ds}{dt}\cdot\frac{d}{ds}\left(\frac{v}{\rho}\right)=v\frac{d}{ds}\left(\frac{v}{\rho}\right)$$

$$=v\left(\frac{1}{\rho}\frac{dv}{ds}-\frac{v}{\rho^2}\frac{d\rho}{ds}\right).$$

Ex. 16. *Prove that if the tangential and normal accelerations of a particle describing a plane curve be constant throughout the motion the angle ψ through which the direction of motion turns in time t is given by*

$$\psi=A\log(1+Bt).$$

Sol Here $\ddot{s}=\text{const.}=a$ say and $\frac{\dot{s}^2}{\rho}=\frac{v^2}{\rho}=\text{const.}=b$, say

$$\therefore \quad \dot{s}=at+c \quad\text{and}\quad \frac{\dot{s}^2}{\frac{ds}{d\psi}}=b \quad\text{or}\quad \dot{s}.\frac{\frac{ds}{dt}}{\frac{ds}{d\psi}}=b \quad\text{or}\quad \dot{s}.\frac{d\psi}{dt}=b$$

$$\therefore \quad (at+c)\frac{d\psi}{dt}=b \qquad \therefore \quad \frac{d\psi}{dt}=\frac{b}{at+c}$$

$$\therefore \quad \psi=A\log(1+Bt).$$

Ex. 17. *A particle moves in a catenary ($s=c\tan\psi$), the direction of its acceleration at any point makes equal angles with the tangent and the normal to the path at the point. If the speed at the vertex where $\psi=0$ be u, show that the velocity and acceleration at any other point are given by ue^{ψ} and $\frac{\sqrt{2}}{c}u^2e^{2\psi}\cos^2\psi$.*

Sol. Since the acc. bisects the angle between the tangent and the normal, the tangential and normal components of acc. are equal.

$$\text{Hence} \qquad v\frac{dv}{ds}=\frac{v^2}{\rho} \quad\text{or}\quad \frac{dv}{ds}=\frac{v}{\rho} \quad\text{or}\quad \frac{dv}{ds}=v\frac{d\psi}{ds}$$

$$\therefore \quad \frac{dv}{v}=d\psi. \quad \text{Integrating } \log v=\psi+\text{const.}$$

$$\therefore \quad v=Ae^{\psi} \text{ but when } \psi=0,\ v=u$$

$$\therefore \quad A=u \text{ hence } v=ue^{\psi}.$$

$$\text{Tang. acc.}=v\frac{dv}{ds}=v\frac{dv}{d\psi}\cdot\frac{d\psi}{ds}=\frac{ue^{\psi}.ue^{\psi}}{c\sec^2\psi}, \qquad \because\ s=c\tan\psi$$

$$=\frac{u^2e^{2\psi}}{c}\cos^2\psi \qquad \therefore\ \frac{ds}{d\psi}=c\sec^2\psi=\rho.$$

$$\text{Normal acc.}=\frac{v^2}{\rho}=\frac{u^2e^{2\psi}}{c\sec^2\psi}=\frac{u^2e^{2\psi}}{c}\cos^2\psi$$

$$\therefore \quad \text{Resultant acc.} = \sqrt{\left(\frac{u^2e^{2\psi}\cos^2\psi}{c}\right)^2 + \left(\frac{u^2e^{2\psi}\cos^2\psi}{c}\right)^2}$$

$$= \frac{\sqrt{2}}{c} u^2 e^{2\psi} \cos^2 \psi.$$

EXAMPLES I (C)

1. If a point moves along a circle, prove that its angular velocity about any point on the circle is half of that about the centre.

2. A point describes a circle of radius a with a uniform speed v; show that the radial and transverse accelerations are $-\frac{v^2}{a}\cos\theta$ and $-\frac{v^2}{a}\sin\theta$ if a diameter is taken as initial line and one end of the diameter as pole.

3. A point describes uniformly a given straight line; show that its angular velocity about a fixed point varies inversely as the square of its distance from the fixed point.

4. A point P describes a curve with a constant velocity and the radius vector joining P to a fixed point O has an angular velocity which is inversely proportional to OP; show that the curve is an equiangular spiral with O as pole and that the acceleration of P along the normal varies inversely as OP.

5. If the velocity of a point moving in a plane curve varies as the radius of curvature, show that the direction of motion revolves with constant angular velocity.

Also, if the angular velocity of the moving point about a fixed origin be constant, show that its transverse acceleration varies as its radial velocity.

6. A point moves in a plane curve so that its tangential acceleration is constant and the magnitudes of the tangential velocity and the normal acceleration are in a constant ratio. Show that the intrinsic equation of the path is of the form $s = A\psi^2 + B\psi + C$.

7. A point moves in a curve so that its tangential and normal accelerations are equal and the tangent rotates with constant angular velocity.

Show that the intrinsic equation of the path is of the form $s = Ae^{\psi} + B$.

8. A particle describes a curve (for which s and ψ vanish simultaneously) with uniform speed v. If the acceleration at any point s be $\frac{v^2c}{s^2+c^2}$, prove that curve is a catenary.

1·7. Relative Motion.

The velocity of a point P relative to a point Q is the rate of displacement of P with respect to Q or, in other words, the velocity of P relative to Q is the velocity which P appears to have when observed from Q, that is, when Q is stationary.

Let the position of P at any time t be given by the co-ordinates (x_1, y_1) relative to two fixed rectangular axes OX, OY and the position of Q be given by its co-ordinates (x_2, y_2) referred to the same axes. Then the position of P relative to Q is given by the co-ordinates (x, y)

where $x=x_1-x_2\ ;\ y=y_1-y_2.$

Then $\dot{x}=\dot{x}_1-\dot{x}_2\ ;\ \dot{y}=\dot{y}_1-\dot{y}_2$

and $\ddot{x}=\ddot{x}_1-\ddot{x}_2\ ;\ \ddot{y}=\ddot{y}_1-\ddot{y}_2.$

But $(\dot{x}_1, \dot{y}_1)$ and $(\dot{x}_2, \dot{y}_2)$ are the actual components of velocities of P and Q, parallel to OX, OY, hence the velocity components of P relative to Q are obtained by adding to the actual velocity components of P, the velocity components which are equal and opposite to those of Q.

The velocity of P relative to Q may, therefore, be regarded as the resultant of two velocities, *viz.*, (*i*) The actual velocity of P and (*ii*) the reversed velocity of Q.

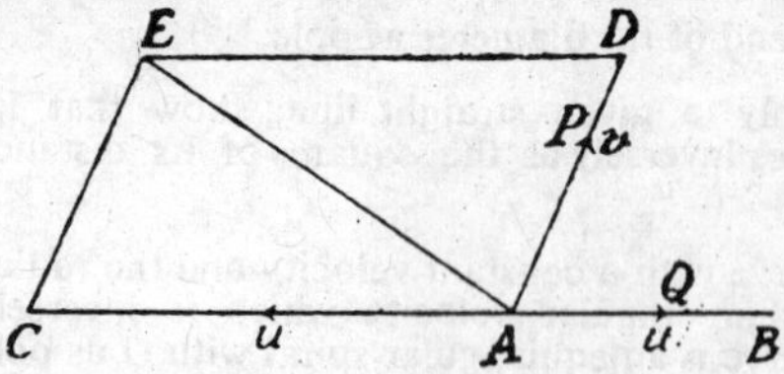

Let the point P move with velocity v along AD and the point Q move with velocity u along AB. To find the relative velocity of P with respect to Q, reduce Q to rest by adding a velocity u in the opposite direction AC. Thus P will have two velocities v along AD and u along AC; the resultant of these two velocities, as obtained by the parallelogram law, is the relative velocity.

If the velocities of P and Q be u and v in directions making angles α and β with x-axis, and if V be the velocity of P relative to Q in a direction making an angle θ with x-axis, then

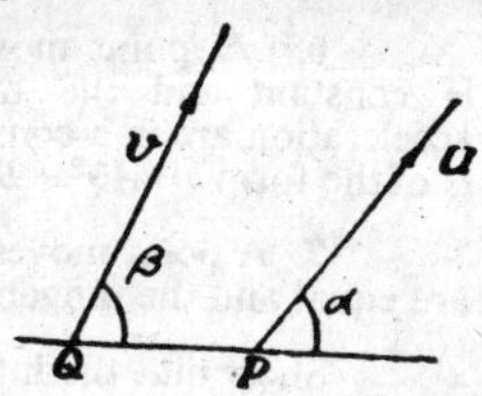

$V\cos\theta$ = velocity of P relative to Q along x-axis

= actual velocity of P along x-axis

− actual velocity of Q along x-axis

$=u\cos\alpha-v\cos\beta$

$V\sin\theta$ = velocity of P relative to Q along y-axis

= actual velocity of P along y-axis

− actual velocity of Q along y-axis

$=u\sin\alpha-v\sin\beta$

Squaring and adding

$$V^2=u^2+v^2-2uv\cos(\alpha-\beta)$$

and dividing

$$\tan\theta=\frac{u\sin\alpha-v\sin\beta}{u\cos\alpha-v\cos\beta}.$$

The angular velocity of P relative to Q

$$=\frac{\text{Velocity of } P \text{ relative to } Q \text{ perp. to } PQ}{PQ}$$

$$=\frac{(\text{actual vel. of } P-\text{actual vel. of } Q) \text{ perp. to } PQ}{PQ}$$

$$=\frac{u \sin \alpha - v \sin \beta}{PQ},$$

The component $(u \cos \alpha - v \cos \beta)$ has no effect in rotating PQ about Q.

A similar proposition is true for relative accelerations.

Ex. 18. *The distance between two moving points at any time is a and their relative velocity is V, u and v being the components of V respectively along and perpendicular to the direction of a. Show that their distance when they are nearest to one another is av/V and the time that elapses before they arrive at their nearest distance is au/V².*

Sol. Let P, Q be their positions when distance between them is a *i.e.*, $QP=a$. At any subsequent time, t, coordinates of P relative to Q are

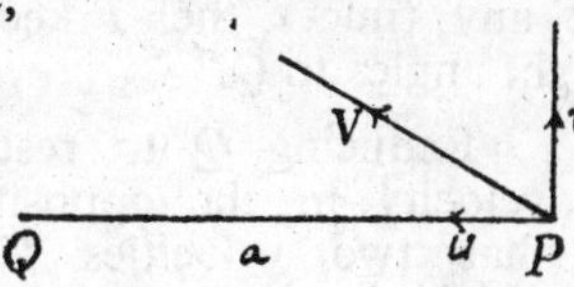

$$x=a-ut, \; y=vt$$

$\therefore$ distance between P, Q is

$$z=\sqrt{x^2+y^2}$$

i.e., $$z^2=(a-ut)^2+v^2t^2.$$

Now z is least when $\frac{dz}{dt}=0$

$$\therefore \qquad 0=2z\frac{dz}{dt}=-2u(a-ut)+2v^2t$$

giving $$t=\frac{au}{u^2+v^2}=\frac{au}{V^2}.$$

Then $$z^2=\left(a-\frac{au^2}{u^2+v^2}\right)^2+v^2\frac{a^2u^2}{(u^2+v^2)^2}=\frac{a^2v^2}{u^2+v^2}$$

$$=\frac{a^2v^2}{V^2}$$

$$\therefore \qquad z=\frac{av}{V}.$$

Ex. 19. *Two particles start simultaneously from the same point and move along two straight lines, one with uniform velocity u and the other from rest with uniform acceleration f. Show that their relative* *velocity is least after a time* $\dfrac{u\cos\alpha}{f}$ *and that the least relative velocity is* $u\sin\alpha$ *where* α *is the angle between the lines.*

Sol. The velocity of the second particle is $v=ft$. If V be the relative velocity,

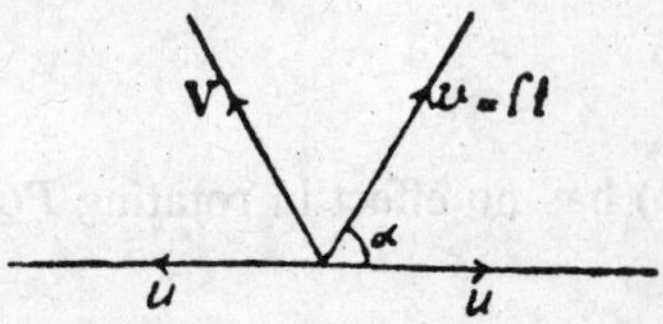

$$\text{Then } V^2=u^2+v^2+2uv\cos(180-\alpha)$$
$$=u^2+f^2t^2-2uft\cos a.$$

V is least when $\dfrac{dV}{dt}=0$, $\quad 0=2V\dfrac{dV}{dt}$

$$=2f^2t-2uf\cos\alpha \qquad \therefore\ t=\frac{u\cos\alpha}{f}$$

and then $V^2=u^2+u^2\cos^2\alpha-2u^2\cos^2 a=u^2\sin^2\alpha$

$\therefore \quad V=u\sin\alpha.$

Ex. 20. *A vessel steams at a constant speed v along a straight line, whilst another vessel steaming at a constant speed V, keeps the first always exactly abeam, show that the path of either vessel relatively to the other is a conic action of eccentricity v/V.*

Sol. If Q be the position of the first vessel and P that of the second at any time t, then P keeps Q abeam means the velocity V of P is at right angles to QP.

Reducing Q to rest by adding a velocity in the opposite direction, P has two velocities V and v as marked.

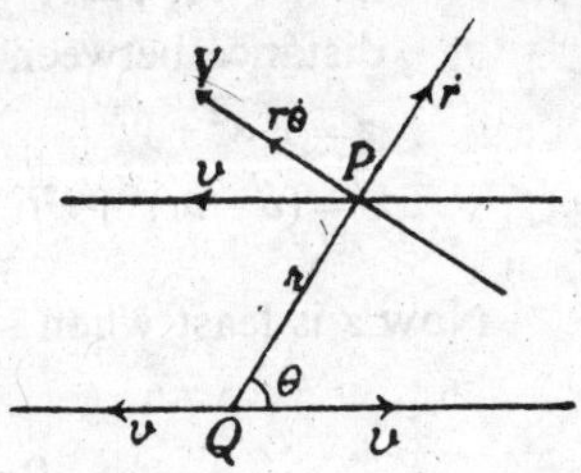

If P be the point (r, θ) with Q as pole and direction of v as initial line then its components of velocity are $\dot r$ and $r\dot\theta$ along and perp. to PQ.

Hence resolving along and perp. to QP, we get

$$\dot r=-v\cos\theta, \qquad r\dot\theta=v\sin\theta+V$$

$$\therefore \quad \text{dividing } \frac{rd\theta}{dr}=-\frac{v\sin\theta+V}{v\cos\theta}$$

$$\therefore \quad \frac{dr}{r}=-\frac{v\cos\theta}{v\sin\theta+V}\,d\theta.$$

Integrating we get $\log r=-\log(v\sin\theta+V)+\text{const.}$

or $\quad r(v\sin\theta+V)=A$

or $$\frac{A}{r}=v\sin\theta+V$$

or $$\frac{A/V}{r}=1+\frac{v}{V}\sin\theta$$

This is a conic section of eccentricity$=\frac{v}{V}$.

Ex. 21. *Two points P and Q start simultaneously from the same point O and moves uniformly in a circle and in a straight line which touches the circle, each with a speed v. Find the path of P with respect to Q and the relative velocity.*

Sol. Let the line OX along which O moves as x-axis and O the starting point. In time t, P has moved through angle θ in the circle so that $a\theta=\text{arc } OP=vt$

and Q has moved through $OQ=vt$

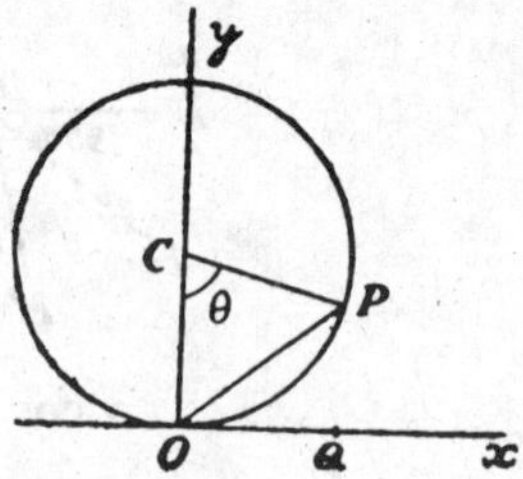

Now $OP=2a\sin\theta/2$ and $\angle POQ=\theta/2$

$\therefore$ coordinates of P are $2a\sin\theta/2\cos\theta/2$, $2a\sin\theta/2\sin\theta/2$

i.e., $a\sin\theta$, $a(1-\cos\theta)$

and co-ordinates of Q are $(vt, 0)$

i.e., $(a\theta, 0)$

$\therefore$ the relative co-ordinates are

$$x=a\sin\theta-a\theta=-a(\theta-\sin\theta)$$

$$y=a(1-\cos\theta)=a(1-\cos\theta)$$

which gives a cycloid.

Also $$\dot{x}=-a(1-\cos\theta)\dot{\theta}=-v(1-\cos\theta)$$

$$\dot{y}=a\sin\theta\dot{\theta}=v\sin\theta$$

$\therefore$ relative velocity$=\sqrt{\dot{x}^2+\dot{y}^2}=2v\sin\theta/2$ at an angle φ with x-axis, where

$$\tan\varphi=\frac{v\sin\theta}{-v(1-\cos\theta)}=-\cot\theta/2$$

$$\therefore\quad \varphi=\frac{\pi}{2}+\frac{\theta}{2}.$$

Ex. 22. *A crank CP rotates about a fixed centre C while the end D (the piston) of the connecting rod PD is constrained to move along CD. If angles PCD and PDC are θ and ϕ, prove that when the angular velocity of P is ω, the linear velocity of D is*

$$-CP.\frac{\sin(\theta+\phi)}{\cos\varphi}\omega.$$

Sol. Let v be the linear velocity of D along CD. Now velocity of P is $r\omega$ along AP which is perp. to CP Reducing P to rest, D has got two velocities, v along DL and $r\omega$ along DM, where DM is parallel to PA; the resultant of these two is the relative velocity of D to P and hence is perp. to PD hence resolving along DP, we have

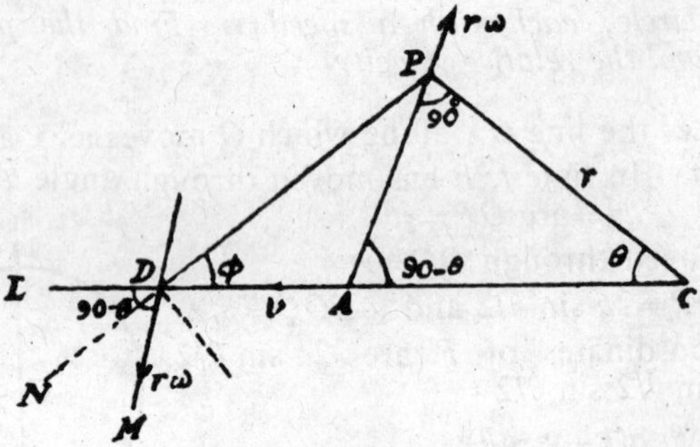

$$v \cos \varphi + r\omega \cos (90-\theta-\varphi)=0$$

$$\therefore \qquad v=-\frac{r\omega \sin (\theta+\varphi)}{\cos \varphi}$$

Ex. 23. *A man runs a race starting with velocity v, the direction of the wind appears to him to make an angle α with his course. It then appears to veer round uniformly through an angle α during the race. If the wind blows uniformly throughout at right angle to the course, and if T be the time of the race, show that the length of the course is*

$$\frac{vT \tan \alpha}{\alpha} \log (2 \cos \alpha)$$

Sol. Let A be the starting point and B the finish of the race and P any point in the course. The actual velocity of wind is V perp. to AB but at A, it appears to make angle α and at B angle 2α, with the course.

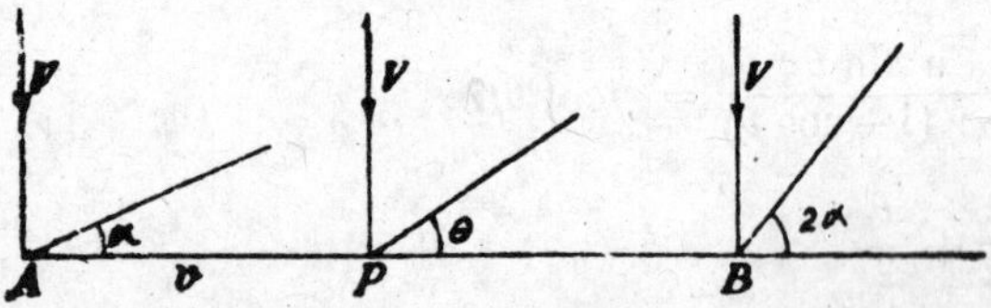

Let it appear to make $\angle\theta$ at P.

Since it veers uniformly

$$\dot{\theta}=\text{const.}=\omega \text{ say}$$

But in time T, change in angle$=2\alpha-\alpha=\alpha$

$$\therefore \qquad \omega T=\alpha$$

i.e. $$\dot{\theta}=\frac{\alpha}{T}$$

At A, $\tan\alpha=\dfrac{V}{v}$;

At P, $\tan\theta=\dfrac{V}{\dot{x}}$

where $AP=x$

$\therefore$ $\dot{x}=V\cot\theta$

or $$\frac{dx}{d\theta}\dot{\theta}=V\cot\theta$$

or $$\frac{dx}{d\theta}=\frac{TV}{\alpha}\cot\theta=\frac{T\,v\tan\alpha}{\alpha}\cot\theta$$

$\therefore$ integrating

$$AB=\frac{Tv\tan\alpha}{\alpha}\int_{\alpha}^{2\alpha}\cot\theta d\theta=\frac{Tv\tan\alpha}{\alpha}\log(2\cos\alpha)$$

Ex. 24. *The line joining two points P and Q is of constant length a and the velocities of P and Q are in directions which make angle α and β with PQ. Prove that the angular velocity of PQ is*

$$\frac{u\sin(\alpha-\beta)}{a\cos\beta}$$

where u is the velocity of P.

Sol. Angular velocity of PQ about Q is

$$\frac{u\sin\alpha-v\sin\beta}{PQ}.$$

The relative velocity along PQ is $u\cos\alpha-v\cos\beta$ and this must be zero since the length of PQ is constant

$\therefore$ $u\cos\alpha-v\cos\beta=0$

i.e., $$v=\frac{u\cos\alpha}{\cos\beta}$$

Hence ang. vel. $$=\frac{u\sin\alpha-\dfrac{u\cos\alpha}{\cos\beta}\sin\beta}{a}=\frac{u\sin(\alpha-\beta)}{a\cos\beta}$$

Ex. 25. *Two particles are describing concentric circles of radii a and b with angular velocities ω and ω' respectively. Prove that the angular velocity of the line joining them when it is of length r is*

$$\frac{(r^2+a^2-b^2)\omega+(r^2-a^2+b^2)\omega'}{2r^2}.$$

Sol. Linear velocities of P and Q are $a\omega$ along the tangent at P and $b\omega'$ along the tangent at Q, to the two circles.

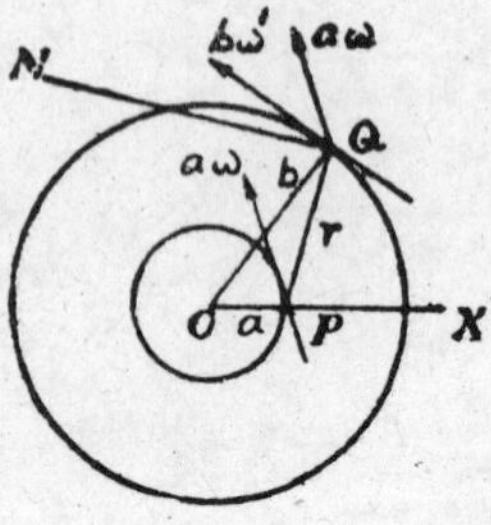

Ang. velocity of QP with respect to P

$$=\frac{\text{(actual vel. of } Q-\text{actual vel. of } P)\text{ perp. to } PQ}{PQ}$$

$$=\frac{bw' \cos OQP - a\omega \cos QPX}{r}$$

$$=\frac{b\omega'\dfrac{b^2+r^2-a^2}{2br}+a\omega\dfrac{a^2+r^2-b^2}{2ar}}{r}=\text{etc.}$$

Ex. 26. *A point P describes a circle of radius a with angular velocity ω′ about its centre O. Another point Q moves so that PQ is of constant length b and it is turning with angular velocity ω in the plane of the first circle. Prove that the angular velocity of OQ when its length is r*

$$=\frac{\omega(r^2+b^2-a^2)+\omega'(r^2-b^2+a^2)}{2r^2}.$$

Sol. Velocity of Q = velocity Q relative to P + velocity of P,

i.e., the resultant of $a\omega$ perp. to OP and $b\omega'$ perp. to PQ.

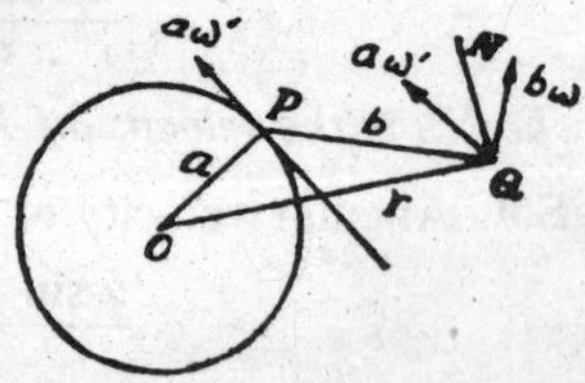

Ang. vel. of OQ

$$=\frac{\text{velocity of } Q \text{ perp. to } OQ}{OQ}$$

$$=\frac{\text{components of } a\omega' \text{ and } b\omega \text{ along } QN}{OQ}$$

$$=\frac{a\omega' \cos POQ+b\omega' \cos OQP}{r}$$

$$=\frac{a\omega\dfrac{a^2+r^2-b^2}{2ar}+b\omega\dfrac{b^2+r^2-a^2}{2br}}{r}=\text{etc.}$$

EXAMPLES I (D)

1. A person travelling towards the north-east finds that the wind appears to blow from the north, but when he doubles his speed it seems to come from a direction inclined at an angle $\cot^{-1} 2$ on the east of north. Prove that the true direction of the wind is towards the east.

2. A battleship A sights another B, a sea-miles away steering in a direction at right angles to AB with a speed of v knots. A alters course so as to pass in front of B and to come to a nearest distance c sea-miles from B. If the new course makes an angle α with AB, prove that A's speed is $v \cos \beta/\sin(\alpha-\beta)$ knots where $\sin \beta=c/a$. [A knot is 1 sea-mile per hour]

3. ABC is right angled triangle with C as a right angle, a particle P starts from A and moves along AC with uniform velocity u ; a second particle Q starts from C at the same instant and moves along CB with uniform velocity v ; show that shortest distance between P and Q will be $\frac{v}{\sqrt{u^2+v^2}} \cdot AC$ after a time $\frac{u}{u^2+v^2} \cdot AC$.

4. Two straight roads OX and OY are inclined to each other at an acute angle α ; one car moves along XO with speed u while a second car moves along OY with speed v ; show that when the cars are at the least distance apart, the ratio of their distances from O is $\frac{v+u \cos \alpha}{u+v \cos \alpha}$.

If the first car is at distance d from O when the second is at O, show that the cars are at their least distance apart after time $\frac{d(u+v \cos \alpha)}{u^2+v^2+2uv \cos \alpha}$.

5. Two straight railways converge to a level crossing at an angle α ; and two trains distant respectively a and b from this point are moving towards it with speed u, v respectively. Find when and where they are nearest each other and prove that their least distant apart is $\frac{(av-bu) \sin \alpha}{\sqrt{u^2+v^2-2uv \cos \alpha}}$.

6. Two ships are steaming along straight courses with such constant velocities that they will collide unless their velocities are altered. Show that to an observer on either ship the other appears to be always moving directly towards him.

7. Two points P and Q are describing concentric circles of radii a and b and centre O with velocities u and v. Find the velocity of P relative to Q when the angle POQ is θ and if the angular velocity of one relative to the other is zero, prove that

$$\cos \theta = \frac{au+bv}{av+bu}.$$

8. If particles P and Q are travelling in straight lines with uniform velocities u and v towards a point O, show that the least distance between them during the motion is $\frac{u \sim v}{v}\ p$, where p is the perpendicular distance of O from the direction of velocity of P relative to Q when P and Q are equidistant from O.

9. If an aviator flies round a triangular course, each side of which is c miles long, while the wind blows at u miles per hour parallel to a side, show that he takes $\frac{c\,(v+\sqrt{4v^2-3u^2})}{v^2-u^2}$ hours to complete the circuit in either direction v being his velocity relative to the air.

B. KINETICS : FORCE AND MOTION

1·8. Newton's Laws of Motion.

We shall now discuss the effects of forces on the motion of a particle. The relation between the force and motion is given by Newton's laws of motion.

Law I. *Everybody continues in its state of rest or of uniform motion in a straight line, except in so far as it is compelled by impressed forces to change that state.*

Law II. *The rate of change of momentum is proportional to the impressed force, and takes place in the direction of the straight line in which the force acts.*

Law III. *To every action there is an equal and opposite reaction.*

1·9. Deductions from Newton's Laws of Motion.

The first law states that a body has no tendency of its own to change its state of rest or of uniform motion and if it is not acted upon by any external force, it will continue to remain in that state of rest or of uniform motion in a straight line. Thus a body in its own nature will remain either at rest or in uniform motion—this is, therefore, the *natural state* of a body and this natural state can be broken only when some external forces are acting on it. Thus the first law gives us the *definition of force.*

Force is something which changes or tends to change the state of rest or of uniform motion in a straight line of a body.

The first law also implies what is called the **Principle of Inertia**. Inertia of rest implies that if a body is initially at rest, it will for ever remain at rest. Inertia of motion implies that if a body is moving with uniform velocity, it will continue to do so for all time to come.

The second law gives us the law governing the measurement of force.

Momentum of a body is the product of its mass and the velocity with which it is moving. Thus if m denotes the mass and v the velocity, mv is the momentum.

The second law says that the impressed force P which causes the motion is proportional to the rate of change of mv.

That is, $$P=k\frac{d}{dt}(mv), \text{ where } k \text{ is a constant,}$$

$$=km\frac{dv}{dt}, \text{ if } m \text{ is constant,}$$

$$=kmf, \text{ where } f \text{ is the acceleration.}$$

If the unit of force be so chosen that it acting on unit mass produces unit acceleration *i.e.*, $P=1$ when $m=1$ and $f=1$, then $k=1$.

Thus $$P=mf.$$

This is the fundamental equation in Dynamics and is known as the working formula. The product of mass and acceleration is called the **effective force.** Thus the above equation can be stated as :

Impressed force in any direction is equal to the effective force in that direction.

As seen above, the second law also gives the definition of the unit of force. *The unit of force is that force which acting on unit mass produces unit acceleration.* In the British (Foot, Pound, Second or F.P.S.) system, the unit of mass is one pound, the unit of acceleration is one foot per second per second and the unit of force is the **Poundal.** Hence *one poundal is that force which acting on a mass of one pound will produce an acceleration of one foot per second per second.* In the French (Centimetre, Gramme, Second or C.G.S.) system, the unit of mass is one gramme, the unit of acceleration is one centimetre per second per second and the unit of force is the **Dyne.** Hence *one dyne is that force which acting on a mass of one gramme will produce an acceleration of one centimetre per second per second.*

The second part of the second law of motion states that the change of momentum produced by a force is in the direction in which the force acts, that is, if two or more forces act on a body at the same time each produces an acceleration proportional to it in its line of action independent of the action of others. This gives us the **Principle of Physical Independence of forces.**

When a body is acted upon by a number of forces, the acceleration due to each force is the same in direction and magnitude as if this is the only force acting on the body and all other forces are absent. Thus each force will produce its effect in direction and magnitude as if it were acting all alone and the resultant effect on the body acted upon by different forces will be the resultant of all these effects produced separately by each force.

The third law of motion states that no force can exist in this universe without the existence at the same time of another force equal and opposite to it. If one body exerts a force on another, the second also exerts an equal force in the opposite direction. The first is called *Action* and the second *Reaction.*

Illustrations :

(1) **Tug-of-war.** *AC* is the ground, *AB* and *CD* the two parties, *LM* is the rope.

The force of tension *T* with which one party pulls the rope is the same as that with which the other party pulls. The parties must push the ground in an inclined position and the ground also gives an equal amount of force to the parties. Let P_1 and P_2 be the forces with which *AB* and *CD* push the ground so that P_1 and P_2 also are the forces acting on *AB* and *CD.* These have components

(R_1 upwards, F_1) and (R F_2). R_1 and R_2, support the weights w_1 and w_2 of the two parties. If $F_1 > F_2$, AB will win over CD.

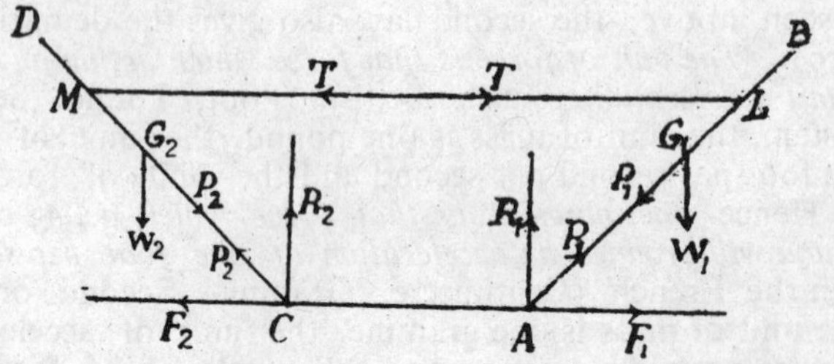

(2) **Horse and Carriage.** AC is the ground, AB is the horse, $CDEF$ the carriage, LM the rope.

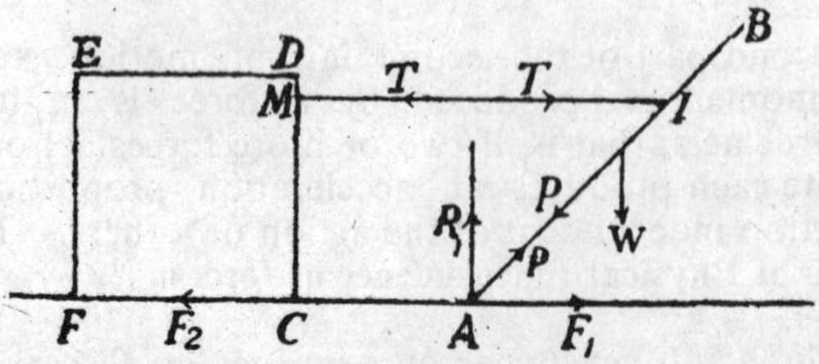

The horse and the carriage pull each other with the same force T, so T does not produce the motion. In trying to go forward, the horse pushes the ground with his feet, say with a force P and the ground gives an equal force P to the horse in the opposite direction. P has two components R_1 and F_1. R_1 will support the weight W of the horse. A force of friction say F_2 also acts on the carriage.

If $F_1 > F_2$, the system will move on.

(3) **Engine and Carriage.** The force of tension in the coupling chain between the engine and carriage with which the engine pulls the carriage is the same as the force with which the carriage pulls the engine.

The driving wheels of the engine are made to rotate by means of a crank. The force of friction, say, F_1 between the wheels and the rails makes the wheels rotate and causes the forward motion of the engine. Again between the other wheels of the engine and wheels of the carriage and the rails, the force of friction acts in the opposite direction and makes the wheels rotate. Let this force of friction be F_2.

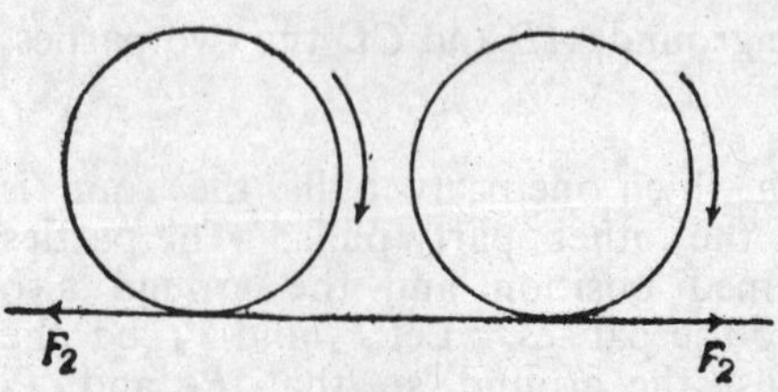

If $F_1 > F_2$, the system will move on.

1·10. Miscellaneous Worked-out Examples.

Ex. 27. *A wheel is rolling, without slipping, in a vertical plane along a horizontal straight line, its centre having the speed u ; to find the velocity of any point on its rim.*

Sol. O is the centre, A the lowest point in contact with the ground, B the highest point at any instant, P any point on the rim where $\angle POB = \theta$. The centre moves in a straight line parallel to the ground with velocity u, velocity of P is equal to the velocity of P relative to O plus the velocity of O.

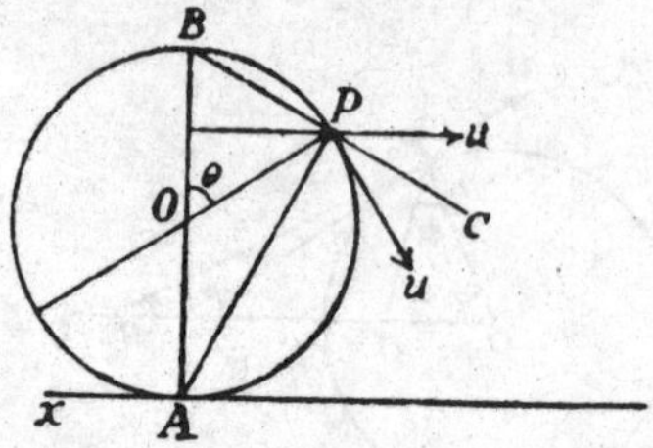

When the wheel has turned through $\angle\theta$, the centre has moved through a distance x, so that $x = a\theta$, if a is the radius.

Hence $\quad u = \dot{x} = a\dot{\theta}$

Again the velocity of P relative to O = that in a circle of radius a $= a\dot{\theta}$, tangentially.

Thus P has two velocities, both equal to u, one horizontally and the other tangentially, the angle between them being also equal to θ.

Hence velocity of $P = 2u \cos \theta/2$ along the bisector PC.

But PC is evidently perpendicular to AP and passes through B.

Also $\qquad AP = 2a \cos \theta/2$

$\therefore$ velocity of $P = u \cdot \dfrac{AP}{a}$, along BP produced.

Ex. 28. *An insect crawls at a constant rate u along the spoke of a cartwheel of radius a, the cart moving with velocity v. Find the accelerations along and perpendicular to the spoke.*

Sol. O is the centre of the wheel, OA the spoke and P the insect at any time t. Let $\angle AOB = \theta$, where B is the highest point. If $OP = r$,

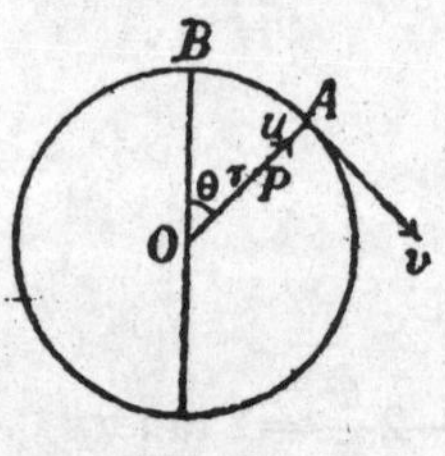

then $\qquad u = \dot{r}$,

Also $\qquad v = a\dot{\theta}$

$\therefore$ Radial acc. of $P = \ddot{r} - r\dot{\theta}^2 = 0 - r\omega^2$

where $\qquad \omega = \dfrac{v}{a}$

$= -r\omega^2 = -ut\omega^2$ for $r = ut$

$\because$ when $t = 0$, $r = 0$.

Transverse acc. $= 2\dot{r}\dot{\theta} + r\ddot{\theta} = 2u\omega + 0 = 2u\omega$.

Ex. 29. *A point is describing a circle of radius a with velocity v. If ω and ω' are its angular velocities about two points which are inverse with respect to the circle, prove that*

$$\omega+\omega'=\frac{v}{a}.$$

Sol. Let A and B be the inverse points. Therefore $OA.OB=a^2$ so that $\frac{OA}{a}=\frac{a}{OB}$. Then the two triangles OAP and OBP are equiangular so that $\angle APO=\angle OBP=\phi$ say.

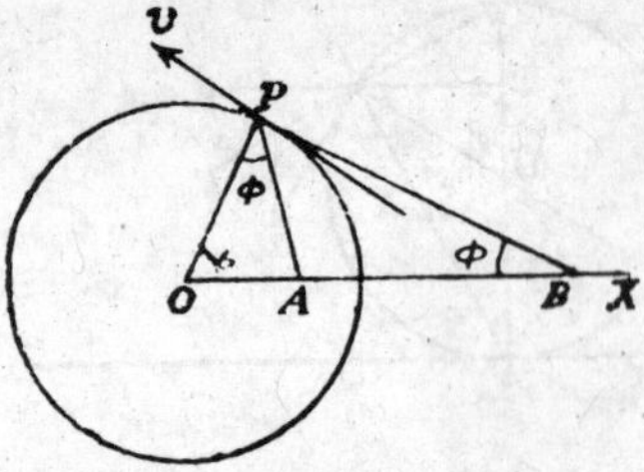

If $\angle AOP=\theta$, then $v=a\dot{\theta}$

ω = angular velocity of P about A

$$=\frac{d}{dt}BAP=\frac{d}{dt}(\theta+\phi)=\dot{\theta}+\dot{\phi}$$

ω' = angular velocity of P about B

$$=\frac{d}{dt}XBP=\frac{d}{dt}(180-\phi)=-\dot{\phi}$$

Hence $\omega+\omega'=\dot{\theta}=\frac{v}{a}$.

Ex. 30. *A particle moves in a semi-circle of which AB is the bounding diameter, in such a manner that AP, where P is any point of the path is increasing at a uniform rate u. Find the acceleration of the particle in the directions PA and PB in terms of u and θ, the angle PAB and show that the resultant acceleration makes with the normal at P an angle* $\cot^{-1}(2\tan\theta)$.

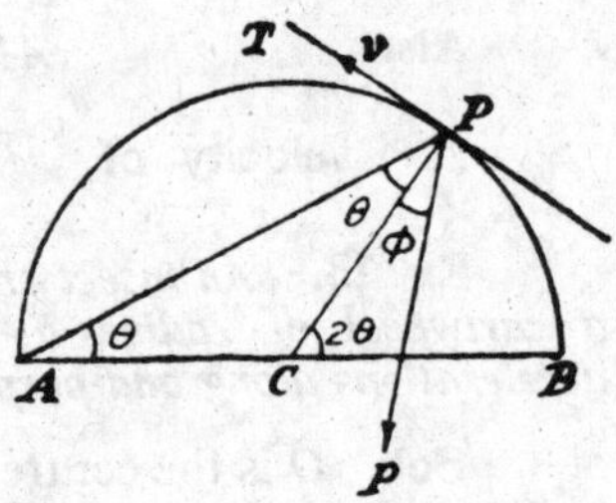

Sol. If $AB=2a$, $AP=2a\cos\theta$

$$\therefore \quad u=\frac{d}{dt}(2a\cos\theta)$$

$$=-2a\sin\theta\dot{\theta}$$

differentiating again

$$O=-2a[\cos\theta\dot{\theta}^2+\sin\theta\ddot{\theta}]$$

$$\therefore \quad \frac{\dot{\theta}^2}{\ddot{\theta}}=-\tan\theta$$

$$\text{Acc. of } P \text{ along } PC=a\left\{\frac{d}{dt}(2\theta)\right\}^2=4a\dot{\theta}^2$$

and $\qquad$ along $PT=a\frac{d^2}{dt^2}(2\theta)=2a\ddot{\theta}$

If the resultant acc. makes an angle ϕ with PC,

$$\cot\phi=\frac{\text{acc. along } PC}{-\text{acc. along } PT}=\frac{4a\dot{\theta}^2}{-2a\ddot{\theta}}=-2\frac{\dot{\theta}^2}{\ddot{\theta}}=2\tan\theta$$

Also acc. along $PA = 4a\dot{\theta}^2 \cos\theta + 2a\ddot{\theta}\sin\theta$

$$= 4a\dot{\theta}^2 \cos\theta - 2a\dot{\theta}^2 \cos\theta = 2a\dot{\theta}^2 \cos\theta$$

$$= \frac{u^2 \cos\theta}{2a \sin^2\theta}$$

acc. along $PB = 4a\dot{\theta}^2 \sin\theta - 2a\ddot{\theta}\cos\theta$

$$= \frac{u^2}{2a} \cdot \frac{1+\sin^2\theta}{\sin^3\theta}.$$

Ex. 31. *A particle moves in the curve $y = a \log \sec \frac{x}{a}$ in such a way that the tangent to the curve rotates uniformly ; prove that the resultant acceleration of the particle varies as the square of the radius of curvature.*

Sol. $y = a \log \sec \dfrac{x}{a}$

$$\frac{dy}{dx} = \frac{1}{\sec \frac{x}{a}} . \sec\frac{x}{a} \tan\frac{x}{a} = \tan\frac{x}{a}$$

$\therefore$ $\tan\psi = \tan\dfrac{x}{a}$

Hence $\psi = \dfrac{x}{a}$

or $x = a\psi$

Given $\dot{\psi} = \text{const.} = c$ say

then $\dot{x} = a\dot{\psi} = ac,\ \ddot{x} = 0$

$y = a \log \sec\psi$

$\therefore$ $\dot{y} = a \tan\psi\dot{\psi} = ac \tan\psi$

$\ddot{y} = ac^2 \sec^2\psi$

Hence acceleration $= ac^2 \sec^2\psi$ since $\ddot{x} = 0$

Now $\rho = \dfrac{(1+y_1^2)^{3/2}}{y_2} = \dfrac{\left(1+\tan^2\frac{x}{a}\right)^{3/2}}{\frac{1}{a}\sec^2\frac{x}{a}}$

$$= a \sec\frac{x}{a} = a \sec\psi$$

$\therefore$ acceleration varies as ρ^2.

Ex. 32. *If a rod which always passes through the origin rotates with uniform angular velocity ω, while one end describes the* *curve $r=a+be^{\theta}$, show that the radial acceleration of any point of the rod is the same at any instant and the radial velocity is the same at any point at a given instant.*

A describes the curve $r=a+be^{\theta}$

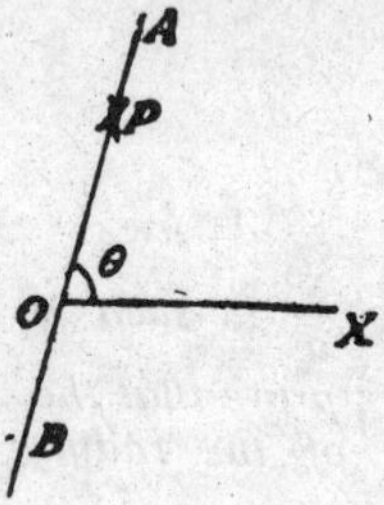

and P be any point such that

$$AP=c$$

Polar co-ordinates of A are r, θ
and those of P are $r-c, \theta$

$$\therefore \text{ radial acc. of } P=\frac{d^2}{dt^2}(r-c)-(r-c)\dot{\theta}^2$$

$$=\ddot{r}-(r-c)\dot{\theta}^2$$

$$=\ddot{r}-(r-c)\omega^2 \text{ if } \dot{\theta}=\omega$$

But $\qquad r=a+be^{\theta}$

$$\therefore \qquad \dot{r}=be^{\theta}\,\dot{\theta}=be^{\theta}\,\omega,\ \ddot{r}=b\omega^2e^{\theta}=(r-a)\omega^2$$

$$\therefore \text{ radial acc. of } P=(r-a)\omega^2-(r-c)\omega^2=(c-a)\omega^2$$

$$=\text{const.}$$

$$\text{Radial velocity of } P=\frac{d}{dt}(r-c)=\dot{r}=b\omega e^{\theta}$$

$$=(r-a)\omega$$

which being independent of c, is the same for every point at a given instant.

Ex. 33. *A particle P moves in an ellipse so that its angular velocity about one focus varies inversely as the square of the radius vector. Show that the angular velocity about the other focus varies inversely as the square of the diameter conjugate to CP, C being the centre of the ellipse.*

Sol. $S\ S'$ are the foci, p, p' are perpendiculars from S, S' on the tangent at P and $SP=r$, $S'P=r'$ so that $\sin\phi=\frac{p}{r}=\frac{p'}{r'}$, tangent being equally inclined to SP and $S'P$.

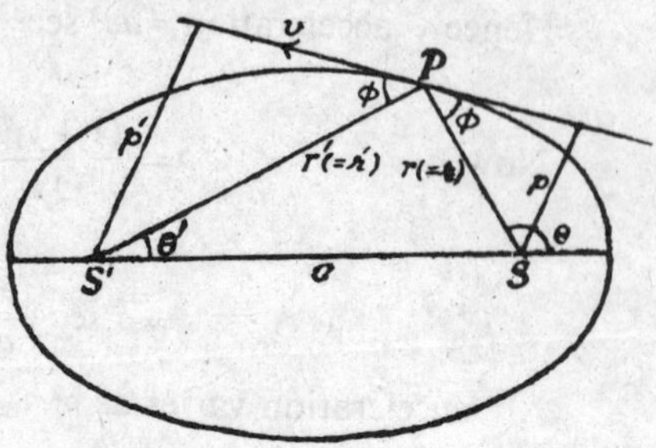

Now $\qquad \dot{\theta}=\frac{vp}{r^2}$,

and $\qquad \dot{\theta}'=\frac{vp'}{r'^2}$

Given that $\dot{\theta}=\frac{c}{r^2}$

$\therefore \qquad vp=c.$

$$\text{Hence} \qquad \dot{\theta}'=\frac{vp'}{r'^2}=\frac{v}{r'}\cdot\frac{p'}{r'}=\frac{v}{r'}\cdot\frac{p}{r}=\frac{vp}{rr'}=\frac{c}{rr'}.$$

But if CD be semi-conjugate diameter to CP,

then $\quad CD^2=rr'$, hence $\theta'=\dfrac{c}{CD^2}$.

Ex. 34. *The acceleration of a point moving in a plane curve is resolved into two components, one parallel to the initial line and the other along the radius vector ; prove that these components are*

$$-\frac{1}{\sin\theta}\frac{d}{dt}(r^2\dot\theta) \text{ and } \frac{\cot\theta}{r}\frac{d}{dt}(r^2\dot\theta)+\ddot r-r\dot\theta^2.$$

Also determine the component accelerations in the directions of the initial line and perpendicular to the radius vector.

Sol. If X, R be the components of acc. parallel to initial line and radially, then resolving radially, and cross-radially, we get

$$X\cos\theta+R=\ddot r-r\dot\theta^2$$

and $\quad X\sin\theta=-\dfrac{1}{r}\dfrac{d}{dt}(r^2\dot\theta).$

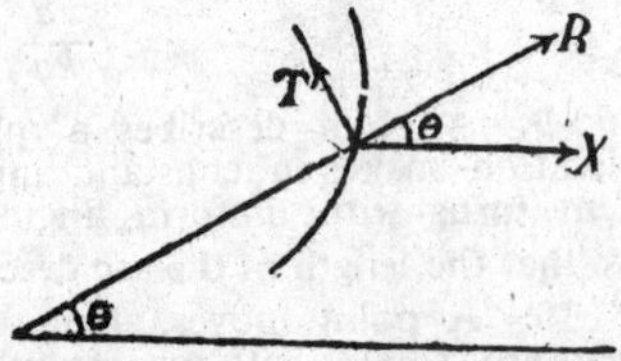

These give X and R.

If X, T be the components of acc. parallel to initial line and cross-radially then resolving radially and cross radially, we get

$$X\cos\theta=\ddot r-\dot\theta^2$$

and $\quad T-X\sin\theta=\dfrac{1}{r}\cdot\dfrac{d}{dt}(r^2\dot\theta).$ These give X and T.

EXAMPLES I (E)

1. A wheel rolls uniformly on the ground without sliding, its centre describing a straight line. Show that the angular velocity of any point on the rim about the point of contact of the wheel with the ground is equal to the angular velocity of the wheel about its centre.

2. Two points are moving with uniform velocities u and v along perpendicular lines OX, OY, the motion being towards O. When $t=0$, they are at distances a, b, respectively from O. Calculate the angular velocity of the line joining them at time t and show that it is the greatest when $t=\dfrac{au+bv}{u^2+v^2}$.

3. A small bead slides with constant speed v on a smooth wire in the shape of the cardioid $r=a(1+\cos\theta)$. Show that the value of $\dot\theta$ is $(v\sec\theta/2)/2a$ and that the radial component of the acceleration is constant.

4. If the curve is the equi-angular spiral $r=ae^{\theta\cot\alpha}$ and if the radius vector to the particle has constant angular velocity, show that the resultant acceleration of the particle makes an angle 2α with the radius of vector and is of magnitude v^2/r where v is the speed of the particle.

5. Two particles P and Q, starting simultaneously from the same point O, move uniformly in a circle and in a straight line which touches the circle respectively, each with a speed u. Prove that the velocity of P relative to Q is $2u \sin \theta$ where $\angle POQ=\theta$, in a direction perp. to OP and the relative path is a cycloid.

6. A man wishes to cross a road breadth c between two cars, each of breadth b, which are moving in a line separated by a distance a along the road with speed u. Show that the least uniform speed of the man must be $\frac{bu}{\sqrt{a^2+b^2}}$ and the time that he takes to cross the road then is $\frac{c}{u}\left(\frac{a}{b}+\frac{b}{a}\right)$.

7. A point starts from the origin in the direction of the initial line with velocity f/ω, and moves with constant angular velocity ω about the origin, and with constant negative radial acceleration $-f$. Show that the rate of growth of the radial velocity is never positive but tends to the limit zero, and prove that the equation of the path is

$$\omega^2 r=f(1-e^{-\theta}).$$

8. A point is describing a circle of radius a in such a way that the tangential acceleration is k times the normal acceleration. If its speed at a certain point is u, prove that it will return to the same point after a time

$$\frac{a}{ku}(1-e^{-2\pi k}).$$

9. A point describes a plane curve in such a way that the resultant acceleration makes a constant angle α with the tangent to the curve and the tangent turns with uniform angular velocity ω. If the initial velocity is $a\omega$, prove that the length of the arc describe in time t is $a \tan \alpha\, (e^{\omega t \cot \alpha}-1)$.

10. A point moves along the arc of a cycloid in such a manner that the tangent at it rotates with a constant angular velocity; show that the acceleration of the moving point is constant in magnitude.

11. The direction of the acceleration of a particle moving in a cycloid makes with the normal an angle equal to that which the tangent to the cycloid at the point makes with the tangent at the vertex and is in the same sense. Prove that the magnitude of the acceleration is constant.

12. One point describes the diameter AB of a circle with uniform velocity and another the semi-circumference AB from rest with uniform tangential acceleration. They start together from A and arrive together at B; show that the velocities at B are as $\pi : 1$.

13. A point describes a parabola in such a manner that the velocity at a distance r from the focus is $\sqrt{\frac{2f}{r}\left(r^2-c^2\right)}$, where f, c are constant; prove that its acceleration is compounded of f parallel to the axis and $f\frac{c^2}{r^2}$ along the radius vector from the focus.

14. A ring which can slide on a thin, long smooth rod, rests at a distance d from one end O. The rod is then set revolving uniformly about O in a horizontal plane; show that in space the ring describes the curve $r=d \cosh \theta$.

15. A straight smooth tube revolves with angular velocity ω in a horizontal plane about one extremity which is fixed. If at zero time a particle inside it be at a distance d from the fixed extremity and moving with velocity V along the tube, show that its distance from the fixed end at time t is

$$d \cosh \omega t+(V/\omega) \sinh \omega t.$$

[Hint. $\ddot{r}-r\dot{\theta}=0$, $\dot{\theta}=\omega$. Initial conditions are $t=0$, $r=d$, $\dot{r}=V$].

16. A straight smooth tube revolves with angular velocity ω in a horizontal plane about one extremity which is fixed, if at zero time a particle inside it be at a distance a from a fixed end and moving with velocity V along the tube, show that its distance at time t is $a\cosh\omega t+(V/\omega) \sinh \omega t$.

2

Rectilinear Motion

2·1. Motion in a straight line with constant acceleration.

A particle moves in a straight line with a constant acceleration f, starting from some point in the line with a velocity u ; to discuss the motion.

Let O be a fixed point on the line, not necessarily the starting point of the particle and let P be the position of the particle at time t and let $OP=s$.

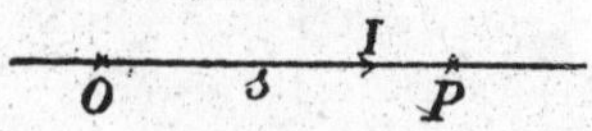

Then the acceleration of the particle is $\ddot{s}$ in the sense OP produced hence the equation is

$$\ddot{s}=f.$$

Integrating this equation with respect to t, we get

$$\dot{s}=ft+C_1 \text{ where } C_1 \text{ is some constant}$$

i.e., $$v=\dot{s}=ft+C_1.$$

Initially when $t=0$, it is given that $v=u$

Hence $\quad u=0+C_1 \quad$ *i.e.,* $C_1=u.$

$\therefore$ we get $\quad v=\dot{s}=u+ft \quad$ ...(1)

Integrating (1) with respect to t, we get

$$s=ut+\tfrac{1}{2}ft^2+C_2 \text{ where } C_2 \text{ is a constant.}$$

If the particle starts from O, say, then when $t=0$, $s=0$

so that $\quad 0=0+C_2 \quad \therefore \quad C_2=0.$

Hence $\quad s=ut+\tfrac{1}{2}ft^2. \quad$ ...(2)

If the particle starts from some other point, say at a distance a from O, then when $t=0$, $s=a$

then $$a=0+C_2, \quad \therefore \quad C_2=a.$$

Hence $$s=a+ut+\tfrac{1}{2}ft^2 \qquad \ldots(3)$$

From equations (1) and (2), we have by eliminating t,

$$v^2=(u+ft)^2=u^2+2uft+f^2t^2$$
$$=u^2+2f(ut+\tfrac{1}{2}ft^2) \qquad =u^2+2fs. \qquad \ldots(4)$$

This equation can also be obtained as

Instead of taking the acceleration as $\ddot{s}$, we can take it as $v\dfrac{dv}{ds}$

hence the equation is $v\dfrac{dv}{ds}=f.$

Integrating the equation with respect to s, we get

$$\tfrac{1}{2}v^2=fs+C \text{ where } C \text{ is a constant.}$$

Initially when $s=0$, $v=u$

$$\therefore \quad \tfrac{1}{2}u^2=0+C \qquad \therefore \quad C=\tfrac{1}{2}u^2$$

Hence $$\tfrac{1}{2}v^2=fs+\tfrac{1}{2}u^2 \qquad \ldots(5)$$

i.e., $$v^2=u^2+2fs.$$

Remark. If f is a continuous function of t_1 then the average value of f in the interval $[0,t_1]$ is $\dfrac{\int_0^{t_1} f.dt}{\int_0^{t_1} dt}$, and is called the time-average of f. Similarly $\dfrac{\int_0^{s_1} f.ds}{\int_0^{s_1} ds}$ is space-average in $[0, s_1]$ if f is a continuous function of s.

Ex. 1. *Two cars start off to race with velocities u and v and travel in a straight line with uniform accelerations α and β; if the race ends in a dead heat prove that the length of the course is*

$$\frac{2(u-v)(u\beta-v\alpha)}{(\alpha-\beta)^2}.$$

Sol. Dead heat means that the cars cover the same distance in the same time. Hence if s be the distance and t the time, we have

$$s=ut+\tfrac{1}{2}\alpha t^2 \quad \text{and} \quad s=vt+\tfrac{1}{2}\beta t^2$$

$$\therefore \quad ut+\tfrac{1}{2}\alpha t^2=vt+\tfrac{1}{2}\beta t^2 \qquad \therefore \quad t=\frac{2(u-v)}{\beta-\alpha}.$$

Hence $$s=\frac{2u(u-v)}{\beta-\alpha}+\frac{1}{2}\cdot\frac{4\alpha(u-v)^2}{(\beta-\alpha)^2}=\frac{2(u-v)(u\beta-v\alpha)}{(\beta-\alpha)^2}.$$

Ex. 2. *A point moving in a straight line with uniform acceleration describes a and b feet in successive intervals of t_1 and t_2 seconds. Prove that the acceleration is*

$$\frac{2(bt_1-at_2)}{t_1t_2(t_1+t_2)}.$$

Also prove that if the point describes successive equal distances in times t_1, t_2, t_3 then

$$\frac{1}{t_1}-\frac{1}{t_2}+\frac{1}{t_3}=\frac{3}{t_1+t_2+t_3}.$$

Sol. $$a=ut_1+\tfrac{1}{2}ft_1^2,\quad v=u+ft_1$$

$$b=vt_2+\tfrac{1}{2}ft_2^2=(u+ft_1)t_2+\tfrac{1}{2}ft_2^2$$

$\therefore$ $bt_1-at_2=\frac{1}{2}ft_1t_2(t_1+t_2)$ hence $f=$etc.

$$s=ut_1+\tfrac{1}{2}ft_1^2 \qquad \therefore \quad \frac{s}{t_1}=u+\tfrac{1}{2}ft_1$$

also $$s=(u+ft_1)t_2+\tfrac{1}{2}ft_2^2 \qquad \therefore \quad \frac{s}{t_2}=u+ft_1+\tfrac{1}{2}ft_2$$

and $$s=[u+f(t_1+t_2)]t_3+\tfrac{1}{2}ft_3^2,$$

$\therefore$ $$\frac{s}{t_3}=u+ft_1+ft_2+\tfrac{1}{2}ft_3$$

$\therefore$ $$\frac{s}{t_1}-\frac{s}{t_2}+\frac{s}{t_3}=u+\tfrac{1}{2}f(t_1+t_2+t_3)$$

also $3s$ is described in time $(t_1+t_2+t_3)$

Hence $$3s=u(t_1+t_2+t_3)+\tfrac{1}{2}f(t_1+t_2+t_3)^2$$

$\therefore$ $$\frac{3s}{t_1+t_2+t_3}=u+\tfrac{1}{2}f(t_1+t_2+t_3)$$

Hence $$\frac{s}{t_1}-\frac{s}{t_2}+\frac{s}{t_3}=\frac{3s}{t_1+t_2+t_3},$$ hence

Ex. 3. *If the co-ordinates of a point moving with constant acceleration be x_1, x_2, x_3 at instants t_1, t_2, t_3 respectively prove that the acceleration is*

$$2\left[\frac{(x_2-x_3)t_1+(x_3-x_1)t_2+(x_1-x_2)t_3}{(t_2-t_3)(t_3-t_1)(t_1-t_2)}\right]$$

Sol. $$x_1=ut_1+\tfrac{1}{2}ft_1^2$$

$$x_2=ut_2+\tfrac{1}{2}ft_2^2$$

$$x_3=ut_3+\tfrac{1}{2}ft_3^2$$

$\therefore$ $$x_1-x_2=u(t_1-t_2)+\tfrac{1}{2}f(t_1^2-t_2^2)$$

$$\therefore \quad (x_1-x_2)t_3=u(t_1t_3-t_2t_3)+\tfrac{1}{2}f(t_1^2t_3-t_2^2t_3)$$

Similarly $(x_2-x_3)t_1$ and $(x_3-x_1)t_2$

$$\therefore \quad (x_2-x_3)t_1+(x_3-x_1)t_2+(x_1-x_2)t_3$$
$$=u(t_2t_1-t_3t_1+t_3t_2-t_1t_2+t_1t_3-t_2t_3)$$
$$+\tfrac{1}{2}f(t_1^2t_3-t_2^2t_3+t_2^2t_1-t_3^2t_1+t_3^2t_2-t_1^2t_2)$$
$$=\tfrac{1}{2}f\,[t_1^2(t_3-t_2)+t_1(t_2^2-t_3^2)+t_2t_3(t_3-t_2)]$$
$$=\tfrac{1}{2}f(t_3-t_2)(t_1-t_2)(t_1-t_3) \qquad \text{hence } f.$$

Ex. 4. *A body is projected vertically upwards with a velocity u ; after time t another body is projected vertically upwards from the same point with a velocity v where $v<u$. If they meet as soon as possible, prove that*

$$t=\frac{u-v+\sqrt{u^2-v^2}}{g}.$$

Sol. Let them meet at a height h after time T of the projection of the second

then
$$h=u(t+T)-\tfrac{1}{2}g(t+T)^2$$
$$=vT-\tfrac{1}{2}gT^2$$
$$\therefore \quad gt^2+2t(gT-u)+2(v-u)T=0 \qquad \ldots(1)$$

T is minimum and therefore h is minimum

when
$$\frac{dh}{dT}=0 \text{ giving } v-g\,T=0 \text{ or } T=\frac{v}{g}$$

$\therefore$ from (1)

$$g^2t^2-2gt(u-v)-2v(u-v)=0$$

Hence
$$t=\frac{(u-v)\pm\sqrt{(u^2-v^2)}}{g}.$$

Negative sign has to be discarded as this gives t negative.

Ex. 5. *A motor starts from rest to go to a distance S and its acceleration t seconds after the start is $K\left(1-\frac{t}{T}\right)$ until it has attained the maximum velocity which occurs when $t=T$, after which the maximum velocity is maintained. If T_1 be the whole time for the whole distance and T_2 for the same motor with a flying start, prove that*

$$T=3\,(T_1-T_2) \text{ and } K=\frac{2S}{3T_2(T_1-T_2)}$$

Sol. Let the motor move through a distance x in time t then

$$\ddot{x}=K\left(1-\frac{t}{T}\right)$$

$$\dot{x}=K\left(t-\frac{t^2}{2T}\right)+A, \text{ but } A=0 \text{ since } t=0=\dot{x}$$

$$x=K\left(\frac{t^2}{2}-\frac{t^3}{6T}\right)+B, \text{ but } B=0 \text{ since } t=0=x.$$

When $t=T$, velocity is maximum. say V

$$\therefore \quad V=K\left(T-\frac{T^2}{2T}\right)=\tfrac{1}{2}KT$$

then
$$x=K\left(\frac{T^2}{2}-\frac{T^3}{6T}\right)=\tfrac{1}{3}KT^2$$

the rest of the distance $=S-x=S-\frac{1}{3}KT^2$ is described with velocity $V=\frac{1}{2}KT$,

$$\text{Time required}=\frac{S-\frac{1}{3}KT^2}{\frac{1}{2}KT}=\frac{2(3S-KT^2)}{3KT}$$

Hence
$$T_1=\text{whole time}=T+\frac{2(3S-KT^2)}{3KT}=\frac{6S+KT^2}{3KT}$$

If the motor is running with flying speed *i.e.*, with maximum velocity $\frac{1}{2}KT$ then

$$T_2=\frac{S}{\frac{1}{2}KT}=\frac{2S}{KT}$$

Hence
$$T_1=\frac{2S}{KT}+\frac{T}{3}=T_2+\frac{T}{3}$$

$$\therefore \quad T=3(T_1-T_2)$$

and
$$K=\frac{2S}{TT_2}=\frac{2S}{3(T_1-T_2)T_2}$$

Ex. 6. *If a point moves with constant acceleration, the space-average of the velocity over any distance is* $\frac{2}{3}\cdot\frac{u_1^2+u_1u_2+u_2^2}{u_1+u_2}$ *where* u_1 *and* u_2 *are the initial and final velocities and the time-average is* $\frac{u_1+u_2}{2}$ *and the difference between is* $\frac{1}{6}\frac{(u_1-u_2)^2}{u_1+u_2}$.

Sol. If v be the velocity at a distance s from the starting point and S be the total space covered, then

$$\text{Space-average of velocity}=\frac{1}{S}\int_0^S v\,ds$$

and if T be the total time, time average of velocity $=\frac{1}{T}\int_0^T v\,dt$.

Now if f be the constant acceleration, $v=u_1+ft$ and $v^2=u_1^2+2fs$

so that space-average $=\frac{1}{S}\int_0^S\sqrt{u_1^2+2fs}\,ds$

$$=\frac{1}{3fS}\left[(u_1^2+2fs)^{\frac{3}{2}}\right]_0^S$$

$$=\frac{1}{3fS}\left[(u_1^2+2fS)^{\frac{3}{2}}-u_1^3\right], \text{ but } u_2^2=u_1^2+2fS$$

$$=\frac{2}{3}\cdot\frac{u_2^3-u_1^3}{u_2^2-u_1^2}=\frac{2}{3}\cdot\frac{u_1^2+u_1u_2+u_2^2}{u_1+u_2}$$

and time-average $$=\frac{1}{T}\int_0^T(u_1+ft)dt=\frac{1}{2fT}\left[(u_1+ft)^2\right]_0^T$$

$$=\frac{1}{2fT}\left[(u_1+fT)^2-u_1^2\right], \text{ but } u_2=u_1+f\,T$$

$$=\frac{1}{2}\cdot\frac{u_2^2-u_1^2}{u_2-u_1}=\frac{u_2+u_1}{2}.$$

Difference is easily found out.

EXAMPLES II (a)

1. A bullet travelling horizontally pierces successively three screens placed at equal distances a apart. If the time from first screen to the second be t_1 and from the second to the third be t_2, prove that the retardation (assumed constant) is

$$\frac{2a(t_2-t_1)}{t_1t_2(t_1+t_2)} \text{ and that the velocity at the middle screen is } \frac{a(t_1^2+t_2^2)}{t_1t_2(t_1+t_2)}.$$

2. Two points move in the same straight line starting at the same moment from the same point in the same direction. The first moves with constant velocity u and the second with constant acceleration f (its initial velocity being zero) Show that the greatest distance between the points before the second catches the first is $\frac{u^2}{2f}$ at the end of time f from the start.

3. A particle moves in a straight line with constant acceleration and its distances from an origin O on the line (not necessarily the position at time $t=0$) at times t_1, t_2 t_3, are x_1, x_2, x_3 respectively. Show that if t_1, t_2, t_3 form an *A.P.* whose common difference is d, and x_1, x_2, x_3 are in *G.P.*, then the acceleration is

$$\frac{(\sqrt{x_1}-\sqrt{x_3})^2}{d^2}.$$

4. A point moves with uniform acceleration and v_1, v_2, v_3 denote the average velocities in three successive intervals t_1, t_2, t_3. Prove that

$$\frac{v_1-v_2}{v_2-v_3}=\frac{t_1+t_2}{t_2+t_3}.$$

5. Two points P and Q move in a straight line AB. The point P starts from A in the direction AB with velocity u and acceleration f, and at the same time Q starts from B in the direction of BA with velocity u_1 and acceleration f_1; if they pass one another at the middle point of AB, and arrive at the other ends of AB with equal velocities, prove that

$$(u+u_1)(f-f_1)=8(fu_1-f_1u).$$

6. One man A pursues another man B along a straight road, their original distance apart being a. A starts with velocity u and moves with constant acceleration α whilst B starts from rest and moves with constant acceleration 2α. Prove that, in order that A may catch B, u must be greater than $\sqrt{2a\alpha}$. Find also how far each will have gone then and when B will overtake A.

2·2. Motion of a train between two stations.

A train starts from one station A with a uniform acceleration f and moves for a time t_1 till it gets its maximum speed V and then moves uniformly with speed V for a time t_2 and at the end of this time, brake is applied and the train gets a uniform retardation r with which it moves for a time t_3 and finally comes to rest at the next station B at a distance d.

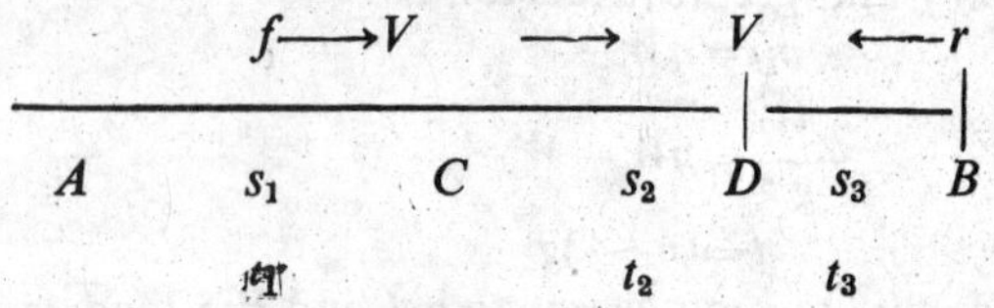

Journey consists of three parts, s_1 during which the motion is accelerated, s_2 during which the motion is uniform and s_3 during which the motion is retarded

$\therefore$ $$d=s_1+s_2+s_3$$

also $$s_1=\tfrac{1}{2}ft_1^2 \text{ and } V=ft_1 \qquad \therefore \quad s_1=\tfrac{1}{2}Vt_1$$

$$s_2=Vt_2$$

$$s_3=Vt_3-\tfrac{1}{2}rt_3^2 \text{ and } 0=V-rt_3$$

$\therefore$ $$V=rt_3 \text{ and } s_3=\tfrac{1}{2}Vt_3.$$

Thus $$d=\tfrac{1}{2}Vt_1+Vt_2+\tfrac{1}{2}Vt_3.$$

Ex. 7. *The speed of a train increases at a constant rate α from 0 to v, then remains constant for an interval and finally decreases to 0 at a constant rate β. If l be the total distance described, prove that the total time occupied is* $\dfrac{l}{v}+\dfrac{v}{2}\left(\dfrac{1}{\alpha}+\dfrac{1}{\beta}\right)$.

Also find the least value of time when $\alpha=\beta$.

Sol. Here $l=\tfrac{1}{2}vt_1+vt_2+\tfrac{1}{2}vt_3$ and $v=\alpha t_1=\beta t_3$

$\therefore$ $$l=v\{t_1+t_2+t_3-\tfrac{1}{2}(t_1+t_3)\}$$

$$=vt-\tfrac{1}{2}v\left(\frac{v}{\alpha}+\frac{v}{\beta}\right)=vt-\frac{v^2}{2}\left(\frac{1}{\alpha}+\frac{1}{\beta}\right)$$

$\therefore$ $$t=\frac{l}{v}+\frac{v}{2}\left(\frac{1}{\alpha}+\frac{1}{\beta}\right)$$

when $\alpha=\beta$, $t=\frac{l}{v}+\frac{v}{\alpha}$. This is least when $\frac{dt}{dv}=0$,

i.e., $$-\frac{l}{v^2}+\frac{1}{\alpha}=0 \text{ i.e., } v^2=l\alpha \quad \text{i.e., } \frac{l}{v}=\frac{v}{\alpha}$$

hence $$t=\frac{l}{v}+\frac{v}{\alpha}=\frac{2v}{\alpha}.$$

Ex. 8. *A load W is to be raised by a rope, from rest to rest, through a height h ; the greatest tension which the rope can safely bear is nW. Show that the least time in which the ascent can be made is*

$$\left[\frac{2nh}{(n-1)g}\right]^{\frac{1}{2}}.$$

Sol. In the first part of the motion, the load moves upwards with an acceleration f and greatest tension nW.

$$\therefore \quad mf=nW-W$$

or $$\frac{W}{g}f=nW-W$$

i.e., $$f=(n-1)g$$

In the second part tension ceases to act and the load is moving under gravity only. Let x_1, x_2 be the distances moved and t_1, t_2 the times taken in the two parts and v the maximum velocity acquired at the end of the first part then

for the first part, $v=ft_1$, $v^2=2fx_1$

for the second part $v=gt_2$, $v^2=2gx_2$

Also $x_1+x_2=h$

$$\therefore \quad \frac{v^2}{2f}+\frac{v^2}{2g}=h$$

or $$\frac{v^2}{2}\left(\frac{1}{f}+\frac{1}{g}\right)=h$$

or $$t_1+t_2=v\left(\frac{1}{f}+\frac{1}{g}\right)=\sqrt{\frac{2h}{\frac{1}{f}+\frac{1}{g}}}\cdot\left(\frac{1}{f}+\frac{1}{g}\right)$$

$$=\sqrt{\left\{2h\left(\frac{1}{f}+\frac{1}{g}\right)\right\}}$$

$$=\sqrt{\left[2h\left\{\frac{1}{(n-1)g}+\frac{1}{g}\right\}\right]}$$

$$=\sqrt{\left[\frac{2nh}{(n-1)g}\right]}$$

EXAMPLES II (B)

1. For $1/m$ of the distance between two stations a train is uniformly accelerated and for $1/n$ of the distance it is uniformly retarded ; it starts from rest at one station and comes to rest at the other. Prove that the ratio of its greatest velocity to its average velocity is $1+\frac{1}{m}+\frac{1}{n}:1$.

2. A body travels a distance s in t seconds. It starts from rest and ends at rest. In the first part of the journey it moves with constant acceleration f and in the second part with constant retardation r. Show that

$$t=\sqrt{2s\left(\frac{1}{f}+\frac{1}{r}\right)}.$$

3. A lift ascends with constant acceleration f then with constant velocity and finally stops under constant retardation f. If the total distance ascended is s and the total time occupied is t, show that the time during which the lift is ascending with constant velocity is

$$\sqrt{\left(t^2-\frac{4s}{f}\right)}.$$

4. Prove that the shortest time from rest to rest in which a steady load of P tons can lift a weight of W tons through a vertical distance h feet is

$$=\sqrt{\frac{2h}{g}\cdot\frac{P}{P-W}}\text{ secs.}$$

5. A stone is dropped from an aeroplane, which is rising with accelera tion f and t seconds after this another stone is dropped. Prove that the distanc between the stones at time T after the second stone is dropped is $\frac{1}{2}(g+f)t$ $(t+2T)$.

[**Hint.** Acceleration of each stone, when dropped relative to the aeroplan $=f+g$. Distance covered by the first stone, relative to the aeroplane in $(t+T)$ sec. $=\frac{1}{2}(f+g)(t+T)^2$ and that by the other one $=\frac{1}{2}(f+g)T^2$.]

2·3. Simple Harmonic Motion.

A particle is said to execute Simple Harmonic Motion if it moves in a straight line such that its acceleration is always directed towards a fixed point in the line and is proportional to the distance of the particle from the fixed point.

Let O be the fixed point on a line BOA and P be the position of the particle at time t where $OP=x$, so that the acceleration of the particle in the sense OP is $\ddot{x}$.

Now the given acceleration is towards O and is proportional to x. Let it be μx, where μ is constant.

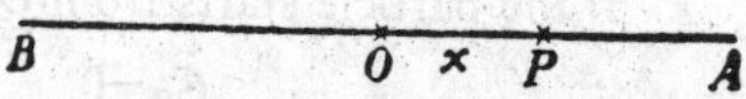

Since $\ddot{x}$ is in the direction of OP produced and μx is towards O, the equation of motion is

$$\ddot{x}=-\mu x.$$

Taking $v\frac{dv}{dx}$ instead of $\ddot{x}$; we can write the above equation as

$$v\frac{dv}{dx}=-\mu x. \qquad ...(1)$$

Integrating with respect to x, we get

$$\frac{v^2}{2}=-\mu\frac{x^2}{2}+\frac{C}{2} \text{ where } C \text{ is a constant}$$

or

$$v^2=-\mu x^2+C.$$

If A be the extreme position of the particle *i.e.*, it is at rest at A *i.e.*, when $x=a$, $v=0$ where $OA=a$, we get

$$0=-\mu a^2+C \quad \therefore \quad C=\mu a^2$$

Hence $$v^2=\mu(a^2-x^2),$$

i.e., $$v=\pm\sqrt{\mu}.\sqrt{a^2-x^2}. \qquad \ldots(2)$$

If the particle moves from A towards O, v is negative

Hence $$\dot{x}=v=-\sqrt{\mu}\sqrt{a^2-x^2}$$

or $$\frac{dx}{dt}=-\sqrt{\mu}\sqrt{a^2-x^2}$$

or $$\sqrt{\mu}.dt=-\frac{dx}{\sqrt{a^2-x^2}}.$$

Integrating we get $\sqrt{\mu}.t=\cos^{-1}\frac{x}{a}+C_1$ where C_1 is a constant. Initially at A, $t=0$, $x=a$ *i.e.*, the particle started from A,

then $$0=\cos^{-1}1+C_1 \quad \therefore \quad C_1=0.$$

Hence $$\sqrt{\mu}.t=\cos^{-1}\frac{x}{a}$$

or $$x=a\cos\sqrt{\mu}.t. \qquad \ldots(3)$$

If the particle moves from O towards A, v is positive

so that $$\dot{x}=\sqrt{\mu}.\sqrt{a^2-x^2}$$

or $$\sqrt{\mu}.dt=\frac{dx}{\sqrt{a^2-x^2}}.$$

Integrating, we get, $\sqrt{\mu}.t=\sin^{-1}\frac{x}{a}+C_2$ where C_2 is a constant.

If the particle starts from O, $t=0$, $x=0$, $0=\sin^{-1}0+C_2$

$$\therefore \quad C_2=0,$$

so that $$x=a\sin\sqrt{\mu}.t. \qquad \ldots(4)$$

Thus the solution of (1) is $x=a\cos\sqrt{\mu}.t$ or $x=a\sin\sqrt{\mu}.t$. according as the starting point is A or O.

From (2), $\quad v=0$ when $x=\pm a$.

Thus if B is a point on the other side of O such that $OB=OA=a$, the particle comes to rest also at B. When $x=0$, $v=\pm\sqrt{\mu}.a$, *i.e.*, at O, the velocity is $\sqrt{\mu}a$.

Consider the solution $x=a\cos\sqrt{\mu}.t$.

The motion starts from A under an attraction towards O. When the particle reaches O, $x=0$. $\therefore \cos\sqrt{\mu}.t=0 \quad \therefore \sqrt{\mu}t=\frac{\pi}{2}$

i.e., $t=\frac{\pi}{2\sqrt{\mu}}$ is the time required in moving from A to O.

As the particle reaches O, the attraction ceases but the particle has a velocity $\sqrt{\mu}.a$ towards the negative side of O hence the particle passes O and moves towards the negative side. As soon as the particle comes to the left side of O, attraction changes direction and becomes towards O ; hence the velocity will go on decreasing as the particle moves towards the left, till at B, the velocity becomes zero so that the particle stops. But the particle is being attracted towards O hence starts moving towards O and reaches O with a velocity $\sqrt{\mu}.a$, due to which it passes O and moves towards A and again stops at A where its velocity becomes zero. The motion is then repeated. Thus the motion is from A to B and back to A and so on. The motion is *oscillatory* Time from O to B is equal to that from A to O hence the **period** *i.e.*, the time from A to B and back to A is $4.\frac{\pi}{2\sqrt{\mu}}=\frac{2\pi}{\sqrt{\mu}}$. The distance $a\,(=OA)$ *i.e.*, the distance of the centre from one of the positions of rest is called the **amplitude.**

Thus the period which is equal to $\frac{2\pi}{\sqrt{\mu}}$ is independent of the amplitude *i.e.*, whatever be the amplitude the period is the same. Thus the simple harmonic motion is oscillatory and periodic, the period being independent of amplitude.

The **frequency** is the number of complete oscillations in one second, so that if n be the frequency and T the periodic time,

$$n=\frac{1}{T}=\frac{\sqrt{\mu}}{2\pi}.$$

Alternative Method. The equation (1), namely $\ddot{x}=-\mu x$, can be solved as a differential equation. The most general solution of this equation is

$$x=A\cos\sqrt{\mu}.t+B\sin\sqrt{\mu}.t \qquad ...(5)$$

A, B are constants to be determined from initial conditions. In the first case when the motion starts from A, the initial conditions are $t=0$, $x=a$, $\dot{x}=0$.

Now $t=0$, $x=a$ gives $a=A$.

Differentiating (5), $\dot{x}=-A\sqrt{\mu}\sin\sqrt{\mu}.t+B\sqrt{\mu}\cos\sqrt{\mu}.t \qquad ...(6)$

The condition $t=0$, $\dot{x}=0$ gives $0=0+B\sqrt{\mu}$ $\therefore$ $B=0$.

Hence the solution is $x=a\cos\sqrt{\mu}t$.

In the second case when the motion starts from O, the first condition is $t=0$, $x=0$. $\therefore$ $0=A$, $A=0$.

Hence $x=B\sin\sqrt{\mu}t$.

To B, we must know the velocity of projection from O.

Let us take the case of a particle, projected from A with velocity V along OA produced, so that the initial conditions are $t=0$, $x=a$, $\dot{x}=V$.

Hence from (5) and (6), we get

$$a=A$$

$$V=B\sqrt{\mu}. \qquad \therefore \quad B=\frac{V}{\sqrt{\mu}}$$

Hence the solution is $x=a\cos\sqrt{\mu}.t+\dfrac{V}{\sqrt{\mu}}\sin\sqrt{\mu}.t.$

Also the general solution of (1) can be written as

$$x=a\cos(\sqrt{\mu}t+\xi)$$

This is periodic with period $\dfrac{2\pi}{\sqrt{\mu}}$.

The quantity ξ is called the **epoch**, the angle $\sqrt{\mu}.t+\xi$ is called the **argument**. The particle is at its maximum distance at time t_0 where $\sqrt{\mu}t_0+\xi=0$ *i.e.*, $t_0=-\dfrac{\xi}{\sqrt{\mu}}$. Hence the time that has elapsed since the particle was at its maximum distance is equal to $t-t_0=t+\dfrac{\xi}{\sqrt{\mu}}=\dfrac{\sqrt{\mu}t+\xi}{\sqrt{\mu}}$. This is the **phase** at time t.

A geometrical representation of the S.H.M.

Let a particle P move on a circle with constant angular velocity ω and let M be the foot of the perpendicular from P on any diameter OA. If a be the radius of the circle, the only acceleration of P is $\omega^2 a$ towards O.

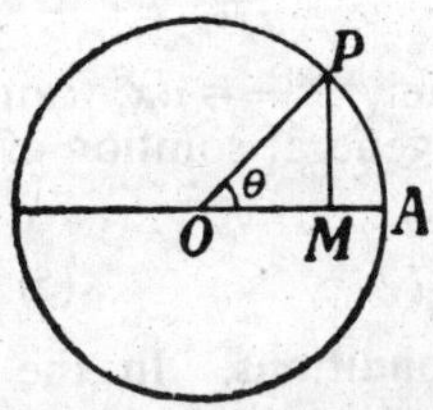

If $\angle AOP=\theta$ and $OM=x$, the component of this acceleration along OA

$$=\omega^2 a.\cos\theta=\omega^2 a.\ \frac{x}{a}=\omega^2 x \text{ towards } O.$$

Hence the equation of motion of the point M is

$$\ddot{x}=-\omega^2 x.$$

This is *S.H.M.*

Thus *if a particle describes a circle with constant angular velocity the foot of the perpendicular from it on any diameter executes a simple harmonic motion.*

Ex. 9. *A particle is moving with S.H.M. and while making an excursion from one position of rest to the other, its distances from the middle point of its path at three consecutive seconds are observed to be* x_1, x_2, x_3, *prove that the time of a complete revolution is*

$$2\pi/\cos^{-1}\left(\frac{x_1+x_3}{2x_2}\right).$$

Sol. From $\qquad x=a \cos \sqrt{\mu}t$

$$x_1=a \cos \sqrt{\mu}t,\ x_2=a \cos \sqrt{\mu}\,(t+1)$$

$$x_3=a \cos \sqrt{\mu}\,(t+2)$$

$$\therefore \quad x_1+x_3=2a \cos \sqrt{\mu}\,(t+1) \cos \sqrt{\mu}$$

$$\therefore \quad \frac{x_1+x_3}{2x_2}=\cos \sqrt{\mu}$$

$$\therefore \quad \sqrt{\mu}=\cos^{-1} \frac{x_1+x_3}{2x_2} \qquad T=\frac{2\pi}{\sqrt{\mu}}$$

Ex. 10. *A particle starts from rest under an acceleration K^2x directed towards a fixed point and after time t another particle starts from the same position under the same acceleration. Show that the particles will collide at time $\frac{\pi}{K}+\frac{t}{2}$ after the start of the first particle provided $t<\frac{2\pi}{K}$.*

Sol. $\qquad \ddot{x}=-K^2x \qquad \therefore \quad \text{period}=\frac{2\pi}{K}$

The condition $t<\frac{2\pi}{K}$ indicates that the second starts before the first has made one complete oscillation. Let them meet after time t' of the start of the second

then $\qquad a \cos K\,(t+t')=a \cos Kt'$

$$\therefore \quad K(t+t')=2\pi-Kt' \quad \therefore \quad t'=\frac{\pi}{K}-\frac{t}{2}.$$

Hence, $\qquad t+t'=\frac{\pi}{K}+\frac{t}{2}.$

Ex. 11. *A horizontal shelf is moved up and down with S.H.M. of period ½ sec. What is the amplitude admissible in order that a weight placed on the shelf may not be jerked off ?*

Sol. $\qquad \frac{2\pi}{\sqrt{\mu}}=\frac{1}{2} \quad \therefore \quad \mu=16\pi^2$

Weight will be jerked off when the max. acc. of *S.H.M.* is greater than g and if it is not to be jerked off, max. acc. of *S.H.M.* must be g

i.e., $\qquad \mu a=g \quad \therefore a=\frac{g}{16\pi^2}$;

Ex. 12. *In a S.H.M. of amplitude a and period T, prove that*

$$\int_0^T v^2 dt = \frac{2\pi^2 a^2}{T}.$$

Sol.
$$x = a\cos\sqrt{\mu}t,\ v = \dot{x} = -a\sqrt{\mu}\sin\sqrt{\mu}t,$$

$$T = \frac{2\pi}{\sqrt{\mu}}$$

$$\int_0^T v^2 dt = a^2\mu\int_0^T \sin^2\sqrt{\mu}t\ dt = a^2\mu\int_0^T \sin^2\frac{2\pi t}{T}dt$$

$$= a^2\mu\int_0^{2\pi}\sin^2 z\ \frac{T}{2\pi}dz \qquad \text{Put } \frac{2\pi t}{T} = z$$

$$= \frac{a^2\mu T}{2\pi}\cdot\pi = \frac{a^2 T}{2}\cdot\frac{4\pi^2}{T^2} = \frac{2\pi^2 a^2}{T}.$$

Ex. 13. *Show that, if a smooth straight tunnel be cut between any two points on the earth's surface; it should be traversed by a particle starting from rest under the action of gravity in about $42\frac{1}{2}$ minutes.*

Sol. Let O be the centre of the earth, AB the tunnel and OC perpendicular from O to AB.

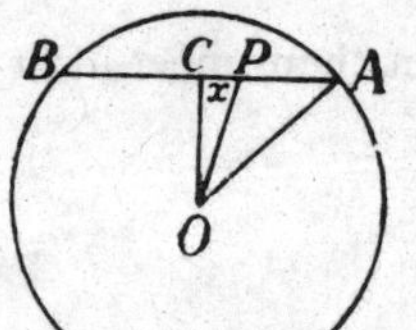

Let P be the position of the particle at time t where $CP = x$.

As the force of attraction inside the earth varies as its distance from the centre, the force on $P = \mu.PO$ along PO.

Its component along $PC = \mu PO\cos OPC$

$$= \mu PO.\frac{x}{PO} = -\mu x$$

Hence the equation of motion of P is $\ddot{x} = -\mu x$.

It is *S.H.M.* of periodic time $T = \dfrac{2\pi}{\sqrt{\mu}}$

$\therefore$ time from A to $B = \frac{1}{2}T = \dfrac{\pi}{\sqrt{\mu}}$.

Now attraction at A is g so that $\mu a = g$, where a = earth's radius

$$\therefore \qquad \mu = \frac{g}{a}$$

$\therefore$ Required time $= \dfrac{\pi}{\sqrt{\mu}} = \pi\sqrt{\dfrac{a}{g}}$

$$=3{\cdot}1416\sqrt{\frac{4000\times1760\times3}{32{\cdot}2}}\text{ secs.}=42\tfrac{1}{2}\text{ minutes nearly.}$$

Ex. 14. *If in a simple harmonic motion u, v, w be the velocities at distances a, b, c from a fixed point on the straight line which is not the centre of force, show that the period T is given by the equation*

$$\frac{4\pi^2}{T^2}(b-c)(c-a)(a-b)=\begin{vmatrix} u^2 & v^2 & w^2 \\ a & b & c \\ 1 & 1 & 1 \end{vmatrix}$$

Sol. Let the intensity of the force be μ and amplitude be λ so that velocity at distance x from centre of force is given by

$$v^2=\mu(\lambda^2-x^2).$$

Let d be the distance of the fixed point from the centre of force.

Hence

$$\begin{aligned} u^2&=\mu[\lambda^2-(a+d)^2] \\ v^2&=\mu[\lambda^2-(b+d)^2] \\ w^2&=\mu(\lambda^2-(c+d^2)] \end{aligned}\quad\text{and}\quad T=\frac{2\pi}{\sqrt{\mu}}\quad\therefore\ \mu=\frac{4\pi^2}{T^2}.$$

Thus

$$\frac{u^2}{\mu}+a^2+2ad+d^2-\lambda^2=0$$

$$\frac{v^2}{\mu}+b^2+2bd+d^2-\lambda^2=0$$

and

$$\frac{w^2}{\mu}+c^2+2cd+d^2-\lambda^2=0$$

Eliminating d and $d^2-\lambda^2$, we get

$$\begin{vmatrix} \dfrac{u^2}{\mu}+a^2 & a & 1 \\ \dfrac{v^2}{\mu}+b^2 & b & 1 \\ \dfrac{w^2}{\mu}+c^2 & c & 1 \end{vmatrix}=0$$

or $$\frac{1}{\mu}\begin{vmatrix} u^2 & a & 1 \\ v^2 & b & 1 \\ w^2 & c & 1 \end{vmatrix}+\begin{vmatrix} a^2 & a & 1 \\ b^2 & b & 1 \\ c^2 & c & 1 \end{vmatrix}=0$$ which gives the answer.

Ex. 15. *A body is attached to one end of an inextensible string and the other end moves in a vertical straight line with n complete oscillations per second. Show that the string will not remain tight during the motion unless*

$$n^2 < \frac{g}{4\pi^2 a}$$

where a is the amplitude of the motion.

Sol. The maximum acceleration of the upper end executing $S.H.M.$ is μa and its period is $T = \frac{\pi}{\sqrt{\mu}}$. Also $T = \frac{1}{n}$.

Hence $\mu = \frac{4\pi^2}{T^2} = 4\pi^2 n^2$

so that the maximum acc. $= 4\pi^2 n^2 a$.

The maximum acc. of the particle is g which is possible when the string is not tight hence the string will not remain tight if the acc. of the upper end is greater than g

i.e., if $4\pi^2 n^2 a > g$

i.e., the string will not remain tight unless $n^2 < \frac{g}{4\pi^2 a}$.

Ex. 16. *A particle P of unit mass free to slide on a straight wire is attracted towards a point O of the wire with a force $\mu.OP$. If the wire be made to revolve about O with constant angular velocity ω in a horizontal plane, show that the motion of the particle on the wire is simple harmonic with a period $\frac{2\pi}{\sqrt{(\mu-\omega^2)}}$ provided $\omega^2 < \mu$; and prove that when $\omega^2 = \frac{1}{2}\mu$, the path of P in space is a circle.*

Sol. When the wire has turned through an angle θ, let P be the position of the particle where $OP = r$ so that attraction on P is μr towards O.

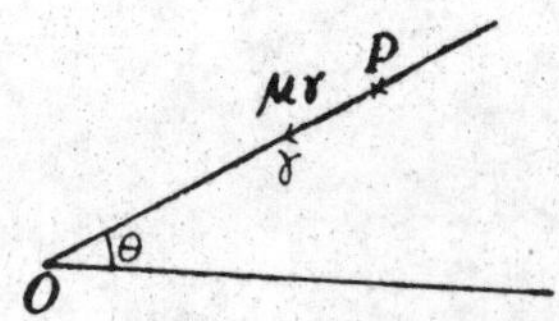

Equation of motion of P is $\ddot{r} - r\dot{\theta}^2 = -\mu r$

But $\dot{\theta} = \text{constant} = \omega \quad \therefore \quad \ddot{r} - r\omega^2 = -\mu r$

or $\ddot{r} = -(\mu - \omega^2) r$

This is a $S.H.M.$ of periodic time

$$\frac{2\pi}{\sqrt{(\mu-\omega^2)}} \text{ if } \mu > \omega^2$$

If $\mu = 2\omega^2$ then the equation becomes $\ddot{r} = -\omega^2 r$

solution of which is $r = a\cos(\omega t + \varepsilon)$

But $\dot{\theta} = \omega \quad \therefore \quad \theta = \omega t + C$. But $C = 0$ since $\theta = 0 = t$,

$\theta = \omega t$,

Hence $r = a\cos(\theta + \varepsilon)$ which is a circle.

EXAMPLES II (C)

1. A particle is performing a $S.H.M.$ of period T about a centre O and it passes through a point $P(OP = b)$ with velocity v in the direction OP; prove that the time which elapses before its return to P is

$$\frac{T}{\pi} \tan^{-1} \frac{vT}{2\pi b}.$$

2. A point in a straight line with *S.H.M.* has velocities v_1 and v_2 when its distances from the centre are x_1 and x_2. Show that the period of motion is

$$2\pi\sqrt{\frac{x_1^2-x_2^2}{v_2^2-v_1^2}}.$$

3. A point executes *S.H.M.* such that in two of its positions the velocities are u, v and the corresponding accelerations are α, β ; show that the distance between the positions is $\frac{v^2-u^2}{\alpha+\beta}$, and the amplitude of the motion is

$$\frac{[(v^2-u^2)(\alpha^2v^2-\beta^2u^2)]^{\frac{1}{2}}}{\beta^2-\alpha^2}.$$

4. Show that in a *S.H.M.* the average speed and the average acceleration (in magnitude) are obtained by multiplying their maximum values by 0·637.

[**Hint.** Average velocity$=\dfrac{\int_0^{T4} \dot{x}\,dt}{T/4}$; average acc.$=\dfrac{\int_0^{T/4} \ddot{x}\,dt}{T/4}$]

5. A body moving in a straight line OAB with *S.H.M.* has zero velocity when at points A and B whose distances from O are a and b respectively and has a velocity v when half-way between them. Show that the complete period is

$$\frac{\pi(b-a)}{v}.$$

6. A particle rests in equilibrium under the attraction of two centres of force which attract directly as the distance, their being μ, μ' ; the particle is slightly displaced towards one of them, show that the time of a small oscillation is

$$\frac{2\pi}{\sqrt{\mu+\mu'}}.$$

7. A particle P moves in a straight line OCP being attracted by a force $m\mu.PC$, always directed towards C whilst C moves along OC with a constant acceleration f. If initially C was at rest at the origin O, and P was at a distance c from O moving with velocity V, prove that the distance of P from O at any time t is

$$\left(\frac{f}{\mu}+c\right)\cos\sqrt{\mu}\,.\,t+\frac{V}{\sqrt{\mu}}\sin\sqrt{\mu}t-\frac{f}{\mu}+\tfrac{1}{2}ft^2.$$

2·4. Hooke's Law

If an elastic string is fixed at one end and pulled at the other, it is found to increase in length. The extension is directly proportional to the product of the tension and the natural length and inversely as the area of the cross-section. Thus if x be the extension, l the natural length, A the area of the cross-section and T the tension, then

$$x=\frac{Tl}{\lambda A} \quad \text{or} \quad T=\lambda\frac{Ax}{l}$$

where λ is a constant depending on the material of the string.

If we take A=unit area, then $T=\lambda\,\dfrac{x}{l}$;

i.e., the tension of the elastic string is proportional to the extension. This is **Hooke's Law** and λ is called the **modulus of elasticity**.

Work done in stretching an elastic string.

Theorem. *The work done against the tension in stretching a light elastic string is equal to the product of its extension and the mean of the initial and final tensions.*

Proof. Let one end of the elastic string of natural length l be fixed at O. Now let the string he extended from A to B, so that $OA=a$ and $OB=b$.

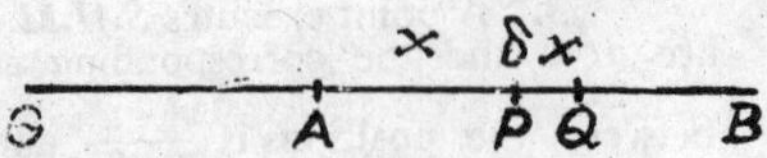

Let $OP=x$ and $PQ=\delta x$. If the end A is stretched upto P, then the tension T in the string is given by $T=\lambda\dfrac{x-l}{l}$ along PO. ......(1)

Now the work done against the tension in stretching the string to a further length δx is $T\delta x$. Hence the total work done in stretching it from length a to length b is given by,

$$\int_a^b T.dx=\frac{\lambda}{l}\int_a^b (x-l)\,dx=(b-a).\tfrac{1}{2}\left[\frac{\lambda}{l}(b-l)+\frac{\lambda}{l}(a-l)\right]$$

$=(b-a).\frac{1}{2}$ [final tension + initial tension]

$=$Extension $\times$ mean of the initial and final tensions.

(*a*) **Horizontal elastic string.** One end of an elastic string is fixed to a point on a smooth horizontal table and to the other end a particle is attached; the particle is pulled out to a distance and then let go; to discuss the motion.

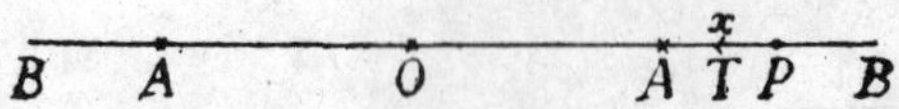

Let O be the fixed point and a the natural length of the string $(=OA)$. Let the particle be pulled out to a point B where $AB=b$ and then released.

Let P be the position of the particle at any subsequent time t, and let $AP=x$, so that x is the extension at time t.

If T be the tension of the string towards A, then by Hooke's Law,

$$T=\lambda\frac{x}{a}, \text{ where } \lambda \text{ is the modulus of elasticity.}$$

If m be the mass of the particle, the equation of motion is

$$m\ddot{x}=-T$$

i.e.,
$$m\ddot{x}=-\lambda\frac{x}{a}$$

or
$$\ddot{x}=-\frac{\lambda}{am}\,x.$$

This shows that the motion is simple harmonic about A.

The period of oscillation $=2\pi\sqrt{\frac{am}{\lambda}}$.

Writing the above equation in the form $v\frac{dv}{dx}=-\frac{\lambda}{am}x$

and integrating with respect to x, we get $\frac{v^2}{2}=-\frac{\lambda}{am}\frac{x^2}{2}+\frac{C}{2}$ where C is a constant

i.e., $$v^2=-\frac{\lambda}{am}x^2+C$$

At the extreme point B where $x=b$, $v=0$

$$\therefore \quad 0=-\frac{\lambda}{am}b^2+C \quad \therefore \quad C=\frac{\lambda}{am}b^2$$

Hence $$v^2=\frac{\lambda}{am}(b^2-x^2).$$

When the particle reaches A, for which $x=0$, $v=\sqrt{\frac{\lambda}{am}}\cdot b$.

As soon as the particle reaches A, the string gets back its natural length, consequently $T=0$ and, therefore, simple harmonic motion ceases but the particle has a velocity $b\sqrt{\frac{\lambda}{am}}$ at A and hence with this velocity the particle moves backwards till it comes to a point A' on the other side of O where $OA'=OA=a$.

This velocity remains constant throughout as the string remains slack and hence the time taken by the particle to move from A to A', a distance $=2a$, is $\frac{2a}{b}\sqrt{\frac{am}{\lambda}}$.

As soon as the particle goes beyond A', the string becomes extended so that the tension comes into play and the motion is again simple harmonic till the particle reaches a point B' where $OB'=OB$, where the motion stops. The particle then retraces its path under $S.H.M.$ till it comes to A' and then under uniform motion till the point A and then again under $S.H.M.$ till the point B where the motion again stops. This motion is repeated.

Thus the motion from B to A, A' to B', B' to A' and A to B is simple harmonic and the total period for which is $2\pi\sqrt{\frac{am}{\lambda}}$

The time from A to A' and A' to A is $\dfrac{4a}{b}\sqrt{\dfrac{am}{\lambda}}$.

Hence the total time for one complete oscillation is

$$2\pi\sqrt{\frac{am}{\lambda}}+\frac{4a}{b}\sqrt{\frac{am}{\lambda}}.$$

Ex. 17. *A particle of mass m executes S.H.M. in the line joining the points A and B on the smooth table and is connected with these points by elastic strings whose tensions in equilibrium are each T; show that the time of an oscillation is* $2\pi\sqrt{\dfrac{mll'}{T(l+l')}}$ *where l, l' are the extensions of the strings beyond their natural lengths.*

Sol. Let a, b be the natural lengths of two strings and O the equilibrium position of the particle.

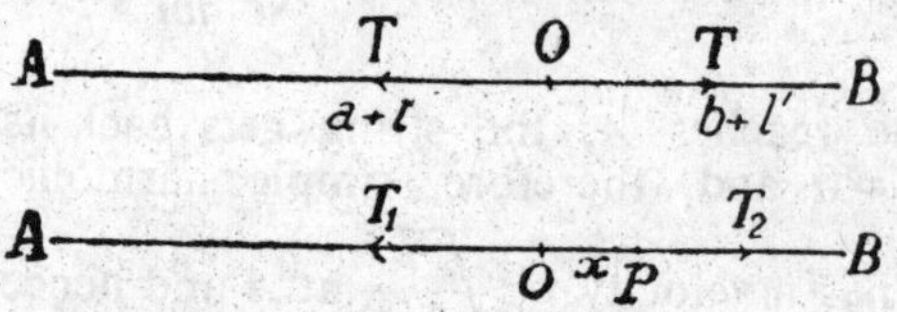

In equilibrium position $T=\lambda_1\dfrac{l}{a}=\lambda_2\dfrac{l'}{b}$ ...(1)

When the particle is slightly displaced from O, let P be its position at time t where $OP=x$.

Then if T_1, T_2 be the tensions in the strings PA and PB, we have

$$T_1=\lambda_1\frac{l+x}{a},\ T_2=\lambda_2\frac{l'-x}{b}.$$

Equation of motion is $m\ddot{x}=T_2-T_1=-x\left(\dfrac{\lambda_1}{a}+\dfrac{\lambda_2}{b}\right)$ from (1)

$$=-Tx\left(\frac{1}{l}+\frac{1}{l'}\right),\text{ also from (1)}$$

i.e., $$\ddot{x}=-\frac{T(l+l')}{mll'}x.$$

This is a $S.H.M.$ of periodic time$=2\pi\sqrt{\dfrac{mll'}{T(l+l')}}$.

Ex. 18 *A particle m is attached to a light wire which is stretched tightly between two fixed points with a tension T. If a, b are the distances of the particle from the two ends, prove that the period of a small transverse oscillation of m is*

$$2\pi\sqrt{\frac{mab}{T(a+b)}}.$$

Sol. When the particle is slightly displaced at right angles to AB and it is at a position P at time t where $OP=x$, the tension T remains practically the same.

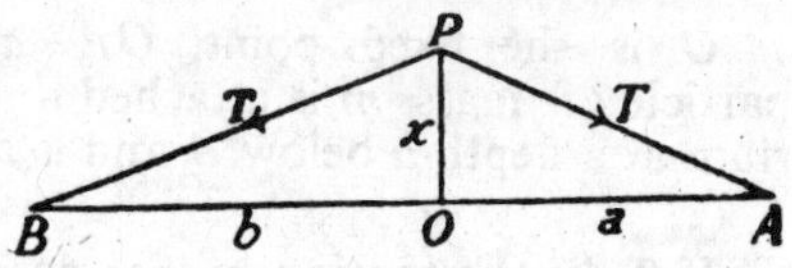

Sum of components of two T's along PO

$$=T.\frac{x}{\sqrt{a^2+x^2}}+T.\frac{x}{\sqrt{b^2+x^2}}$$

$$=T\left(\frac{x}{a}+\frac{x}{b}\right) \text{ neglecting other terms.}$$

Hence the equation of motion is

$$m\ddot{x}=-Tx\left(\frac{1}{a}+\frac{1}{b}\right)=-T\frac{a+b}{ab}x$$

i.e.,
$$\ddot{x}=-\frac{T(a+b)}{mab}x.$$

This is a $S.H.M.$ of periodic time$=2\pi\sqrt{\frac{mab}{T(a+b)}}$.

If the length of the wire be given, *i.e.*, $a+b$ is given, then time period is maximum when ab is maximum, *i.e.*, when $a=b$. Hence for a wire of given length the period is longest when the particle is attached to the middle point.

EXAMPLES II (D)

1. A light elastic string of modulus λ is stretched to double its length and is tied to two fixed points distant $2a$ apart. A particle of mass, m, tied to its middle point, is displaced in the line of the string through a distance equal to half its distance from the fixed points and released. Prove that the time of a complete oscillation is $\pi\sqrt{\frac{am}{\lambda}}$ and the maximum velocity is $\sqrt{\frac{a\lambda}{m}}$ where λ is the modulus of elasticity.

2. A particle of mass m resting on a smooth horizontal plane is attached to two fixed points A and B on the plane by elastic string of unstretched lengths a and b respectively $(a>b)$, the points A and B being $(a+b)$ apart. The particle is held at B and then released. Prove that the particle will oscillate to and fro through a distance $b(\sqrt{a}+\sqrt{b})/\sqrt{a}$ in a periodic time $\pi(\sqrt{a}+\sqrt{b})\sqrt{m/\lambda}$, where λ is the modulus of each string.

3. The particles of masses m_1 and m_2 are tied to the ends of an elastic string of natural length a and modulus λ. They are placed on a smooth table so that the string is just taut and m_2 is projected with any velocity directly away from m_1. Prove that the string will become slack after the lapse of time $\pi[am_1m_2/\lambda(m_1+m_2)]^{\frac{1}{2}}$.

(*b*) **Vertical elastic string.** A particle is suspended from a fixed point by an elastic string ; to discuss the motion.

O is the fixed point, $OA=a$ the natural length of the string. A particle of mass m is attached to the free end and hangs in equilibrium at a depth b below A and is at B so that $AB=b$.

If T_0 be the tension in that position then $T_0=mg$

i.e., $$\lambda\frac{b}{a}=mg \qquad ...(1)$$

The particle is then taken to a depth c below B to the position C so that $BC=c$ and then released.

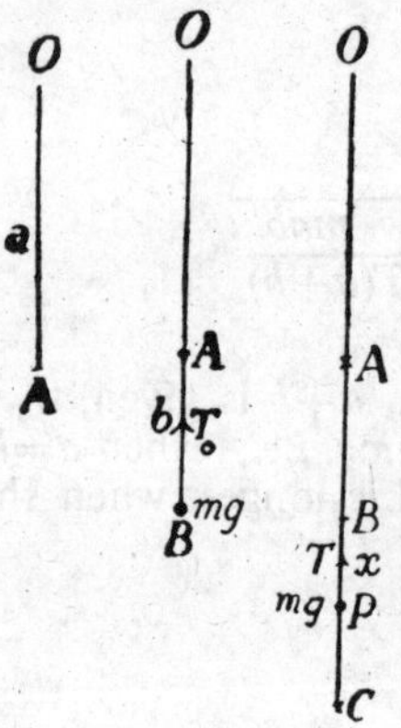

Let P be the position of the particle at any subsequent time t when $BP=x$; so that if T be the tension in that position

$$T=\lambda\frac{b+x}{a}=mg+\frac{\lambda x}{a}, \text{ from (1)}$$

The equation of motion is

$$m\ddot{x}=mg-T=mg-mg-\frac{\lambda x}{a}=-\frac{\lambda x}{a}$$

i.e., $$\ddot{x}=-\frac{\lambda}{am}x. \qquad ...(2)$$

This is a $S.H.M.$ having B, the position of equilibrium, as the centre of oscillation.

$$\text{The periodic time}=2\pi\sqrt{\frac{am}{\lambda}},$$

$$=2\pi\sqrt{\frac{b}{g}} \text{ from (1).}$$

The amplitude of this $S.H.M.$ is c and if $BC<AB$, i.e., $c<b$, the motion is completely simple harmonic about the point B.

If, however, $c>b$, the particle in its upward motion goes above A, but at A the string gets its natural length and so the tension becomes zero hence the $S.H.M.$ ceases, but the particle has a velocity at A with which it rises against gravity.

Now integrating (2), we get

$$v^2=C-\frac{\lambda}{am}x^2=C-\frac{g}{b}x^2 \text{ ; where } C \text{ is a constant}$$

But when $x=c$, $v=0$ $\therefore$ $0=C-\frac{g}{b}c^2$ $\therefore$ $C=\frac{gc^2}{b}$

Hence $$v^2=\frac{g}{b}(c^2-x^2).$$

Hence at A, where $x=-b$, $v^2=\frac{g}{b}(c^2-b^2)$.

This is the velocity at A.

Hence the height through which the particle rises above A is

$$\frac{v^2}{2g}=\frac{c^2-b^2}{2b}.$$

This will be the case provided the height risen above O does not exceed a, the natural length of the string, otherwise the motion again becomes *S.H.M.*

Hence the condition is $\frac{c^2-b^2}{2b}<2a$, *i.e.*, $c^2<b^2+4ab$.

Ex 19. *A heavy particle is attached at one point of a uniform light elastic string. The ends of the string are attached to two points in a vertical line. Show that the period of vertical oscillation in which the string remains taut is $2\pi\sqrt{(mh/2\lambda)}$, where λ is the co-efficient of elasticity of the string and h the harmonic mean of the unstretched lengths of the two parts of the string.*

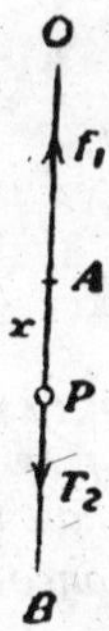

Sol. O, B are the points where the ends are attached and A the point where the particle is attached. Let $OA=a$, $AB=b$ and let l_1, l_2 be the natural lengths of these two portions.

Therefore in equilibrium

$$\lambda\frac{a-l_1}{l_1}=mg+\lambda\frac{b-l_2}{l_2} \qquad \ldots(1)$$

Let the particle be slightly displaced. When it is at P where $AP=x$,

the lengths of the two portions are $a+x$, $b-x$

and T_1 and T_2 be their tensions

$$T_1=\lambda\frac{a+x-l_1}{l_1},\ T_2=\lambda\frac{b-x-l_2}{l_2}$$

Therefore the equation of motion is

$$m\ddot{x}=mg+T_2-T_1=mg+\lambda\frac{b-x-l_2}{l_2}-\lambda\frac{a+x-l_1}{l_1}$$

$$=-\lambda x\left(\frac{1}{l_1}+\frac{1}{l_2}\right)=-\lambda x\frac{l_1+l_2}{l_1l_2}$$

$$=-\frac{2\lambda}{h}x$$

$\therefore \quad \text{period} = 2\pi\sqrt{\dfrac{mh}{2\lambda}}.$

Ex. 20. *Two bodies M and M′ are attached to the lower end of an elastic string whose upper end is fixed and are hung at rest, M′ falls off; show that the distance of M from the upper end of the string at time t is*

$$a+b+c\cos\sqrt{\left(\frac{g}{b}\right)}\,t$$

where a is unstretched length of the string, b and c the distances by which it would be extended when supporting M and M′ respectively.

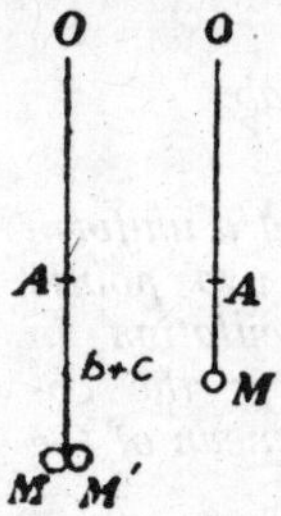

Sol. $Mg=\lambda\dfrac{b}{a},\ M'g=\lambda\dfrac{c}{a}$

$$M\ddot{x}=Mg-T=Mg-\lambda\frac{x}{a}=Mg-\frac{Mg}{b}x$$

$$\therefore \quad \ddot{x}=-\frac{g}{b}(x-b)$$

$$\therefore \quad x-b=A\cos\sqrt{\frac{g}{b}}\,t+B\sin\sqrt{\frac{g}{b}}\,t$$

Initially $\quad x=b+c,\ t=0,\ \dot{x}=0$

$\therefore \quad B=0, \quad A=c$

Hence $\quad x=b+c\cos\sqrt{\dfrac{g}{b}}\,t$

Reqd. depth $\quad =a+x=$etc.

Ex. 21. *A tight elastic string of natural length l has one extremity fixed at a point A and the other attached to a stone the weight of which in equilibrium, would extend the string to a length l_1; show that if the stone be dropped from rest at A, it will come to instantaneous rest at a depth $\sqrt{l_1^2-l^2}$ below the equilibrium position and this depth is attained in time* $\sqrt{\dfrac{2l}{g}}+\sqrt{\dfrac{l_1-l}{g}}\left\{\pi-\cos^{-1}\sqrt{\dfrac{l_1-l}{l_1+l}}\right\}$

Prove also that if the greatest depth below A be $l\cot^2\dfrac{\theta}{2}$, the modulus of elasticity is $\frac{1}{2}mg\tan^2\theta$ and the above time is

$$\sqrt{\frac{2l}{g}}\left[1+(\pi-\theta)\cot\theta\right].$$

Sol. Let P be the position of the particle at time t, where $AP=x$. The equation of motion is

$$m\ddot{x}=mg-T$$

$$=mg-\lambda\frac{x-l}{l}.$$

In equilibrium position, $\ddot{x}=0$, $x=l_1$

$$\therefore \quad 0=mg-\lambda\,\frac{l_1-l}{l}.$$

Hence $$m\ddot{x}=mg-\frac{mg}{l_1-l}(x-l)=-\frac{mg}{l_1-l}(x-l_1)$$

i.e., $$\ddot{x}=-\frac{g}{l_1-l}(x-l_1).$$

Integrating this equation we get

$$\dot{x}^2=-\frac{g}{l_1-l}(x-l_1)^2+C$$

where C is constant.

Now the simple harmonic motion begins when the particle has fallen freely through a distance l so that when $x=l$, $\dot{x}^2=2gl$.

$$\therefore \quad 2gl=-\frac{g}{l_1-l}(l-l_1)^2+C$$

or $$C=2gl-g(l-l_1)=g(l+l_1)$$

Hence $$\dot{x}^2=g(l+l_1)-\frac{g}{l_1-l}(x-l_1)^2 \qquad ...(1)$$

The particle will come to rest when $\dot{x}=0$

$$\therefore \quad (x-l_1)^2=(l_1^2-l^2)$$

i.e., $x-l_1=\sqrt{l_1^2-l^2}$, which is the answer.

Time for free fall is given by $l=\frac{1}{2}gt^2$

i.e., $$t=\sqrt{\frac{2l}{g}}$$

From (1) $$\dot{x}^2=\frac{g}{(l_1-l)}\left\{l_1^2-l^2-(x-l_1)^2\right\}$$

Time from B to the lowest point is given by

$$\therefore \quad \sqrt{\frac{g}{l_1-l}}\,t$$

$$=\int_l^{l_1+\sqrt{l_1^2-l^2}}\frac{dx}{\sqrt{\{l_1^2-l^2-(x-l_1)^2\}}}=\sin^{-1}\frac{x-l_1}{\sqrt{l_1^2-l^2}}\Bigg]_l^{l_1+\sqrt{l_1^2-l^2}}$$

$$=\sin^{-1}1+\sin^{-1}\sqrt{\frac{l_1-l}{l_1+l}}$$

$$=\frac{\pi}{2}+\frac{\pi}{2}-\cos^{-1}\sqrt{\frac{l_1-l}{l_1+l}}=\pi-\cos^{-1}\sqrt{\frac{l_1-l}{l_1+l}}$$

$\therefore$ total time of fall

$$=\sqrt{\frac{2l}{g}}+\sqrt{\frac{l_1-l}{g}}\left\{\pi-\cos^{-1}\sqrt{\frac{l_1-l}{l_1+l}}\right\}$$

Further if the greatest depth below the centre be $l \cot^2 \frac{\theta}{2}$ we have

$$l \cot^2 \frac{\theta}{2} = l_1 + \sqrt{l_1^2 - l^2}$$

or

$$l_1^2 - l^2 = l^2 \cot^4 \frac{\theta}{2} + l_1^2 - 2ll_1 \cot^2 \frac{\theta}{2}$$

$$\therefore \qquad l_1 = \frac{l}{2} \cdot \frac{1 + \cot^4 \frac{\theta}{2}}{\cot^2 \frac{\theta}{2}}.$$

Hence

$$\lambda = \frac{mgl}{l_1 - l} = \frac{2mg \cot^2 \frac{\theta}{2}}{\left(1 - \cot^2 \frac{\theta}{2}\right)^2} = \tfrac{1}{2} mg \tan^2 \theta.$$

Also time of fall $= \sqrt{\frac{2l}{g}} \left[1 + \sqrt{\frac{l_1 - l}{2l}} \left\{ \pi - \cos^{-1} \sqrt{\frac{l_1 - l}{l_1 + l}} \right\} \right]$

Now

$$\sqrt{\frac{l_1 - l}{2l}} = \frac{\cot^2 \theta/2 - 1}{2 \cot \theta/2} = \cot \theta$$

and

$$\sqrt{\frac{l_1 - l}{l_1 + l}} = \left(\frac{\cot^2 \theta/2 - 1}{\cot^2 \theta/2 + 1} \right)^2 = \cos \theta$$

Hence time $= \sqrt{\frac{2l}{g}} \left[1 + (\pi - \theta) \cot \theta \right]$.

Ex. 22. *A particle is attached to the mid-point of a light elastic string of natural length a. The ends of the string are attached to fixed points A and B, A being at a height 2a vertically above B, and in equilibrium the particle rests at a depth $\frac{5a}{4}$ below A. The particle is projected vertically downwards from this position with velocity $\sqrt{ga}$. Prove that the lower string slackens after a time $\frac{\pi}{g} \sqrt{\frac{a}{12}}$ and that the particle comes to rest after a further time $\beta \sqrt{\frac{a}{2g}}$ where β is the acute angle defined by the equation $\tan \beta = \sqrt{\frac{3}{2}}$.*

Sol. In equilibrium position C_0,

$$T_0 = mg + T_0'$$

$$\lambda \frac{\frac{5a}{4} - \frac{a}{2}}{\frac{a}{2}} = mg + \lambda \frac{\frac{3a}{4} - \frac{a}{2}}{\frac{a}{2}}$$

giving $$\lambda = mg$$

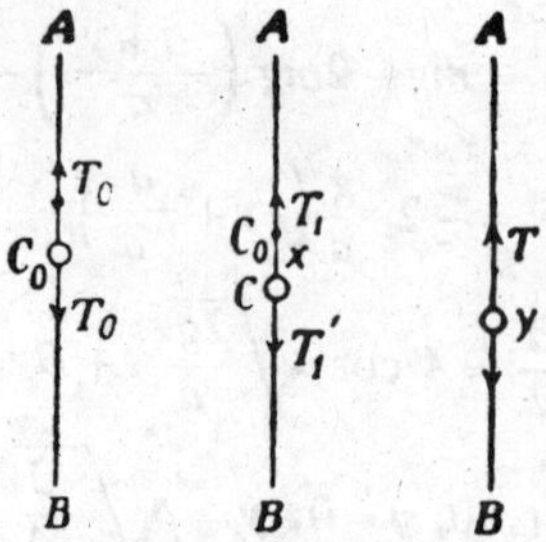

When the particle has gone through x below C_0,

$$m\ddot{x} = mg + T_1' - T_1 = mg + \lambda \frac{\frac{3a}{4} - x - \frac{a}{2}}{\frac{a}{2}} - \lambda \frac{\frac{5a}{4} + x - \frac{a}{2}}{\frac{a}{2}}$$

$$= -4 \frac{mg}{a} x \text{ using the value of } \lambda.$$

$$\therefore \quad \ddot{x} = -\frac{4g}{a} x \qquad \ldots(1)$$

Solution is $\quad x = A \cos 2\sqrt{\frac{g}{a}}\, t + B \sin 2\sqrt{\frac{g}{a}}\, t$

Initially $\quad t = 0,\ x = 0,\ \dot{x} = \sqrt{ga}$

$$\therefore \quad x = \frac{a}{2} \sin 2\sqrt{\frac{g}{a}}\, t,$$

Lower slackens when $T_1' = 0 \quad \therefore \quad x = \frac{a}{4}$

$$\therefore \quad \frac{a}{4} = \frac{a}{2} \sin 2\sqrt{\frac{g}{a}}\, t \text{ hence } t = \frac{\pi}{12}\sqrt{\frac{a}{g}}.$$

Integrating (1) $\dot{x}^2 = \frac{g}{a}(a^2 - 4x^2)$ since

when $\quad x = 0,\ \dot{x} = \sqrt{ga}.$

Therefore when $x = \frac{a}{4}, \quad \dot{x}^2 = \frac{3}{4} ga.$

At the instant the string slackens, its depth

$$=\frac{5a}{4}+\frac{a}{4}=\frac{3}{2}a.$$

After this instant, let the particle go down through y.

then $$m\ddot{y}=mg-T=mg-\lambda\frac{\frac{3}{2}a+y-\frac{a}{2}}{\frac{a}{2}}$$

$$=mg-2mg\left(\frac{a+y}{a}\right)=-mg-2\frac{mg}{a}y$$

$\therefore$ $$\ddot{y}=-2\frac{g}{a}\left(y+\frac{a}{2}\right)$$

Solution is $y+\dfrac{a}{2}=A\cos\sqrt{\dfrac{2g}{a}}\,t+B\sin\sqrt{\dfrac{2g}{a}}\,t.$

Initially $$t=0,\ y=0,\ \dot{y}=\sqrt{\frac{3}{4}ga}$$

$\therefore$ $$A=\frac{a}{2},\ B=\sqrt{\frac{3}{8}}\,a$$

Hence $$\dot{y}=0 \text{ when } \tan\sqrt{\frac{2g}{a}}\,t=\sqrt{\frac{3}{2}}$$

i.e., $$\tan\beta=\sqrt{\frac{3}{2}} \text{ where } \beta=\sqrt{\frac{2g}{a}}\,t \text{ i.e., } t=\beta\sqrt{\frac{a}{2g}}.$$

Ex. 23. *A heavy particle is attached to one end of a fine elastic string, the other end of which is fixed. The unstretched length of the string is a and its modulus of elasticity is n times the weight of the particle. The particle is pulled vertically downwards till the length of the string is a′ and is then let go from rest. Show that the time it returns to this position is* $2\left(\dfrac{a}{ng}\right)^{\frac{1}{2}}(\pi-\theta+\theta'+\tan\theta-\tan\theta')$ *where and* θ' *are positive acute angles given by*

$$\sec\theta=\frac{na'}{a}-n-1,\ \sec^2\theta'=\sec^2\theta-4n.$$

Sol. When the depth of the particle below Q is x, we have

$$m\ddot{x}=mg-T=mg-nmg\,\frac{x-a}{a}$$

i.e., $$\ddot{x}=-\frac{ng}{a}\left(x-\frac{n+1}{n}a\right) \quad \ldots(1)$$

$\therefore$ $$x-\frac{n+1}{n}a=A\cos\sqrt{\frac{ng}{a}}\,t+B\sin\sqrt{\frac{ng}{a}}\,t.$$

When $t=0$, $x=a'$ and $\dot{x}=0$

giving

$$x=\frac{n+1}{n}a+\left(a'-\frac{n+1}{n}a\right)\cos\sqrt{\frac{ng}{a}}t$$

When $x=a$, we have

$$a=\frac{n+1}{n}a+\left(a'-\frac{n+1}{n}a\right)\cos\sqrt{\frac{ng}{a}}.t$$

i.e. $$-1=\left\{\frac{na'}{a}-(n+1)\right\}\cos\sqrt{\frac{ng}{a}}.t. \quad (2)$$

$$=\sec\theta\cos\sqrt{\frac{ng}{a}}.t$$

$$\therefore \quad \cos\sqrt{\frac{ng}{a}}t=-\cos\theta \quad \therefore \quad \sqrt{\frac{ng}{a}}.t=\pi-\theta$$

$$t=\sqrt{\frac{a}{ng}}(\pi-\theta).$$

Integrating (1) with the condition $\dot{x}=0$ when $x=a'$, we have

$$\dot{x}^2=\frac{ng}{a}\left\{\left(a'-\frac{n+1}{n}a\right)^2-\left(x-\frac{n+1}{n}a\right)^2\right\}$$

$\therefore$ when $x=a$, $\dot{x}=u$ say

$$u^2=\frac{ng}{a}\left\{\left(a'-\frac{n+1}{n}a\right)^2-\frac{a^2}{n^2}\right\}=\frac{ag}{n}\tan^2\theta.$$

The particle with the velocity u goes against gravity till a height a above O. Then velocity v is

$$v^2=u^2-4ag=\frac{ag}{n}\tan^2\theta-4ag=\frac{ag}{n}\tan^2\theta'$$

Time to this point is

$$t=\frac{u-v}{g}=\sqrt{\frac{a}{ng}}(\tan\theta-\tan\theta') \quad \ldots(3)$$

The motion thereafter is given by

$$m\ddot{x}=-mg-T=-mg-nmg\frac{x-a}{a}$$

$$\therefore \quad \ddot{x}=-\frac{ng}{a}\left(x-a+\frac{a}{n}\right)$$

$$\therefore \quad x-a+\frac{a}{n}=A\cos\sqrt{\frac{ng}{a}}.t+B\sin\sqrt{\frac{ng}{a}}.t$$

with conditions $t=0,\ x=a,\ \dot{x}=v=\sqrt{\frac{ag}{n}}.\tan\theta'$

$$x-a+\frac{a}{n}=\frac{a}{n}\cos\sqrt{\frac{ng}{a}}t+\frac{a}{n}\tan\theta'\sin\sqrt{\frac{ng}{a}}.t.$$

Hence $\dot{x}=0$ gives

$$\tan\sqrt{\frac{ng}{a}}t=\tan\theta' \ \therefore\ t=\sqrt{\frac{a}{ng}}.\theta'. \quad (4)$$

Hence from (2), (3) and (4), we get the answer being twice the sum of three.

EXAMPLES II (E)

1. An elastic string without weight of which the unstretched length is l and the modulus of elasticity is the weight of n oz. is suspended by one end and a man of m oz. is attached to the other. Show that the time of a small vertical oscillation is $2\pi\sqrt{\dfrac{ml}{ng}}$.

2. One end of an elastic string is fixed and to the other end is fastened a particle heavy enough to stretch the string to double its natural length a. The string is drawn vertically down till it is four times its natural length and then let go. Show that the particle returns to this point in time

$$\sqrt{\frac{a}{g}}\left(2\sqrt{3}+\frac{4\pi}{3}\right).$$

3. A light elastic string of natural length l is hung by one end and to the other end are tied successively particles of masses m and m'. If T and T' be the periods of small oscillations corresponding to these weights and a, a' the statical extensions prove that

$$g(T^2-T'^2)=4\pi^2(a-a').$$

4. A mass m hangs from a fixed point by a light string and is given a small vertical oscillation. If l is the length of the string in equilibrium position and n the number of oscillations per second show that the natural length of the string is

$$\left(l-\frac{g}{4\pi^2n^2}\right).$$

5. A smooth light pulley is suspended from a fixed point by a string of natural length l and modulus of elasticity ng. If masses m_1 and m_2 hang at the ends of a light inextensible string passing round the pulley, show that the pulley executes *S.H.M.*, about a centre whose depth below the point of suspension is

$$l\left(1+\frac{2M}{n}\right)$$

where M is the harmonic mean between m_1 and m_2.

6. Two particles of masses M and $2M$ are connected by an inextensible string passing over a smooth peg. From the particle M another equal particle hangs by an elastic string of natural length a and modulus Mg. The system is initially supported with the strings vertical, the first being taut and the second at its natural length and then released. Show that the motion is *S.H.M.* with period $\pi\ (3a/g)^{\frac{1}{2}}$ and the extension of the second string at time t is $a[1-\cos\{2t\sqrt{g/3a}\}]$.

7. One end of a light elastic string of natural length a and modulus $2mg$ is attached to a point O and the other end to a particle of mass m. The particle initially held at rest at O, is allowed to fall. Find the greatest extension of the string and show that the particle will reach O again after a time t $(\pi+2-\tan^{-1}2)\sqrt{2a/g}$.

2·5. Repulsion from a fixed point varying as the distance from that point.

O is the fixed point on a line and a particle is under a force of repulsion from O. If P be the position of the particle at time t and

$OP = x$, then the force of repulsion varies as x.

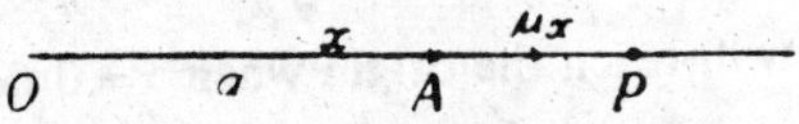

The equation of motion is $\ddot{x} = \mu x$

or $$v\frac{dv}{dx} = \mu x.$$

Integrating we get, $v^2 = \mu x^2 + C$ where C is a constant.

Let the particle start from a point A where $OA = a$

then when $x = a, v = 0$ $\therefore$ $C = -\mu a^2$.

Hence $$v^2 = \mu(x^2 - a^2)$$

or $$v = \frac{dx}{dt} = \sqrt{\mu}\sqrt{x^2 - a^2}$$

or $$\sqrt{\mu}\, dt = \frac{dx}{\sqrt{x^2 - a^2}}.$$

Integrating we get, $\sqrt{\mu}.t = \log \dfrac{x + \sqrt{x^2 - a^2}}{a} + A$

where A is constant

Initially when $t = 0$, $x = a$ $\therefore$ $A = 0$,

Hence $$\log \frac{x + \sqrt{x^2 - a^2}}{a} = \sqrt{\mu}.t$$

or $$x + \sqrt{x^2 - a^2} = ae^{\sqrt{\mu}.t}$$

and, therefore, $$x - \sqrt{x^2 - a^2} = ae^{-\sqrt{\mu}.t}$$

so that $$x = \frac{a}{2}\left(e^{\sqrt{\mu}.t} + e^{-\sqrt{\mu}.t}\right) = a \cosh \sqrt{\mu}.t.$$

Suppose the particle be projected from the point A with velocity V towards O.

Now the general solution of the equation $\ddot{x} = \mu x$ is

$$x = A \cosh \sqrt{\mu}t + B \sinh \sqrt{\mu}.t,$$

where A, B are two constants.

The conditions are when $t = 0$, $x = a$ and $\dot{x} = -V$.

The first condition $t = 0$, $x = a$ gives $a = A$.

Differentiating we get $\dot{x} = A\sqrt{\mu} \sinh \sqrt{\mu}.t + B\sqrt{\mu} \cosh \sqrt{\mu}.t$.

The condition $t = 0$, $\dot{x} = -V$ gives $-V = B\sqrt{\mu}$ $\therefore$ $B = -\dfrac{V}{\sqrt{\mu}}$

Hence $$x=a\cosh\sqrt{\mu}.t-\frac{V}{\sqrt{\mu}}\sinh\sqrt{\mu}.t.$$

The particle will reach the origin when $x=0$ *i.e.*, when

$$\tanh\sqrt{\mu}.t=\frac{a\sqrt{\mu}}{V}.$$

In particular if $V=a\sqrt{\mu}$, $\tanh\sqrt{\mu}.t=1$ gives $t=\infty$.

Thus the particle will always be moving towards the origin but will take an infinite time to reach it, *i.e.*, will never reach it.

Ex. 24. *A particle moves along the axis of x starting from rest at $x=a$; for an interval t_1 from the beginning of the motion the acceleration is $-\mu x$, for a subsequent time t_2 the acceleration is μx, and at the end of this interval the particle is at the origin again; prove that*

$$\tan(t_1\sqrt{\mu})\tanh(t_2\sqrt{\mu})=1$$

Sol.

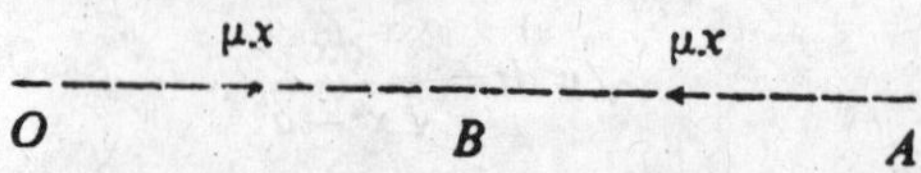

Let A be the starting point, O the origin, $OA=a$; B the point such that time from A to B is t_1 and time from B to O is t_2 and let $OB=b$.

The motion from A to B is given by $x=a\cos\sqrt{\mu}.t$ so that time A to B is given by $b=a\cos\sqrt{\mu}.t_1$, and the velocity at B is V where

$$V^2=\mu(a^2-b^2).$$

The motion from B to O is given by

$$x=b\cosh\sqrt{\mu}t-\frac{V}{\sqrt{\mu}}\sinh\sqrt{\mu}.t$$

Hence time from B to O is given by

$$0=b\cosh\sqrt{\mu}.t_2-\frac{V}{\sqrt{\mu}}\sinh\sqrt{\mu}.t_2$$

$$\therefore\qquad \tanh\sqrt{\mu}.t_2=\frac{b\sqrt{\mu}}{V}=\frac{b}{\sqrt{a^2-b^2}}.$$

Also since $\cos\sqrt{\mu}.t_1=\dfrac{b}{a}$, $\tan\sqrt{\mu}.t_1=\dfrac{\sqrt{a^2-b^2}}{b}$.

Hence $$\tan(\sqrt{\mu}.t_1)\tanh(\sqrt{\mu}.t_2)=\frac{\sqrt{a^2-b^2}}{b}\times\frac{b}{\sqrt{a^2-b^2}}=1.$$

2·6. Motion under Inverse square law. A particle moves in a straight line under an attraction towards a fixed point on the line, varying inversely as the square of the distance from the fixed point ; to discuss the motion.

Let O be the fixed point on the line and P the position of the particle at time t, where $OP=x$. If μ/x^2 be the acceleration of the force of attraction on P, so the equation of motion is

$$\ddot{x}=-\frac{\mu}{x^2} \text{ or } v\frac{dv}{dx}=-\frac{\mu}{x^2}$$

Integrating with respect to x, we get

$$v^2=\frac{2\mu}{x}+C \text{ where } C \text{ is a constant}$$

If the particle starts from rest from a point A where $OA=a$;

then when $x=a$, $v=0$, $\therefore\ C=-\frac{2\mu}{a}$

Hence $$v^2=2\mu\left(\frac{1}{x}-\frac{1}{a}\right)$$

or $$v=\pm\sqrt{2\mu}\sqrt{\frac{1}{x}-\frac{1}{a}}, \qquad \ldots(1)$$

Since the particle is proceeding towards O and $\dot{x}$ is positive in the direction OP, we have $\dot{x}=-\sqrt{2\mu}\sqrt{\frac{1}{x}-\frac{1}{a}}$.

i.e., $$\frac{dx}{dt}=-\sqrt{2\mu}\sqrt{\frac{a-x}{ax}}$$

or $$\sqrt{\frac{2\mu}{a}}\,.\,dt=-\frac{\sqrt{x}}{\sqrt{a-x}}\,dx$$

To integrate this, put $x=a\cos^2\theta$

so that $$dx=-2a\cos\theta\sin\theta d\theta$$

Hence $$\sqrt{\frac{2\mu}{a}}\,dt=\frac{\sqrt{a}\cos\theta}{\sqrt{a}\sin\theta}.\,2a\cos\theta\sin\theta d\theta=2a\cos^2\theta d\theta$$

$$=a(1+\cos 2\theta)d\theta.$$

Integrating we get

$$\sqrt{\frac{2\mu}{a}}\,t=a\left\{\theta+\frac{\sin 2\theta}{2}\right\}+C_1, \text{ where } C_1 \text{ is a constant.}$$

Initially when $t=0$, $x=a$, $\quad\therefore\quad \theta=0$, $\quad\therefore\quad C_1=0$.

Hence $$t=\sqrt{\frac{a}{2\mu}}\left[a\cos^{-1}\sqrt{\frac{x}{a}}+\sqrt{ax-x^2}\right] \qquad \ldots(2)$$

Therefore, the time taken by the particle to reach O is given by putting $x=0$, $\quad t=\sqrt{\dfrac{a}{2\mu}}\cdot a\dfrac{\pi}{2}=\dfrac{\pi a^{3/2}}{2\sqrt{2\mu}}.$

From (1), the velocity at $x=0$ is infinity. Thus the particle reaches the origin with an infinite velocity. As soon as the particle passes O and goes to the left of it, the attraction is directed towards O and the velocity goes on decreasing as the particle moves away from O and ultimately the velocity becomes zero again at a distance a from O to the left where the particle comes to rest. It then retraces its path, passes through O and comes to rest at A. The motion is repeated. The motion is thus oscillatory about O and the period of one complete oscillation is

$$4.\frac{\pi a^{\frac{3}{2}}}{2\sqrt{2\mu}}=\frac{2\pi a^{\frac{3}{2}}}{\sqrt{2\mu}}.$$

Illustration. Earth attracts a body outside its surface with a force varying inversely as the square of the distance from the centre and a body inside its surface with a force varying directly as the distance from the centre.

From equation (1) above, the velocity at a distance x from centre is given by $\quad v^2=2\mu\left(\dfrac{1}{x}-\dfrac{1}{a}\right).$

If the particle comes from infinity, a is infinite $\quad\therefore\quad v^2=\dfrac{2\mu}{x}.$

On the earth's surface, attraction is $g \quad\therefore\quad \dfrac{\mu}{r^2}=g \quad\therefore\quad \mu=r^2g$

where r is the earth's radius.

Hence $\qquad v^2=\dfrac{2r^2g}{x}.$

Therefore, on reaching the earth's surface, the velocity of the particle is given by $\qquad v^2=\dfrac{2r^2g}{r}=2gr.$

i.e., $\qquad v=\sqrt{(2gr)}=\sqrt{(2\times 32\times 4000\times 1760\times 3)}$ ft. per sec.

$\qquad =7$ miles per second nearly.

Thus a particle coming from infinity will reach the earth's surface with a velocity nearly equal to 7 miles per sec. If the problem be inverted then we can say that a particle thrown away from the earth's surface with a velocity of 7 miles per second will never return to the surface of the earth.

Suppose the particle goes inside the earth's surface through a hole. As soon as the particle goes inside, the attraction becomes λx.

But on the earth's surface, attraction is $g \therefore \lambda r=g$, hence $\lambda=\dfrac{g}{r}$

$\therefore$ Attraction on the particle at distance x from the centre is

$$\frac{gx}{r}.$$

Hence the equation of motion is $\ddot{x}=-\frac{gx}{r}$

or $$v\frac{dv}{dx}=-\frac{gx}{r}.$$

Integrating we get $v^2=-\frac{gx^2}{r}+C$ ...(3)

where C is a constant. But for a particle coming from infinity, velocity when it reaches the earth is given by $v^2=2gr$.

Hence when $\quad x=r,\ v^2=2gr$

$\therefore \quad 2gr=-gr+C \quad \therefore \quad C=3gr$

$\therefore \quad v^2=3gr-\frac{gx^2}{r}.$

When the particle reaches the centre of the earth, $x=0$

$\therefore \quad v=\sqrt{3gr}.$

If the particle comes from a distance a outside the earth's surface its velocity is given by $v^2=2\mu\left(\frac{1}{x}-\frac{1}{a}\right)$ where $\mu=r^2g$.

$\therefore$ On reaching the earth's surface, $v^2=2r^2g\left(\frac{1}{r}-\frac{1}{a}\right)$

Hence for the particle going inside the earth's surface we have from (3) $2r^2g\left(\frac{1}{r}-\frac{1}{a}\right)=-\frac{gr^2}{r}+C \quad \therefore \quad C=gr\left(3-\frac{2r}{a}\right)$

$$\therefore \quad v^2=-\frac{gx^2}{r}+gr\left(3-\frac{2r}{a}\right)$$

On reaching the centre of the earth, $v^2=gr\left(3-\frac{2r}{a}\right)$

$$v=\sqrt{gr\left(3-\frac{2r}{a}\right)}.$$

Ex. 25. *If h be the height due to the velocity v at the earth's surface, supposing its attraction constant and H the corresponding height when the variation of gravity is taken into account, prove that $\frac{1}{h}-\frac{1}{H}=\frac{1}{r}$ where r is the earth's radius.*

Sol. When the attraction is constant and is equal to g, $v^2=2gh$.

When the attraction is $\frac{\mu}{x^2}$ where $\frac{\mu}{r^2}=g$

the equation of motion is $v\frac{dv}{dx}=-\frac{r^2g}{x^2}$

so that $$v^2=\frac{2r^2g}{x}+C.$$

If the particle starts from a height H so that when $x=r+H$, $v=0$.

$$\therefore \quad C=-\frac{2r^2g}{r+H}$$

$$\therefore \quad v^2=2r^2g\left(\frac{1}{x}-\frac{1}{r+H}\right).$$

On reaching the earth's surface $v^2=2r^2g\left(\frac{1}{r}-\frac{1}{r+H}\right)$

$$\therefore \quad 2gh=2r^2g\left(\frac{1}{r}-\frac{1}{r+H}\right)$$ which gives the answer.

Ex. 26. *Show that the time of descent to the centre of force, the force varying inversely as the square of the distance from the centre, through the first half of its initial distance is to that through the last half as* $\pi+2 : \pi-2$.

Sol. $$v\frac{dv}{dx}=-\frac{\mu}{x^2}$$

$$\therefore \quad v^2=\frac{2\mu}{x}+C$$

when $$x=a,\ v=0 \quad \therefore \quad C=-\frac{2\mu}{a}$$

Hence $$v^2=2\mu\left(\frac{1}{x}-\frac{1}{a}\right)$$

or $$\frac{dx}{dt}=-\sqrt{2\mu}\sqrt{\frac{1}{x}-\frac{1}{a}}=-\sqrt{2\mu}\sqrt{\frac{a-x}{ax}}$$

or $$\sqrt{\frac{2\mu}{a}}\,dt=-\sqrt{\frac{x}{a-x}}\,dx.$$

If t_1 be the time from a to $\frac{a}{2}$ and t_2 from a to 0, then

$$\sqrt{\frac{2\mu}{a}}\cdot t_1=-\int_a^{a/2}\sqrt{\frac{x}{a-x}}\cdot dx=\int_{a/2}^{a}\sqrt{\frac{x}{a-x}}dx$$

$$=\frac{a}{4}(\pi+2), \text{ putting } x=a\sin^2\theta,$$

and $\sqrt{\frac{2\mu}{a}}\cdot t_2=-\int_a^0\sqrt{\frac{x}{a-x}}\cdot dx=\int_0^a\sqrt{\frac{x}{a-x}}\,dx=a.\frac{\pi}{2}$

$\therefore$ time from $\frac{a}{2}$ to $0=t_2-t_1=a\sqrt{\frac{a}{2\mu}}\left(\frac{\pi}{2}-\frac{\pi}{4}-\frac{1}{2}\right)$

$$=a\sqrt{\frac{a}{2\mu}}\left(\frac{\pi}{4}-\frac{1}{2}\right)=\frac{a}{4}\sqrt{\frac{a}{2\mu}}(\pi-2).$$

Hence the required ratio $=\frac{\pi+2}{\pi-2}$.

EXAMPLES II (F)

1. A particle is projected vertically upwards from the earth's surface with a velocity just sufficient to carry it to infinity. Prove that the time it takes in reaching a height h is

$$\frac{1}{3}\sqrt{\frac{2a}{g}}\left\{\left(1+\frac{h}{a}\right)^{\frac{3}{2}}-1\right\},$$ where a is the earth's radius.

2. A body is falling from a height h to the earth's surface. Find its velocity on reaching the surface and show that the corresponding time is

$$\sqrt{\frac{a+h}{2a}}\left[\frac{a+h}{a}\sin^{-1}\sqrt{\frac{h}{a+h}}+\sqrt{\frac{h}{a}}\right].$$

3. A particle falls towards the earth from infinity; show that its velocity, on reaching the surface of the earth, is the same as that which it would have acquired in falling with constant acceleration through a distance equal to the earth's radius.

2·7. Motion under other laws of force.

Ex. 27. *A particle moves in a straight line from a distance a towards the centre of force, the force varying inversely as the cube of the distance; show that the time of descent to the centre is*

$$\frac{a^2}{\sqrt{\mu}}.$$

Sol. The equation of motion is $\ddot{x}=-\frac{\mu}{x^3}$

or $$v\frac{dv}{dx}=-\frac{\mu}{x^3}$$

Integrating $$v^2=\frac{\mu}{x^2}+C$$

when $x=a$, $v=0$ $$C=-\frac{\mu}{a^2}.$$

Hence $v^2=\mu\left(\frac{1}{x^2}-\frac{1}{a^2}\right)=\frac{\mu(a^2-x^2)}{a^2x^2}.$

$$\therefore \quad \frac{dx}{dt}=-\sqrt{\mu}.\frac{\sqrt{a^2-x^2}}{ax}$$

since the particle is going towards the origin.

$$\therefore \quad \sqrt{\mu}.dt=-\frac{ax}{\sqrt{a^2-x^2}}dx.$$

Integrating, if T be the time from $x=a$ to $x=b$, we have

$$\sqrt{\mu}.T=-\int_a^b \frac{ax}{\sqrt{a^2-x^2}}\,dx=\left[a\sqrt{a^2-x^2}\right]_a^b=a\sqrt{a^2-b^2}$$

$$\therefore \qquad T=\frac{a\sqrt{a^2-b^2}}{\sqrt{\mu}}.$$

If $b=0$, *i.e.*, time to the origin$=\dfrac{a^2}{\sqrt{\mu}}$.

Ex. 28. *If the acceleration be* $\dfrac{\mu}{x^{5/3}}$, *find the time of descent to the centre.*

Here $\quad v\dfrac{dv}{dx}=-\dfrac{\mu}{x^{5/3}}$.

Integrating $\quad v^2=\dfrac{3\mu}{x^{2/3}}+C,$

when $x=a$, $v=0$ $\quad \therefore \; C=-\dfrac{3\mu}{a^{2/3}}$

Hence $\quad v^2=3\mu\left(\dfrac{1}{x^{2/3}}-\dfrac{1}{a^{2/3}}\right)=3\mu\,\dfrac{a^{2/3}-x^{2/3}}{a^{2/3}\,x^{2/3}}$

$$\therefore \qquad \frac{dx}{dt}=-\sqrt{3\mu}\frac{\sqrt{a^{2/3}-x^{2/3}}}{a^{1/3}\,x^{1/3}}$$

or $\quad \sqrt{3\mu}.dt=-\dfrac{a^{1/3}\,x^{1/3}\,dx}{\sqrt{(a^{2/3}-x^{2/3})}}$.

If T be the time a to 0, then

$$\sqrt{3\mu}.T=-\int_a^0 \frac{a^{1/3}\,x^{1/3}}{\sqrt{a^{2/3}-x^{2/3}}}\,dx=\int_0^a \frac{a^{1/3}\,x^{1/3}}{\sqrt{a^{2/3}-x^{2/3}}}\,dx$$

Put $x=a\cos^3\theta$ so that $dx=-3a\cos^2\theta\sin\theta d\theta$

we get $\sqrt{3\mu}T=\displaystyle\int_0^{\pi/2} 3a^{4/3}\cos^3\theta\,d\theta=2a^{4/3}$ $\quad \therefore \quad T=\dfrac{2a^{4/3}}{\sqrt{3\mu}}$.

Ex. 29. *If the law of attraction be inverse distance, find the time of descent to the centre.*

$$v\frac{dv}{dx}=-\frac{\mu}{x}.$$

Integrating $\quad v^2=-2\mu\log x+C$

when $\quad x=a, v=0$

$\therefore \quad C=2\mu\log a \quad \therefore \; v^2=2\mu\log\dfrac{a}{x}$

$$\therefore \qquad \frac{dx}{dt}=-\sqrt{2\mu}\sqrt{\log\frac{a}{x}}$$

or
$$\sqrt{2\mu}.dt=-\frac{dx}{\sqrt{\log\frac{a}{x}}}.$$

If T be the time from a to 0, then

$$\sqrt{2\mu}.T=-\int_a^0\frac{dx}{\sqrt{\log\frac{a}{x}}}.$$

Put $\log\frac{a}{x}=y^2$ or $x=ae^{-y^2}$ $\therefore$ $dx=-2ay\,e^{-y^2}dy$

so that $\sqrt{2\mu}.T=\int_0^\infty 2ae^{-y^2}dy=2a.\frac{\sqrt{\pi}}{2}=a\sqrt{\pi}$

$$\therefore \qquad T=a\sqrt{\frac{\pi}{2\mu}}.$$

Ex. 30. *If the law of attraction be* $\mu\left(x+\frac{a^4}{x^3}\right)$, *show that the time of descent to the centre is* $\pi/4\sqrt{\mu}$.

then Here $v\frac{dv}{dx}=-\mu\left(x+\frac{a^4}{x^3}\right).$

so that $v^2=\mu\left(-x^2+\frac{a^4}{x^2}\right)+C$

When $x=a,\ v=0$ $\therefore$ $C=0$

Hence $\frac{dx}{dt}=-\sqrt{\mu}\sqrt{\frac{a^4}{x^2}-x^2}$

$$\therefore \quad \sqrt{\mu}.T=-\int_a^0\frac{x\,dx}{\sqrt{a^4-x^4}}=\tfrac{1}{2}\cos^{-1}\frac{x^2}{a^2}\Big]_a^0=\frac{1}{2}\cdot\frac{\pi}{2}$$

$$\therefore \qquad T=\frac{\pi}{4\sqrt{\mu}}.$$

EXAMPLES II (G)

1. Prove that it is impossible for a particle to move from rest so that its velocity varies as the distance described from the commencement of the motion. If the velocity as (distance)^n show that n cannot be greater than $\frac{1}{2}$.

2. A particle moves in a straight line under a force to a point in it varying as (distance)$^{-\frac{4}{3}}$; show that the velocity in falling from rest at infinity to a distance a is equal to that acquired in falling from rest at a distance a to a distance $a/8$.

3. A particle moves in a straight line ; its acceleration is directed towards a fixed point O in the line and is always equal to $\mu\left(\frac{a^5}{x^2}\right)^{\frac{1}{3}}$, when it is at a distance x from O. If it starts from rest at a distance a from O, show that it will arrive at O with velocity $a\sqrt{6\mu}$ after time

$$\frac{8}{15}\sqrt{\frac{6}{\mu}}.$$

4. A particle is attracted by a force to a fixed point varying inversely as the (distance)^n. If the velocity acquired in falling from the infinite distance to a distance a from the centre be equal to the velocity acquired in falling from rest at a distance a to a distance $\frac{a}{4}$, then $n=\frac{3}{2}$.

5. A particle moves in a straight line with an acceleration towards a fixed point in a straight line which is equal to $\frac{\mu}{x^2}-\frac{\lambda}{x^3}$ at a distance x from the given point ; the particle starts from rest at a distance a ; show that it oscillates between that distance and the distance $\frac{\lambda}{2\mu a-\lambda}$ and the periodic time is

$$\frac{2\pi\mu a^3}{(2a\mu-\lambda)^{\frac{3}{2}}}.$$

6. Show that the curve which is such that a particle will slide down each of its tangents to the horizontal axis in a given time is a cycloid whose axis is vertical.

7. A parabola is placed with its axis vertical and vertex downwards ; prove that the time of falling down any chord to the vertex is equal to the time of falling vertically through a space equal to the parallel focal chord.

8. A particle of mass m moving in a straight line acted upon by an attractive force which is expressed by $\frac{m\mu x}{a}$ for $x<a$ and by $\frac{m\mu a^2}{x^2}$ for $x>a$, from a fixed origin in the line. If the particle starts from rest at a distance $2a$ from the origin, prove that it will reach the origin with a velocity $\sqrt{2\mu a}$ and the time taken is $\sqrt{\frac{a}{\mu}}\left(\frac{3\pi}{4}+1\right)$.

3

Uniplanar Motion

PROJECTILE

3·1. The position of a particle in a plane is known if we know its co-ordinates referred to two fixed axes in the plane. Thus two equations of motion will be required in order to obtain the two co-ordinates which define the position of the particle in the plane. The two equations of motion can be obtained by resolving the forces in any two convenient directions at right angles to one another.

If the two directions are taken parallel to the co-ordinate axes, the equations of motion, as deduced from the second law of motion will be of the form

$$m\ddot{x}=X, \text{ and } m\ddot{y}=Y,$$

where X, Y are sums of the resolved parts of the forces acting on the particle of mass m, parallel to the axes of x and y.

Integrating each of these equations twice, we have two equations containing four constants which can be determined from the initial conditions of the problem. From these two equations we then eliminate t and obtain a relation between x and y which is the equation to the path of the particle.

Ex. 1. *A particle moves freely in a semi-circle under a force perpendicular to the bounding diameter ; show that the force varies inversely as the cube of the ordinate to the diameter.*

Sol. Take the given diameter as x-axis and the perpendicular diameter as y axis. The equations of motion are

$$\ddot{x}=0, \text{ and } \ddot{y}=P$$

if P be the force per unit mass.

Equation to the circle is

$$x^2+y^2=a^2$$

where a is the radius.

Differentiating the equation with respect to t, we get

$$x\dot{x}+y\dot{y}=0.$$

But $\ddot{x}=0 \quad \therefore \quad \dot{x}=\text{const.}=c$ say hence $\dot{y}=-\dfrac{cx}{y}$.

Differentiating again with respect to t, we get

$$x\ddot{x}+\dot{x}^2+y\ddot{y}+\dot{y}^2=0$$

$$\therefore \quad 0+c^2+yP+\frac{c^2x^2}{y^2}=0$$

$$\therefore \quad P=-\frac{c^2\left(1+\dfrac{x^2}{y^2}\right)}{y}=-\frac{c^2(x^2+y^2)}{y^3}=-\frac{c^2a^2}{y^3}.$$

The negative sign shows that the force is directed towards the bounding diameter.

Ex. 2. *A particle is acted on by a force parallel to the axis of y whose acceleration is λy and is initially projected with a velocity $a\sqrt{\lambda}$ parallel to the axis of x at the point where $y=a$; prove that it will describe a catenary.*

Sol. The equations of motion are

$$\ddot{x}=0 \text{ and } \ddot{y}=\lambda y.$$

Integrating the first, we get $\dot{x}=\text{const}$.

Initially when $t=0$, $\dot{x}=a\sqrt{\lambda}$ $\therefore$ $\dot{x}=a\sqrt{\lambda}$

Again integrating, we get, $x=a\sqrt{\lambda}.\, t+C$.

Initially when $t=0$, $x=0$, $\therefore$ $C=0$ hence $x=a\sqrt{\lambda}\,.\,t$.

The solution of the second equation $\ddot{y}=\lambda y$ is

$$y=A\cosh\sqrt{\lambda}.t+B\sinh\sqrt{\lambda}.t$$

where A, B are constants.

Initially when $t=0$, $y=a$, $\dot{y}=0$ $\therefore$ $A=a$, $B=0$.

Hence $y = a \cosh \sqrt{\lambda} \,.\, t.$

Eliminating t, we get $y = a \cosh \dfrac{x}{a}$

which is the required path. This is a catenary.

Ex. 3. *A straight smooth tube revolves with angular velocity ω in a horizontal plane about one extremity which is fixed ; if at zero time a particle inside it be at a distance a from the fixed end and moving with a velocity V along the tube, show that its distance at time t is a $\cosh \omega t + \dfrac{V}{\omega} \sinh \omega t$.*

$$\ddot{r} - r\dot{\theta}^2 = 0$$

i.e., $$\ddot{r} = r\omega^2$$

its solution is $$r = A \cosh \omega t + \sinh \omega t.$$

Initially $t = 0$, $$r = a, \ \dot{r} = V$$

$\therefore$ $$A = a, \ B = \frac{V}{\omega} \text{ hence.}$$

EXAMPLES III (A)

1. Prove that the parabola $y^2 = 4ax$ can be described under a constant force parallel to the axis of y and a force proportional to y parallel to the axis of x.

2. A particle is acted upon by a force parallel to the axis of y whose acceleration (always towards the axis of x) is $\dfrac{\mu}{y^2}$; and when $y = a$, it is projected parallel to the axis of x with velocity $\sqrt{\dfrac{2\mu}{a}}$, prove that it will describe a cycloid.

3. A particle moves in a plane with an acceleration which is parallel to the axis of y and varies as the distance from the axis of x ; show that the equation to the path is of the form $y = Aa^x + Ba^{-x}$ when the acceleration is a repulsion.

If the acceleration is attraction then the equation is of the form

$$y = A \cos (ax + B).$$

4. If a rectangular hyperbola be described under a force parallel to an asymptote the magnitude of the force varies inversely as the cube of the distance from the other asymptote.

Projectile

3·2. If a particle is thrown into vacuum then the particle which is thrown or projected is called the *projectile* ; the angle which the direction of projection makes in the plane of projection with the horizontal plane through the point of projection is called the *angle of projection* ; the path which the projectile describes is called the *trajectory* ; the time taken by the particle to come back to the horizontal plane through the point of projection again is called the *time of flight* and the distance between the point of projection and the point where the particle strikes the horizontal plane again is called the *range*.

To find the position of such a particle at any time t, we take the point of projection as origin, the horizontal and vertical lines through the point of projection as axes of x and y, and let P be the position of the particle at time t so that its co-ordinates referred to those axes are x and y.

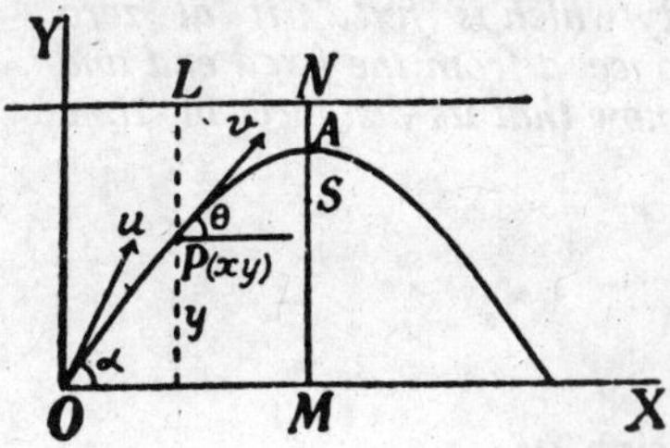

If u is the velocity of projection and α the angle of projection, then initially components of velocity are $u \cos \alpha$ horizontally and $u \sin \alpha$ vertically.

During the course of motion of the particle, it is being acted on by its weight only which is a downward force equal to mg, where m is the mass of the particle. Components of acceleration of the particle are $\ddot{x}$ horizontally and $\ddot{y}$ vertically upwards hence the equations of motion are

$$m\ddot{x}=0 \text{ and } m\ddot{y}=-mg$$

i.e., $$\ddot{x}=0 \qquad ...(1)$$

and $$\ddot{y}=-g. \qquad ...(2)$$

To find the path of the particle we will have to find x and y from these equations. Now integrating these equations with respect to t, we can find x and y both in terms of t ; so eliminating t between these equations we can get an equation involving x and y only. That will, therefore, be the path of the particle.

From (1) we have on integration

$$\dot{x}=\text{constant}=c_1 \text{ say.}$$

To find c_1, we use the initial condition namely when it was projected, we had

$$t=0,\ \dot{x}=\text{horizontal component of velocity}=u\cos\alpha$$

Hence $c_1=u\cos\alpha \quad \therefore \quad \dot{x}=u\cos\alpha \qquad ...(3)$

If we again integrate this equation with respect to t, we get

$$x=u\cos\alpha.\, t+c_2 \text{ where } c_2 \text{ is a constant.}$$

To find c_2 we again use initial condition namely when $t=0$, $x=0$. Putting these values in the above equation we find $c_2=0$.

Hence $$x=u\cos\alpha.t. \qquad ...(4)$$

Similarly from (2), we have by integrating with respect to t

$$\dot{y}=-gt+c_3 \text{ where } c_3 \text{ is a constant.}$$

To find c_3, we again use initial condition namely when $t=0$, $\dot{y}$ = vertical component of velocity $=u \sin \alpha$.

$\therefore$ Using these in the above equation we get

$$c_3=u \sin \alpha \text{ so that}$$

$$\dot{y}=u \sin \alpha-gt \qquad \ldots(5)$$

To find y, we again integrate with respect to t so that

$$y=u \sin \alpha\, t-\tfrac{1}{2}gt^2+c_4 \text{ where } c_4 \text{ is a constant.}$$

Again to find c_4, we use initial conditions namely $t=0$, $y=0$ so using these in the above equation we find $c_4=0$.

Hence $$y=u \sin \alpha t-\tfrac{1}{2}gt^2 \qquad \ldots(6)$$

The equations (3) and (5) give the components of velocity of the particle at time t, we notice that the horizontal velocity $\dot{x}$ is always constant and equal to $u \cos \alpha$.

The equations (4) and (6) give the values of x and y in terms of t so that the path can be found by eliminating t from these equations so that we have from (4),

$$t=\frac{x}{u \cos \alpha}.$$

Hence from (6), $y=u \sin \alpha \dfrac{x}{u \cos \alpha}-\tfrac{1}{2}g\left(\dfrac{x}{u \cos \alpha}\right)^2$

i.e., $$y=x \tan \alpha-\tfrac{1}{2}g \frac{x^2}{u^2 \cos^2 \alpha} \qquad \ldots(7)$$

This is, therefore, the equation to the path.

This is seen to be a parabola with axis vertical and vertex upwards. *Thus the path of the projectile in vacuum i.e., the trajectory is a parabola.*

The equation (7) can be written as

$$x^2-\frac{2u^2 \sin \alpha \cos \alpha}{g}x=-\frac{2u^2 \cos^2 \alpha}{g}y$$

$$\left(x-\frac{u^2 \sin \alpha \cos \alpha}{g}\right)^2=-\frac{2u^2 \cos^2 \alpha}{g}\left(y-\frac{u^2 \sin^2 \alpha}{2g}\right) \qquad \ldots(8)$$

Thus the vertex is the point $x=\dfrac{u^2 \sin \alpha \cos \alpha}{g}$, $y=\dfrac{u^2 \sin^2 \alpha}{2g}$

and **latus rectum** $=\dfrac{2u^2 \cos^2 \alpha}{g}$.

The negative sign on the right hand side of (8) shows that the concavity of the parabola is downwards.

The focus and the directrix. The vertex A and the focus S are on the same vertical line ASM such that

$$AS=\frac{1}{4}\text{(latus rectum)}=\frac{1}{4}\cdot\frac{2u^2\cos^2\alpha}{g}=\frac{u^2\cos^2\alpha}{2g}.$$

Hence the co-ordinates of the focus S are

$$x=\text{same as that of } A=\frac{u^2\sin\alpha\cos\alpha}{g}=\frac{u^2\sin 2\alpha}{2g}$$

$$y=SM=AM-AS=\frac{u^2\sin^2\alpha}{2g}-\frac{u^2\cos^2\alpha}{2g}=-\frac{u^2\cos 2\alpha}{2g}.$$

Also the directrix of the parabola is a line LN, parallel to OMB at a distance AN from A, equal to AS on the other side of S, hence, its equation is

$$y=NM=AM+AN=\frac{u^2\sin^2\alpha}{2g}+\frac{u^2\cos^2\alpha}{2g}=\frac{u^2}{2g}.$$

This is independent of α hence so long as the point of projection is the same and u the same, height of the directrix is the same, that is, *if particles are projected from the same point with the same velocity the directrix of each path will be the same.*

If S be the focus, then OS=distance of O from the directrix $=\frac{u^2}{2g}$. Also the tangent at O which is the direction of u, bisects the angle between OS and OY hence

$$\angle SOX=\alpha-\left(\frac{\pi}{2}-\alpha\right)=2\alpha-\frac{\pi}{2}.$$

Time to the highest point. In the vertical direction the particle is moving under a retardation g so that the vertical velocity goes on decreasing. After a certain time t_1 say, it must be zero so that at that instant the particle has no vertical velocity, only velocity is horizontal and equal to $u\cos\alpha$. At that instant the particle is at the highest point of its path. Hence putting

$$\dot{y}=0 \text{ in } \dot{y}=u\sin\alpha-gt \text{ and } t=t_1$$

we get
$$t_1=\frac{u\sin\alpha}{g}.$$

This is the time to the highest point.

Time of flight. By definition time of flight is the time taken by the particle to come back to the horizontal plane again, *i.e.*, if T

be the time of flight then when $t=T$, $y=0$. Putting these in the equation $y=u \sin \alpha t-\frac{1}{2}gt^2$

we get
$$T=\frac{2u \sin \alpha}{g}.$$

Evidently $T=2t_1$ proving thereby that the time taken by the particle from the point of projection to the highest point is equal to that taken from highest point to the horizontal plane again.

Range. By definition it is the distance between the point of projection and the point where the particle strikes the plane again hence it is the distance the particle moves horizontally in time T. Since the horizontal velocity is constant and equal to $u \cos \alpha$, we have
$$\text{range}=u \cos \alpha . T=u \cos \alpha \frac{2u \sin \alpha}{g}=\frac{u^2 \sin 2\alpha}{g}.$$

This is maximum when $2\alpha=\pi/2$ or $\alpha=\frac{\pi}{4}$.

That is, the *range is maximum* when the angle of projection is 45° *i.e.*, the direction of projection bisects the angle between the horizontal and the vertical and the maximum range is $\frac{u^2}{g}$.

Also when $\alpha=\frac{\pi}{4}$, the co-ordinates of the focus S are
$$x=\frac{u^2}{2g},\ y=0$$
i.e., the focus lies on the horizontal plane.

To show that for a given velocity of projection, there are, for a given horizontal range, in general, two directions of projection which are equally inclined to the direction of maximum range.

For given u and α, the horizontal range is $\frac{u^2 \sin 2\alpha}{g}$. If u remains the same but the angle of projection becomes $\frac{\pi}{2}-\alpha$ then the horizontal range will be $\frac{u^2}{g} \sin 2\left(\frac{\pi}{2}-\alpha\right)=\frac{u^2}{g} \sin 2\alpha$.

Hence for the same range we have two angles of projection, α and $\frac{\pi}{2}-\alpha$. These two angles are obviously equally inclined to the angle $\frac{\pi}{4}$ *i.e.*, the angle of maximum range.

Greatest height. The greatest height H is obtained at that point where the vertical velocity is zero, evidently after time

$$t_1 = \frac{u \sin \alpha}{g}.$$

Hence $H = u \sin \alpha t_1 - \frac{1}{2} g t_1^2 = \dfrac{u^2 \sin^2 \alpha}{2g}$.

Velocity at any height. Let v be the velocity at angle θ with the horizon. It is evidently along the tangent to the path at the point. Then

$$v \cos \theta = \dot{x} = u \cos \alpha$$

$$v \sin \theta = \dot{y} = u \sin \alpha - gt$$

$\therefore$ $$\tan \theta = \frac{\dot{y}}{\dot{x}} = \frac{u \sin \alpha - gt}{u \cos \alpha}$$

and $$v^2 = u^2 - 2u\, gt \sin \alpha + g^2 t^2$$

$$= u^2 - 2g(u \sin \alpha t - \tfrac{1}{2} g t^2) = u^2 - 2gy.$$

To show that the velocity of a particle at any point of its path is equal in magnitude to that which would be acquired by it in falling freely through the height from the directrix to that point i.e., the velocity is due to a fall from the directrix.

We have $$v^2 = u^2 - 2gy = u^2 - 2g(MN - PL)$$

$$= u^2 - 2g\left(\frac{u^2}{2g} - PL\right) = 2g.PL$$

Ex. 4. *If particles are projected from the same point and in the same plane so as to describe equal parabolas, show that the vertices of their paths lie on a parabola.*

Sol. For equal parabolas, latus rectum is the same; hence for all paths

$$u \cos \alpha = \text{constant} = k \text{ say.} \qquad \ldots(i)$$

Now vertex of one path is

$$x = \frac{u^2 \sin \alpha \cos \alpha}{g}, \quad y = \frac{u^2 \sin^2 \alpha}{2g}$$

Hence from (i)

$$x = \frac{ku \sin \alpha}{g} \text{ so that } y = \frac{g}{2k^2} x^2$$

which is a parabola.

Ex. 5. *If at any instant the velocity of the projectile be u and its direction of motion α to horizon, then it will be moving at right angles to this direction after the time* $\frac{u}{g}$ *cosec α.*

Sol. Let the velocity at any point be v at an angle θ to the horizontal then $v \cos \theta = u \cos \alpha$

$$v \sin \theta = u \sin \alpha - gt$$

$$\therefore \quad \tan \theta = \frac{u \sin \alpha - gt}{u \cos \alpha}.$$

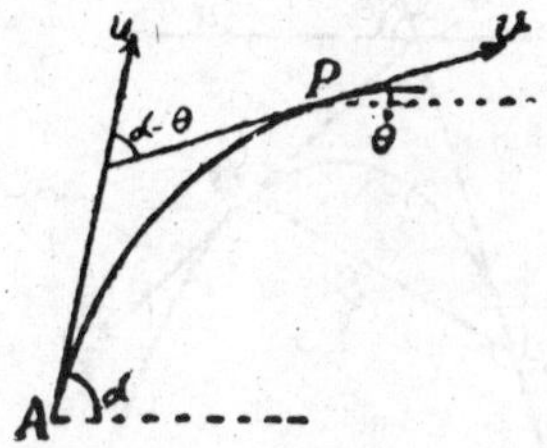

The angle between the two directions is $(\alpha - \theta)$ $\therefore$ the condition of moving at right angles is $\alpha - \theta = \pi/2$ or $\theta = \alpha - \pi/2$.

$$\text{Hence} \quad -\cot \alpha = \frac{u \sin \alpha - gt}{u \cos \alpha} = \tan \alpha - \frac{gt}{u \cos \alpha}$$

$$\therefore \quad \frac{gt}{u \cos \alpha} = \tan \alpha + \cot \alpha = \frac{1 + \tan^2 \alpha}{\tan \alpha} = \frac{\sec^2 \alpha}{\tan \alpha}$$

$$\therefore \quad t = \frac{u}{g} \cdot \frac{\sec \alpha}{\tan \alpha} = \frac{u}{g \sin \alpha}.$$

Ex. 6. *If α be the angle between the tangents at the extremities of any arc of a parabolic path, v and v′ the velocities at those extremities and u the velocity at the vertex of the path, show that the time of describing the arc is* $\frac{vv' \sin \alpha}{gu}$.

Sol. If θ, θ' be the angles v, v' make with the horizontal.

$$v' \cos \theta' = v \cos \theta = u$$

$$v' \sin \theta' = v \sin \theta - gt$$

$\therefore \quad vv' \cos \theta \sin \theta' = v' \cos \theta' (v \sin \theta - gt) = vv' \cos \theta' \sin \theta - v' \cos \theta' gt$

or $\quad vv' (\sin \theta \cos \theta' - \cos \theta \sin \theta') = v' \cos \theta' gt = u\, gt$

i.e., $\quad vv' \sin (\theta - \theta') = ugt$

i.e., $\quad vv' \sin \alpha = ugt$, hence $t = \frac{vv' \sin \alpha}{ug}$.

Ex. 7. *If* v_1, v_2 *be the velocities at the ends of a focal chord of a projectile's path and u the horizontal component of the velocity show*

that
$$\frac{1}{v_1^2}+\frac{1}{v_2^2}=\frac{1}{u^2}.$$

Sol. Let PQ be a focal chord. The tangents at P, Q meet on the directrix at right angles. If θ be the angle v_1 makes with the horizontal then v_2 makes $\angle(90-\theta)$ with the horizontal hence resolving horizontally we have

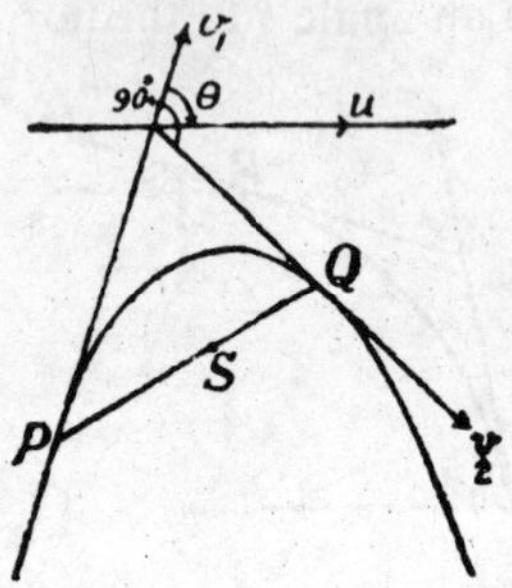

$$v_1 \cos \theta = u$$
$$v_2 \sin \theta = u$$
$$\therefore \quad \frac{u^2}{v_1^2}+\frac{u^2}{v_2^2}=\cos^2\theta+\sin^2\theta=1$$
i.e.,
$$\frac{1}{v_1^2}+\frac{1}{v_2^2}=\frac{1}{u^2}.$$

Ex. 8. *Three particles are projected simultaneously and in the same vertical plane from a point with velocities u, v, w in directions making angles α, β, γ with the horizontal. Show that the area of the triangle formed by the particles at any time t is proportional to the square of the time elapsed from the instant of projection, and that the three particles will always lie in the same straight line if*
$$\frac{\sin(\beta-\gamma)}{u}+\frac{\sin(\gamma-\alpha)}{v}+\frac{\sin(\alpha-\beta)}{w}=0.$$

Sol. If (x_2, y_1), (x_2, y_2), (x_3, y_3) be the positions of the particles at time t then
$$x_1=u\cos\alpha t,\ y_1=u\sin\alpha t-\tfrac{1}{2}gt^2$$
$$x_2=v\cos\beta t,\ y_2=v\sin\beta t-\tfrac{1}{2}gt^2$$
$$x_3=w\cos\gamma t,\ y_3=w\sin\gamma t-\tfrac{1}{2}gt^2.$$

Area of the triangle $=\tfrac{1}{2}[x_1(y_2-y_3)+x_2(y_3-y_1)+x_3(y_1-y_2)]$
$$=\tfrac{1}{2}[u\cos\alpha t(v\sin\beta-w\sin\gamma)t+v\cos\beta t(w\sin\gamma-u\sin\alpha)t+w\cos\gamma t(u\sin\alpha-v\sin\beta)t]$$
$$=\tfrac{1}{2}t^2[u\,v\sin(\gamma-\beta)+\ldots\ldots]$$

If the particles are in a straight line, the area vanishes hence the result.

Ex. 9. *Three tangents to a projectile's path form a triangle ABC the velocities are u along BC, v along CA and w along AB. Prove that*
$$\frac{BC}{u}+\frac{CA}{v}+\frac{AB}{w}=0$$

Sol. Let BC, CA, AB make angles α, β, γ with the forward direction of BD, the horizontal line through B. Since the horizontal components of the velocity is always the same

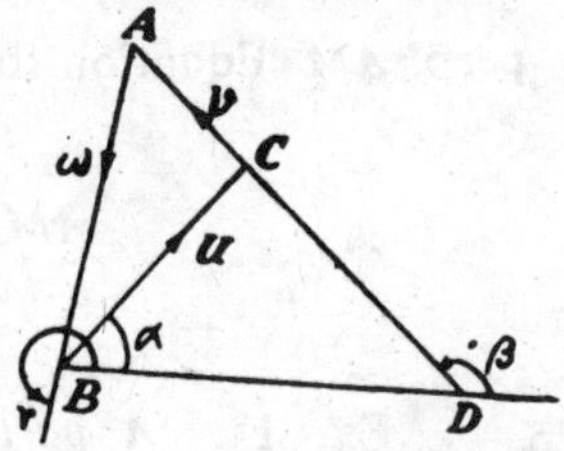

$$u \cos \alpha = v \cos \beta = w \cos \gamma = K \text{ say} \quad ...(1)$$

Projecting BC, CA, AB along BD, we have

$$BC \cos \alpha + CA \cos \beta + AB \cos \gamma = 0$$

hence from (1)

$$\frac{BC}{u} + \frac{CA}{v} + \frac{AB}{w} = 0$$

Ex. 10. *A and B are projectiles describing parabolic paths under gravity. Prove that relative to A, B appears to describe a straight line uniformly.*

Two particles are projected at different times from the same point with the same velocity of projection and in the same direction. The first is above the level of the point of projection when the second is projected. Prove that the two particles are closest together when the line joining them is horizontal.

If A and B are observed at one instant to be in a vertical line at a distance a apart and at another instant they are seen to be in a horizontal line at a distance b apart, prove that the shortest distance that occurred between them during this interval was $\dfrac{ab}{\sqrt{a^2+b^2}}$, *A and B moving in the same vertical plane.*

Sol The horizontal velocities of A and B are constant throughout. Both A and B have the same vertical acceleration namely g hence their relative vertical acceleration is zero that is, their relative vertical velocity is constant. Hence relative to A, B always has a constant velocity hence relative to A, B moves in a straight line.

When A and B are projected from the same point with the same velocity and the same direction but at different times their horizontal velocities are the same hence B has no relative horizontal velocity with respect to A. Its vertical velocity relative to A is constant hence B moves in a vertical line relative to A. The shortest distance between them is the perpendicular distance from A on the path of B which is a vertical line i.e., A and B are nearest when they are horizontal.

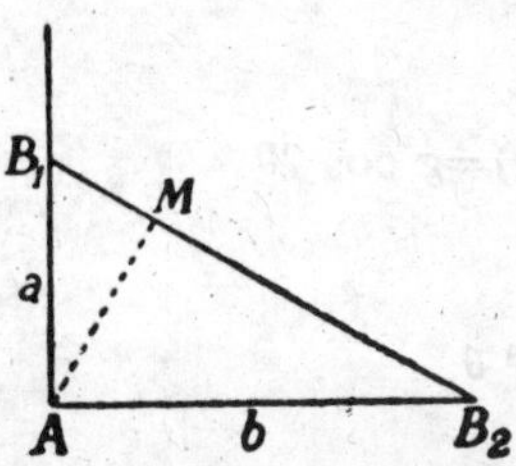

B_1 and B_2 are the positions of B relative to A when in a vertical line and in a horizontal line so that

$AB_1=a$, $AB_2=b$. Thus relative to A, B describes the hypotenuse of the right angled triangle $AB_1\,B_2$ and the shortest distance is the perp. AM. Equation to B_1B_2 with A as origin is $\frac{x}{b}+\frac{y}{a}=1$.

$$\therefore \qquad AM=\frac{1}{\sqrt{\frac{1}{a^2}+\frac{1}{b^2}}}=\frac{ab}{\sqrt{a^2+b^2}}.$$

Ex. 11. *A particle moving under gravity describes a parabola whose focus is S. Prove that when the particle is at P where $SP=r$, the components of its velocity along and perpendicular to SP are respectively equal to the vertical and horizontal components of its velocity. Show also that the component in the direction SP of the acceleration of the particle is $\left(g-\frac{u^2}{r}\right)$ where u is the velocity at the vertex.*

Sol. v is the velocity along the tangent PT, its components are v_1, v_2 along PX (horizontally) and PY (vertically) respectively and u_1, u_2 along SPP' and perpendicularly.

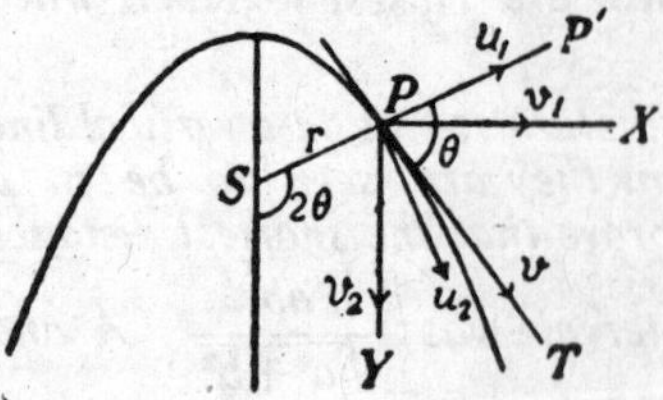

Resolving v along PP' and perpendicularly

$$v\cos\theta=u_1,$$
$$v\sin\theta=u_2$$

Now $\quad \angle\theta=\angle TPP'=TPY$
$$=90°-\angle TPX$$

Resolving v along PX and PY,

$$v\cos TPX=v_1 \text{ and } v\sin TPX=v_2$$

i.e., $\qquad v\sin\theta=v_1$

and $\qquad v\cos\theta=v_2$

Hence $\qquad u_2=v_1$

and $\qquad u_1=v_2$

Acceleration along SP

$$=-g\cos(180-2\theta)=g\cos 2\theta$$
$$=g(1-2\sin^2\theta)$$

but $\qquad \frac{2a}{r}=1-\cos 2\theta=2\sin^2\theta$

and $\qquad u^2=2ga$

Hence acc. along $SP=g\left(1-\frac{2a}{r}\right)$

$$=g-\frac{2ag}{r}=g-\frac{u^2}{r}.$$

EXAMPLES III (B)

1. If t be the time in which a projectile reaches a point P in its path and t' the time from P till it reaches the horizontal plane through the point of projection, show that the height of P above a horizontal plane is $\frac{1}{2}gtt'$.

2. If the focus of a trajectory lies as much below the horizontal plane through the point of projection as the vertex is above, prove that the angle of projection is given by $\sin\alpha=1/\sqrt{3}$.

3. Particles are projected simultaneously in the same vertical plane from the same point. Show that the locus of the foci of all the trajectories is a parabola when (*i*), the horizontal component of the velocity for each is the same, (*ii*) the initial vertical component of the velocity is the same, and (*iii*) the time of flight of each is the same.

4. A number of particles is projected simultaneously from the same point with equal velocities in the same vertical plane, in different directions. Prove that the foci of all the paths lie on a circle with the point of projection as centre.

5. Particles are projected from a point O in a vertical plane with velocity $\sqrt{2gk}$; prove that the locus of the vertices of their paths is the ellipse $x^2+4y(y-k)=0$.

6. A number of particles start simultaneously from the same point in all directions in a vertical plane with the same speed u; show that after time t they will all lie on a circle of radius ut. Show also that the radius increases uniformly with the time and that the centre of the circle descends with acceleration g.

7. A particle is projected with velocity u and elevation $\alpha\left(<\frac{\pi}{4}\right)$ from a point O situated at a height h above the horizontal plane. If the particle is moving at right angles to its direction of projection when it hits the plane after time t, show that $t=\frac{u}{g\sin\alpha}$ and $h=\frac{u^2\cos 2\alpha}{2g\sin^2\alpha}$.

8. Prove that the time of describing any portion PQ of the parabolic path of a projectile is proportional to the difference of the tangents of the angle which the linear tangents at P and Q make with the horizon.

9. Prove that if lines are drawn from a point representing in magnitude and direction the velocity of projectile at successive points of its paths, their extremities lie upon a vertical straight line.

10. Show that the velocity at any point of a trajectory is that due to a fall from a height equal to one quarter of the focal chord parallel to the tangent at the point.

11. Prove that the angular velocity of a projectile about the focus of its path varies inversely as its distance from the focus.

12. A gun is fired from the sea level out to sea. It is then mounted in a battery h feet higher up and fired at the same elevation α. Show that the range is increased by

$$\frac{1}{2}\left\{\left(1+\frac{2gh}{v^2\sin^2\alpha}\right)^{\frac{1}{2}}-1\right\}$$ of itself, v being the velocity of projection.

13. Three particles are projected from the same point in the same vertical plane with velocities v_1, v_2, v_3 at elevations $\beta_1, \beta_2, \beta_3$. Prove that the foci of their paths will be in a straight line if

$$\frac{\sin 2(\beta_1-\beta_2)}{v_1^2}+\frac{\sin 2(\beta_3-\beta_1)}{v_2^2}+\frac{\sin 2(\beta_1-\beta_2)}{v_3^2}=0.$$

14. Show that the greatest height which a particle with initial velocity v can reach in a vertical wall at a distance a from the point of projection is

$$\frac{v^2}{2g}-\frac{a^2g}{2v^2}.$$

15. A shell bursts on contact with the ground and fragments fly in all directions with all speed upto 80 ft/sec. Show that a man 100 feet away is in danger for $\frac{5\sqrt{2}}{2}$ seconds.

3·3. Projection to pass through a given point.

Let a particle be projected from a point O so as to pass through a given point A. If the co-ordinates of A referred to O as origin and horizontal and vertical lines as axes be (a, b) then these co-ordinates satisfy the equation

$$y=x\tan\alpha-\tfrac{1}{2}\frac{gx^2}{u^2\cos^2\alpha}$$

$$\therefore \quad b=a\tan\alpha-\tfrac{1}{2}\frac{ga^2}{u^2\cos^2\alpha}$$

i.e.,
$$b=a\tan\alpha-\tfrac{1}{2}\frac{ga^2}{u^2}(1+\tan^2\alpha)$$

i.e.,
$$\tan^2\alpha-\frac{2u^2}{ga}\tan\alpha+1+\frac{2u^2b}{ga^2}=0.$$

This is a quadratic equation in $\tan\alpha$ hence in general may have two roots say $\tan\alpha_1$ and $\tan\alpha_2$, provided these roots are real.

Thus we may have two angles α_1 and α_2, such that when the particle is projected from O at these angles, it may pass through the point A. The condition for real roots of this equation is

$$\frac{4u^4}{g^2a^2}-4\left(1+\frac{2u^2b}{ga^2}\right)\geqslant 0$$

which gives

$$u^4-(g^2a^2+2u^2g\,b)\geqslant 0$$

or
$$(u^2-gb)^2-(g^2a^2+g^2b^2)\geqslant 0$$

or
$$u^2-gb\geqslant g(a^2+b^2)^{\frac{1}{2}}$$

or
$$u^2\geqslant g(b+\sqrt{a^2+b^2}).$$

This shows that when u is greater than $[g(b+\sqrt{a^2+b^2})]^{\frac{1}{2}}$, the particle will pass through A and when u is just equal to this value

Ex. 16. *A particle is projected with a velocity $2\sqrt{ag}$ so that it just clears two walls of equal height a which are at a distance 2a from each other. Show that the latus rectum of the path is equal to 2a and that the time of passing between the wall is* $2\sqrt{\dfrac{a}{g}}$.

Sol. Height of directrix above O

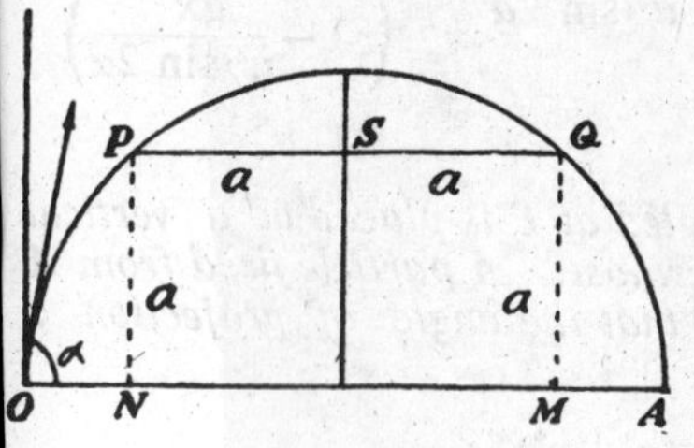

$$=\frac{u^2}{2g}=\frac{4ag}{2g}=2a$$

hence height of directrix above P

$$=2a-a=a$$

but the distances of a point on the parabola from the directrix and focus are same hence if S be the focus,

$$PS=a \quad \therefore \quad S \text{ lies on } PQ$$

Hence $PQ=$latus rectum$=2a$

Also latus rectum$=\dfrac{2u^2\cos^2\alpha}{g}=2a$

$$\therefore \quad u\cos\alpha=\sqrt{ag}$$

If t be the time from P to Q

$$u\cos\alpha.t=PQ=2a \quad \therefore \quad t=\frac{2a}{\sqrt{ag}}=2\sqrt{\frac{a}{g}}.$$

Ex. 17. *A body is projected at an angle α to the horizon so as to clear two walls of equal height a at a distance 2a from each other.*

Show that the range is equal to $2a\cot\dfrac{\alpha}{2}$.

$$R=\frac{2u^2\sin\alpha\cos\alpha}{g}.$$

If P be the point (x, a) then from the equation

$$y=x\tan\alpha-\tfrac{1}{2}\frac{gx^2}{u^2\cos^2\alpha},$$

we have

$$a=x\tan\alpha-\tfrac{1}{2}\frac{gx^2}{u^2\cos^2\alpha}$$

$$gx^2-2u^2\sin\alpha\cos\alpha+2au^2\cos^2\alpha=0$$

it may pass through A, that is, therefore, the *least value of the velocity* of projection from O in order that the particle may pass through A.

If t be the time taken by the particle from O to A, we have from the equation

$$x=u\cos\alpha\ t$$

and

$$y=u\sin\alpha\ t-\tfrac{1}{2}gt^2$$

$$a=u\cos\alpha.t$$

$$b+\tfrac{1}{2}gt^2=u\sin\alpha t$$

Squaring and adding we get

$$a^2+(b+\tfrac{1}{2}gt^2)^2=u^2t^2\cos^2\alpha+u^2t^2\sin^2\alpha=u^2t^2$$

or $$a^2+b^2+\tfrac{1}{4}g^2t^4+gbt^2=u^2t^2$$

or $\tfrac{1}{4}g^2t^4+(bg-u^2)t^2+a^2+b^2=0.$

This is a quadratic equation in t^2 so that if t_1^2, t_2^2 are the roots, we have

$$t_1^2+t_2^2=\frac{4(u^2-bg)}{g^2}$$

and

$$t_1^2t_2^2=\frac{4(a^2+b^2)}{g^2}.$$

If r be the distance of A from O then

$$r^2=a^2+b^2$$

$$t_1^2\ t_2^2=\frac{4r^2}{g^2}$$

or

$$t_1\ t_2=\frac{2r}{g}.$$

Ex. 12. *A projectile is fired from a point O with velocity* $\sqrt{2gh}$ *and hits a tangent at the point P (xy), the axes being horizontal and downward vertical lines through O. Show that if the two possible directions of projection be at right angles then* $x^2=2hy$ *and then one of the possible directions of projection bisects the angle POX.*

Sol. $$x=\sqrt{2gh}\cos\theta.\ t$$

$$y=\sqrt{2gh}\sin\theta t+\tfrac{1}{2}gt^2$$

$$=x\tan\theta+\frac{x^2\sec^2\theta}{4h}$$

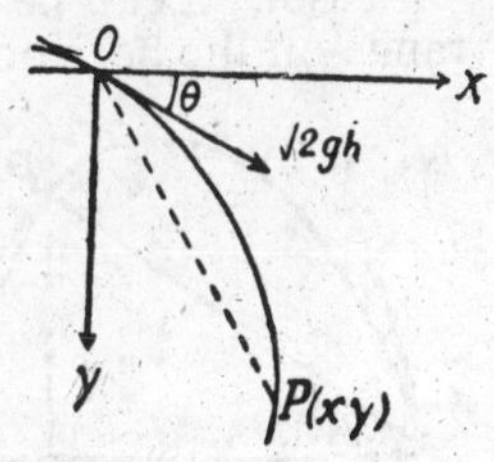

or $\tan^2\theta+\dfrac{4h}{x}\tan\theta+1-\dfrac{4hy}{x^2}=0$

If the directions are at right angles

$\tan\theta_1\tan\theta_2=-1$

$$\therefore \quad 1-\frac{4hy}{x^2}=-1$$

$$\therefore \qquad \frac{2hy}{x^2}=1$$

Hence $\quad \tan^2\theta+\frac{4h}{x}\tan\theta-1=0$

or $\quad \frac{2\tan\theta}{1-\tan^2\theta}=\frac{x}{2h}=\frac{y}{x}$

i.e., $\quad \tan 2\theta=\frac{y}{x}=\tan\phi$ where $\angle POX=\phi$,

$$\theta=\frac{\phi}{2}.$$

Ex. 13. *Two particles are projected under gravity from O with the same velocity in the same vertical plane through O at angles of elevation α and β. If the trajectories meet at the point (h, k) referred to horizontal and upward vertical axes through O,*

prove that $\quad k=h\tan\left(\alpha+\beta-\frac{\pi}{2}\right).$

$$\tan^2\alpha-\frac{2u^2}{gh}\tan\alpha+1+\frac{2u^2k}{gh^2}=0$$

$$\tan(\alpha+\beta)=\frac{\tan\alpha+\tan\beta}{1-\tan\alpha\tan\beta}=\frac{\frac{2u^2}{gh}}{1-\left(1+\frac{2u^2k}{gh^2}\right)}$$

$$=-\frac{h}{k}$$

$$\therefore \qquad \frac{k}{h}=-\cot(\alpha+\beta)=\tan(\alpha+\beta-\pi/2)$$

Ex. 14. *A particle is thrown over a triangle from one end of a horizontal base and grazing the vertex falls on the other end of the base.*

If A, B be the base angles and α the angle of projection, prove that

$$\tan\alpha=\tan A+\tan B.$$

Sol. Let u be the velocity of projection. Evidently AB is the range on the horizontal plane

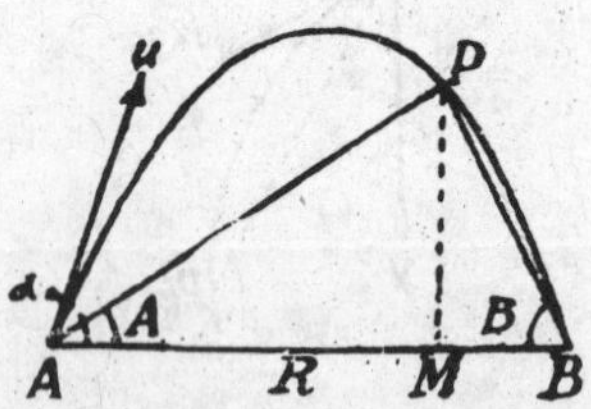

i.e., $\quad AB=R=\frac{u^2\sin 2\alpha}{g}.$

If P be the point (x, y),

$$\tan A+\tan B=\frac{y}{x}+\frac{y}{R-x}$$

$$=\frac{y}{x}\cdot\frac{R-x+x}{R-x}=\frac{y}{x}\cdot\frac{R}{R-x}.$$

But $\quad y=x\tan\alpha-\frac{1}{2}\frac{gx^2}{u^2\cos^2\alpha}.$

Hence $\quad \tan A+\tan B=\left(\tan\alpha-\frac{1}{2}\frac{gx}{\cos^2 x}\right)\frac{\frac{u^2\sin 2\alpha}{g}}{\frac{u^2\sin 2\alpha}{g}-x}$

$$=\tan\alpha\left(1-\frac{gx}{u^2\sin 2a}\right)\cdot\frac{1}{\left(1-\frac{g}{u^2 s\ldots}\right.}$$

$$=\tan\alpha.$$

Ex. 15. *A triangle APB right angled at P is placed in a plane with AB horizontal and P uppermost. A particle fire… passed exactly through P and B. Show that the angle of pro… and velocity u are given by*

$$\tan\alpha=\frac{c}{p},\quad u^2=\frac{g(c^2+p^2)}{2p}$$

where AB=c and p is the altitude of the triangle.

Sol. From the previous example,

$$\angle P=90^\circ \quad \therefore \quad \angle B=90-\angle A$$

Hence $\quad \tan\alpha=\tan A+\cot A=\frac{1}{\sin A\cos A}$

$$=\operatorname{cosec} A\sec A.$$

From triangle APM, $\quad \operatorname{cosec} A=\frac{AP}{PM}$

From triangle BPM, $\sec A=\frac{BP}{PM}$

$$\therefore \qquad \tan\alpha=\frac{AP\ BP}{PM^2}$$

Also area of the triangle $=\frac{1}{2}AB.PM=\frac{1}{2}AP.BP$

$$\therefore \qquad \tan\alpha=\frac{AB.PM}{PM^2}=\frac{c}{p}$$

Also $\quad c=\text{Range}=\frac{u^2\sin 2\alpha}{g}$

$$=\frac{u^2}{g}\cdot\frac{2\tan\alpha}{1+\tan^2\alpha}=\frac{u^2}{g}\cdot\frac{2cp}{c^2+p\ldots}$$

Hence $\quad u^2=$etc.

If $ON=x_1$, $OM=x_2$, x_1, x_2 are roots of the above equation so that

$$x_1+x_2=\frac{2u^2 \sin \alpha \cos \alpha}{g}=R$$

$$x_1x_2=\frac{2au^2 \cos^2 \alpha}{g}$$

Also $\quad 2a=x_2-x_1$

$$\therefore \quad 4a^2=(x_2+x_1)^2-4x_1x_2=R^2-\frac{8au^2 \cos^2 \alpha}{g}$$

$$=R^2-4a\,R \cot \alpha$$

or $\quad R^2-4a\,R \cot \alpha-4a^2=0$

$$R=\frac{4a \cot \alpha+\sqrt{16a^2 \cot^2 \alpha+16a^2}}{2},$$

discarding the negative sign

$$=2a \cot \alpha+2a \operatorname{cosec} \alpha=2a \cot\frac{\alpha}{2}.$$

Ex. 18. *A ball is projected so as to just clear two walls, the first of height a at a distance b from the point of projection and the second of height b and at a distance a from the point of projection. Show that the range on the horizontal plane is* $\frac{a^2+ab+b^2}{a+b}$ *and the angle of projection exceeds* $\tan^{-1} 3$.

Sol. The points (b, a) and (a, b) lie on

$$y=x \tan \alpha-\tfrac{1}{2}g\,\frac{x^2}{u^2 \cos^2 \alpha}$$

$$\therefore \quad a=b \tan \alpha-\tfrac{1}{2}g\,\frac{b^2}{u^2 \cos^2 \alpha}$$

$$b=a \tan \alpha-\tfrac{1}{2}g\,\frac{a^2}{u^2 \cos^2 \alpha}$$

Multiplying first by b and second by a and subtracting

$$0=(a^2-b^2) \tan \alpha-\frac{g}{2u^2 \cos^2 \alpha}(a^3-b^3)$$

or $$\frac{2u^2 \sin \alpha \cos \alpha}{g}=\frac{a^2+ab+b^2}{a+b}=R$$

Also $$b-a \tan \alpha=-\frac{ga^2}{2u^2 \cos^2 \alpha}$$

and $$a-b\tan\alpha=-\frac{gb^2}{2u^2\cos^2\alpha}$$

Hence dividing $$\frac{b-a\tan\alpha}{a-b\tan\alpha}=\frac{a^2}{b^2}$$

or $$(a^2b-ab^2)\tan\alpha=a^3-b^3$$

$$\therefore\quad \tan\alpha=\frac{a^3-b^3}{ab(a-b)}=\frac{(a-b)^3+3ab(a-b)}{ab(a-b)}$$

$$=3+\frac{(a-b)^2}{ab}\text{ hence }\tan\alpha>3.$$

Ex. 19. *A stone is projected with velocity u from a height h to hit a point in the level at a horizontal distance R from the point of projection. Show that the angle of projection is given by*

$$R^2\tan^2\alpha-\frac{2u^2}{g}R\tan\alpha+R^2-\frac{2hu^2}{g}=0.$$

Hence deduce that the maximum range on the level for this velocity is $\sqrt{\left(\frac{u^4}{g^2}+\frac{2hu^2}{g}\right)}$ *and that if R′ is this maximum range and α the angle of projection to give the maximum range then* $\tan\alpha=\frac{u^2}{gR'}$ *and* $\tan 2\alpha=\frac{R'}{h}$.

Further if $u=\sqrt{2ag}$, *prove that* $R'=2\sqrt{a(a+h)}$ *and* $\tan\alpha=\sqrt{\frac{a}{a+h}}$.

Sol. The point $(R, -h)$ is on

$$y=x\tan\alpha-\tfrac{1}{2}\frac{gx^2}{u^2\cos^2 a}$$

$$\therefore\quad -h=R\tan\alpha-\tfrac{1}{2}\frac{gR^2}{u^2}(1+\tan^2\alpha)$$

or $$R^2\tan^2\alpha-\frac{2u^2}{g}R\tan\alpha+R^2-\frac{2hu^2}{g}=0 \qquad \text{...(1)}$$

R is maximum when $\frac{dR}{d\alpha}=0$

$$2R^2\tan\alpha\sec^2\alpha-\frac{2u^2}{g}R\sec^2\alpha=0$$

giving $R=\frac{u^2}{g}\cot\alpha$ or $\tan\alpha=\frac{u^2}{gR}$.

Substituting this value of R in (1) we get

$$R^2=\frac{2u^2h}{g}+\frac{u^4}{g^2}$$

i.e.,
$$\tan\alpha=\frac{u^2}{gR'} \text{ and } R'=\sqrt{\left(\frac{2u^2h}{g}+\frac{u^4}{g^2}\right)}$$

$$\tan 2\alpha=\frac{2\tan\alpha}{1-\tan^2\alpha}=\frac{2u^2gR'}{g^2R'^2-u^4}$$

$$=\frac{2u^2g\,R'}{g^2\left(\frac{2u^2h}{g}+\frac{u^4}{g^2}\right)-u^4}=\frac{2u^2g\,R'}{2u^2gh}=\frac{R'}{h}.$$

If
$$u=\sqrt{2ag}$$

then
$$R'=\sqrt{\left\{\frac{4ag\,h}{g}+\frac{4a^2g^2}{g^2}\right\}}=\sqrt{4a(a+h)}$$

and
$$\tan\alpha=\frac{2ag}{g\,.\,2\sqrt{(a+h)}}=\frac{a}{\sqrt{a(a+h)}}=\sqrt{\frac{a}{a+h}}$$

Ex. 20. *A ship is under fire from the guns of a fort h feet above sea level. Assuming that two guns of the ship and the fort fire with a velocity* $\sqrt{2ag}$, *prove that*

(1) *if* d_1 *and* d_2 *are the greatest ranges at which the fort and the ship can engage*

$$\frac{d_1}{d_2}=\sqrt{\frac{a+h}{a-h}}.$$

(2) *the width of the zone under fire from the fort which the ship has to cross before being able to reply is*

$$2\sqrt{a}\left[\sqrt{a+h}-\sqrt{a-h}\right]$$

(3) *the angular region in which the fort is out of range of the ship but the ship is not out of range of the fort is* $8\pi ah$.

Sol. As in previous example

$$d_1=2\sqrt{a(a+h)}$$

For the gun in the ship, putting $-h$ for h

$$d_2=2\sqrt{a(a-h)}$$

Width of the zone $=d_1-d_2=$etc.

and the annular area $=\pi d_1^2-\pi d_2^2$

$$=4\pi[a(a+h)-a(a-h)]=8\pi ah.$$

Ex. 21. *Two particles are simultaneously projected in the same vertical plane from the same point with velocities u and v at angles α and β to the horizontal. Prove that*

(*i*) *the line joining them moves always parallel to itself*

(*ii*) *the time that elapses when their velocities are parallel is*

$$\frac{uv \sin(\alpha-\beta)}{g(v\cos\beta-u\cos\alpha)}$$

and (*iii*) *the time that elapses between their transits through the other common point is*

$$\frac{2uv\sin(\alpha-\beta)}{g(u\cos\alpha+v\cos\beta)}.$$

Sol. Let O be the point of projection, $P(x_1, y_1)$ and $Q(x_2, y_2)$ be the two particles at time t.

Then
$$x_1=u\cos\alpha.t, \qquad y_1=u\sin\alpha.t-\tfrac{1}{2}gt^2$$
$$x_2=v\cos\beta.t, \qquad y_2=v\sin\beta.t-\tfrac{1}{2}gt^2$$

$\therefore$ slope of $PQ=\dfrac{y_1-y_2}{x_1-x_2}=\dfrac{u\sin\alpha-v\sin\beta}{u\cos\alpha-v\cos\beta}$ independent of t, hence same for all times *i.e.*, PQ remains parallel to itself. If θ_1, θ_2 be the angles directions of motion make at time t,

$$\tan\theta_1=\frac{u\sin\alpha-gt}{u\cos\alpha} \text{ and } \tan\theta_2=\frac{v\sin\beta-gt}{v\cos\beta}$$

Hence when $\theta_1=\theta_2$ we have $\dfrac{u\sin\alpha-gt}{u\cos\alpha}=\dfrac{v\sin\beta-gt}{v\cos\beta}$

$$\therefore \qquad t=\frac{uv\sin(\alpha-\beta)}{g(v\cos\beta-u\cos\alpha)}.$$

The paths of the two particles are

$$y=x\tan\alpha-\frac{1}{2}\frac{gx^2}{u^2\cos^2\alpha}$$

and
$$y=x\tan\beta-\frac{1}{2}\frac{gx^2}{u^2\cos^2\beta}.$$

If (a, b) be the common point then $b=a\tan\alpha-\dfrac{1}{2}\dfrac{ga^2}{u^2\cos^2\alpha}$

and
$$b=a\tan\beta-\frac{1}{2}\frac{ga^2}{v^2\cos^2\beta}$$

Hence $a\tan\alpha-\dfrac{1}{2}\dfrac{ga^2}{u^2\cos^2\alpha}=a\tan\beta-\dfrac{1}{2}\dfrac{ga^2}{v^2\cos^2\beta}$

i.e., $$\tan\alpha-\tan\beta=\tfrac{1}{2}ga\left(\frac{1}{u^2\cos^2\alpha}-\frac{1}{v^2\cos^2\beta}\right). \quad ...(1)$$

If t_1, t_2 be their times to pass through the common point then $a=u\cos\alpha.t_1$ also $a=v\cos\beta.t_2$

$$\therefore \quad t=t_1-t_2=a\left(\frac{1}{u\cos\alpha}-\frac{1}{v\cos\beta}\right). \quad ...(2)$$

Hence dividing (1) by (2), we get

$$\frac{\tan\alpha-\tan\beta}{t}=\tfrac{1}{2}g.\left(\frac{1}{u\cos\alpha}+\frac{1}{v\cos\beta}\right)$$

$$=\frac{g(u\cos\alpha+v\cos\beta)}{2uv\cos\alpha\cos\beta}$$

Hence $$t=\frac{2uv\sin(\alpha-\beta)}{g(u\cos\alpha+v\cos\beta)}.$$

Ex. 22. *Particles are simultaneously fired from the same point in different directions with such velocities as just to clear a vertical wall of height b and to fall into a narrow ditch parallel to the wall; show that the particles will fall into the ditch at the same moment.*

Sol. Let u and α be the velocity and angle of projection from A.

PN is the wall and B the point in the ditch.

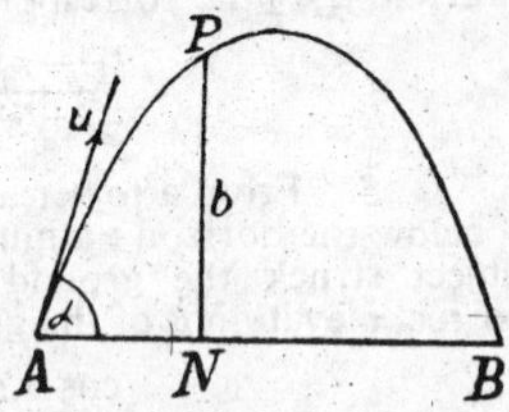

$AN=a$, $AB=c$, say.

Let t be the time from A to P and t' the time from A to B.

Then $\quad a=u\cos\alpha t$

and $\quad b=u\sin\alpha t-\tfrac{1}{2}gt^2$

also $\quad t'=\dfrac{2u\sin\alpha}{g}$ and $c=u\cos\alpha t'$

$$\frac{a}{c}=\frac{t}{t'}.$$

Now $$b=\tfrac{1}{2}gt'.t-\tfrac{1}{2}gt^2=\tfrac{1}{2}gt^2\left(\frac{t'}{t}-1\right)=\tfrac{1}{2}gt^2\left(\frac{c}{a}-1\right)$$

$$=\tfrac{1}{2}g\left(\frac{c}{a}-1\right)\left(\frac{a}{c}\right)^2t'^2$$

$$t'^2=\frac{2b}{g\left(\dfrac{a}{c}\right)^2\left(\dfrac{c}{a}-1\right)}.$$

Thus t' depends only on the ratio $\frac{c}{a}$. But the ratio $\frac{AN}{AB}$ *i.e.*, $\frac{a}{c}$ is the same hence t' is the same for all trajectories.

EXAMPLES III (C)

1. Shots fired simultaneously from the top and bottom of a vertical cliff with elevations α and β respectively, strike an object simultaneously at the same point. Show that if a is the horizontal distance of the object from the cliff, the height of the cliff is a $(\tan\beta-\tan\alpha)$.

2. From a point on the ground at a distance x from the foot of a vertical wall a ball is thrown at an angle of 45°, which just clears the top of the wall and afterwards strikes the ground at a distance y on the other side. Show that the height of the wall is $\frac{xy}{x+y}$.

3 A particle is projected in a direction making an angle α with the horizon. If it passes through the point (x_1, y_1) show that

$$\tan\alpha=\frac{y_1}{x_1}\cdot\frac{R}{R-x_1}$$

and that if it also passes through the point (x_2, y_2) then

$$\tan\alpha=\frac{x_2^2y_1-x_1^2y_2}{x_1x_2(x_2-x_1)}$$

where R denotes the horizontal range.

4. A particle is projected so as to have a range R on the horizontal plane through the point of projection. If α, β are the possible angles of projection and t_1, t_2 the corresponding times of flights, show that

$$\frac{t_1^2-t_2^2}{t_1^2+t_2^2}=\frac{\sin(\alpha-\beta)}{\sin(\alpha+\beta)}.$$

5. From a tower an object on the ground was observed at a depression φ below the horizon; a gun was fired at an elevation α but the shot missing the object struck the ground at a point whose depression was ψ. Prove that the correct elevation θ of the gun is given by

$$\frac{\cos\theta\sin(\theta+\varphi)}{\cos\alpha\sin(\alpha+\psi)}=\frac{\cos^2\varphi\sin\psi}{\cos^2\psi\sin\varphi}.$$

6. A gun fires a shell with a muzzle velocity u so that the farthest horizontal distance at which an aeroplane at a height h can be hit is $\frac{u}{g}\sqrt{u^2-2gh}$ and the gun's elevation is then $\tan^{-1}\{u/\sqrt{(u^2-2gh)}\}$.

7. The angular elevation of an enemy's position on a hill h feet high is β; show that in order to shell it, the initial velocity of the projectile must not be less than $\sqrt{gh(1+\operatorname{cosec}\beta)}$.

8. Prove that the minimum velocity required to project a particle from a height h to fall at a distance d from the point of projection is

$$\sqrt{g(d-h)}.$$

9. Prove that if t_1, t_2 be the times of flight from a point A to a point B and α the inclination of AB to the horizontal, then $t_1^2+2t_1t_2\sin a+t_2^2$ is independent of α.

10. Two particles are projected from the same point in the same vertical plane with equal velocities. If t, t' be the times taken to reach the other common point of their paths and T, T' the times to the highest point, show that $tT+t'T'$ is independent of the directions of projection.

11. The line joining C to D is inclined at an angle α to the horizon. Show that the least velocity required to shoot from C to D is $\tan\left(\frac{\pi}{4}+\frac{\alpha}{2}\right)$ times the least velocity required to shoot from D to C.

12. A shot fired at an elevation α is observed to strike the foot of a tower which rises above a horizontal plane through the point of projection. If θ be the angle subtended by the tower at this point, show that the elevation required to make the shot strike the top of the tower is

$$\tfrac{1}{2}[\theta+\sin^{-1}(\sin\theta+\sin 2\alpha\cos\theta)].$$

13. A particle is projected from P so as to pass through Q the initial speed being given. If S be the focus of either possible trajectory show that the two possible times of flight are

$$\{(SP+SQ+PQ)^{\frac{1}{2}} \pm (SP+SQ-PQ)^{\frac{1}{2}}\}\sqrt{g}.$$

3·4. Range on an inclined plane. *A particle is projected with velocity u at an inclination α to the horizon from the foot of the plane inclined at an angle β to the horizon. If the particle is projected in the plane passing through the normal to the inclined plane and the line of greatest slope, to find the range on the inclined plane.*

Let OX and OA be respectively the horizontal and the inclined plane through the point of projection O. The components of the initial velocity u along OA, and perpendicularly are $u\cos(\alpha-\beta)$ and $u\sin(\alpha-\beta)$ and the components of g along these lines are $-g\sin\beta$ and $-g\cos\beta$ respectively. Take O as OA as x-axis and the perp. line through origin, O as y-axis.

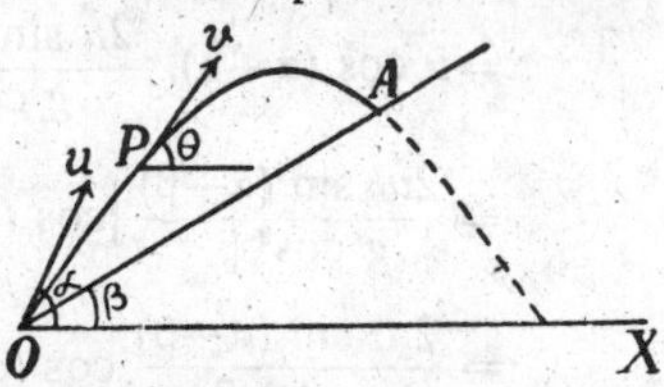

Let P be the position of the particle at time t, co-ordinates x, y; the components of acceleration in those directions are $\ddot{x}$ and $\ddot{y}$ and the components of the external force namely the weight of the particle are $-mg\sin\beta$ and $-mg\cos\beta$.

Hence the equations of motions are

$$m\ddot{x}=-mg\sin\beta,\quad m\ddot{y}=-mg\cos\beta$$

or $$\ddot{x}=-g\sin\beta,\qquad \ddot{y}=-g\cos\beta \qquad \text{...(1)}$$

Integrating the first equation with respect to t, we have

$$\dot{x}=-gt\sin\beta+c_1 \text{ where } c_1 \text{ is a constant.}$$

Initially $\quad t=0,\ \dot{x}=u\cos(\alpha-\beta) \quad \therefore \quad c_1=u\cos(\alpha-\beta)$

Hence $\quad \dot{x}=u\cos(\alpha-\beta)-gt\sin\beta. \qquad \text{...(2)}$

Again integrating we get

$$x = ut\cos(\alpha - \beta) - \tfrac{1}{2}gt^2 \sin\beta + c_2 \text{ where } c_2 \text{ is a constant.}$$

From the conditions $t=0$ $x=0$ we have $c_2=0$

$$\therefore \qquad x = ut\cos(\alpha-\beta) - \tfrac{1}{2}gt^2\sin\beta. \qquad ...(3)$$

Similarly from the second equation of (1) we have on integration

$$\dot{y} = -gt\cos\beta + c_3 \text{ where } c_3 \text{ is a constant.}$$

Using conditions $t=0$, $\dot{y}=u\sin(\alpha-\beta)$, we get $c_3 = u\sin(\alpha-\beta)$

Hence $$\dot{y} = u\sin(\alpha-\beta) - gt\cos\beta. \qquad ...(4)$$

Again integrating and using $t=0$, $y=0$, we get

$$y = ut\sin(\alpha-\beta) - \tfrac{1}{2}gt^2\cos\beta. \qquad ...(5)$$

If the particle strikes the plane at A after time T then when

$$t=T,\ y=0$$

$$\therefore \qquad T = \frac{2u\sin(\alpha-\beta)}{g\cos\beta} \qquad ...(6)$$

This is evidently the time of flight over the inclined plane.

Substituting this value of t in (3), we have

$$x = u\cos(\alpha-\beta).\frac{2u\sin(\alpha-\beta)}{g\cos\beta} - \tfrac{1}{2}g\sin\beta.\frac{4u^2\sin^2(\alpha-\beta)}{g^2\cos^2\beta}$$

$$= \frac{2u^2\sin(\alpha-\beta)}{g\cos^2\beta}[\cos(\alpha-\beta)\cos\beta - \sin(\alpha-\beta)\sin\beta]$$

$$= \frac{2u^2\sin(\alpha-\beta)}{g\cos^2\beta}\cos(\alpha-\beta+\beta) = \frac{2u^2\sin(\alpha-\beta)\cos\alpha}{g\cos^2\beta}$$

This is the required *range*.

To find the maximum, we note that g is constant and u, β are given so that the range is maximum when

$$\frac{d}{d\alpha}\{\sin(\alpha-\beta)\cos\alpha\} = 0$$

giving $$\cos(\alpha-\beta)\cos\alpha - \sin(\alpha-\beta)\sin\alpha = 0$$

or $$\cos(2\alpha-\beta) = 0 \quad \text{or} \quad 2\alpha - \beta = \frac{\pi}{2}$$

or $$\alpha = \frac{\pi}{4} + \frac{\beta}{2}$$

and the maximum range $=\frac{2u^2}{g}\cdot\frac{\cos^2\alpha}{\cos^2\beta}=\frac{u^2}{g\cos^2\beta}(1+\cos 2\alpha)$

$$=\frac{u^2}{g\cos^2\beta}(1-\sin\beta)=\frac{u^2}{g(1+\sin\beta)}$$

For the maximum range, $\alpha-\beta=\frac{\pi}{2}-\alpha$

i.e., the direction for the max. range bisects the angle between the vertical and the inclined plane. Now if S be the focus then the tangent at O also bisects the angle between the vertical and OS hence S must lie on the inclined plane. *Thus when the range on the inclined plane is max., the focus lies on it.* Further in that case OA becomes a focal chord and the tangents at the extremities of a focal chord meet at right angles hence the tangent at A is at right angles to the tangent at O. *Thus in the case of max. range the particle strikes the plane at right angles to the direction of projection.*

Again if the range with the angle of projection α_1 be the same as that with the angle α_2 then

$$\sin(\alpha_1-\beta)\cos\alpha_1=\sin(\alpha_2-\beta)\cos\alpha_2$$

or $$\sin(2\alpha_1-\beta)-\sin\beta=\sin(2\alpha_2-\beta)-\sin\beta$$

or $$\sin(2\alpha_1-\beta)=\sin(2\alpha_2-\beta)$$

$$\therefore\ 2\alpha_1-\beta=\pi-(2\alpha_2-\beta) \text{ or } \alpha_1=\frac{\pi}{2}+\beta-\alpha_2$$

or $$\alpha_1-\left(\frac{\pi}{4}+\frac{\beta}{2}\right)=\left(\frac{\pi}{4}+\frac{\beta}{2}\right)-\alpha_2.$$

But $\left(\frac{\pi}{4}+\frac{\beta}{2}\right)$ is the angle for the max. range. Hence it follows that *for a given range with a given velocity, there are two angles of projection which are equally inclined to the angle of max. projection on the inclined plane.*

The equations (2) and (4) give the components of velocity at any point $P(x, y)$ parallel and perp. to the inclined plane.

If the particle strikes the inclined plane at right angles

then evidently $\dot{x}=0$ when $t=T=\frac{2u\sin(\alpha-\beta)}{g\cos\beta}$

$$\therefore\quad u\cos(\alpha-\beta)=g\sin\beta.\frac{2u\sin(\alpha-\beta)}{g\cos\beta}$$

giving $$\cot\beta=2\tan(\alpha-\beta).$$

Simplifying we get $\cot\beta=2\dfrac{\tan\alpha-\tan\beta}{1+\tan\alpha\tan\beta}$

or $\cot\beta+\tan\alpha=2\tan\alpha-2\tan\beta$ or $\tan\alpha=2\tan\beta+\cot\beta$.

If *the particle strikes the plane horizontally*, then the vertical component of velocity is zero in time

$$T=\frac{2u\sin(\alpha-\beta)}{g\cos\beta}$$

i.e.,
$$O=u\sin\alpha-gT'$$

$\therefore$
$$u\sin\alpha=\frac{2u\sin(\alpha-\beta)}{\cos\beta}$$

or
$$\sin\alpha\cos\beta=2(\sin\alpha\cos\beta-\cos\alpha\sin\beta)$$

or
$$\tan\alpha=2\tan\beta.$$

If the tangent at P makes an angle θ with the horizontal and v be the velocity there

then
$$v\cos\theta=\text{horizontal velocity}=u\cos\alpha$$
$$v\sin\theta=\text{vertical velocity}=u\sin\alpha-gt.$$

At the point A where the particle strikes the inclined plane

$$t=T=\frac{2u\sin(\alpha-\beta)}{g\cos\beta}$$

$$\therefore\quad \tan\theta=\frac{u\sin\alpha-gT}{u\cos\alpha}=\frac{u\sin\alpha-\dfrac{2u\sin(\alpha-\beta)}{\cos\beta}}{u\cos\alpha}.$$

$$=\frac{\sin\alpha\cos\beta-2\sin(\alpha-\beta)}{\cos\alpha\cos\beta}=\frac{2\cos\alpha\sin\beta-\sin\alpha\cos\beta}{\cos\alpha\cos\beta}$$

$$=2\tan\beta-\tan\alpha.$$

Ex. 23. *Show that if a gun be situated on an inclined plane, the maximum range in a direction at right angles to the line of greatest slope is a harmonic mean between the maximum ranges up and down the plane respectively.*

Sol. Let u be the velocity of projection and β the inclination of the plane.

$$\text{Max. range up the plane}=\frac{u^2}{g(1+\sin\beta)}=R_1 \text{ say}$$

$$\text{Max. range down the plane}=\frac{u^2}{g(1-\sin\beta)}=R_2 \text{ say}$$

Harmonic mean of R_1 and R_2

$$=\frac{2R_1R_2}{R_1+R_2}=\frac{\dfrac{2u^4}{g^2(1-\sin^2\beta)}}{\dfrac{u^2}{g}\left[\dfrac{1}{1+\sin\beta}+\dfrac{1}{1-\sin\beta}\right]}=\frac{u^2}{g}$$

=max. horizontal range

=max. range in a direction at right angles to the line of greatest slope.

Ex. 24. *A perfectly elastic ball is thrown from the foot of a plane inclined at angle β to the horizon. If after striking the plane at a distance l from the point of projection it rebounds and retraces its former path, show that the velocity of projection is*

$$\sqrt{\frac{gl(1+3\sin^2\beta)}{2\sin\beta}}.$$

Sol The ball will retrace its path only if it strikes the plane at right angles, the condition for which is

$$\tan\alpha=\cot\beta+2\tan\beta \text{ or } \tan(\alpha-\beta)=\tfrac{1}{2}\cot\beta.$$

If u is the velocity of projection then the velocity at the point where it strikes the plane has no component along the plane

Hence $$u^2\cos^2(\alpha-\beta)-2g\sin\beta l=0$$

$$\therefore\ u^2=2gl\sin\beta[1+\tan^2(\alpha-\beta)]=2gl\sin\beta\left[1+\frac{\cot^2\beta}{4}\right]$$

$$=\frac{gl}{2}\ \frac{1+3\sin^2\beta}{\sin\beta}.$$

Ex. 25. *A particle is projected with a velocity u from the foot of a plane inclined to the horizon at angle β. If it strikes the plane at right angles show that the range on the plane is*

$$\frac{2u^2\sin\beta}{g(1+3\sin^2\beta)},$$

the vertical height of the point struck is $\dfrac{2u^2\sin^2\beta}{g(1+3\sin^2\beta)}$ *and the time of flight is* $\dfrac{2u}{g\sqrt{(1+3\sin^2\beta)}}$.

Sol As in the previous example, range is l and vertical height $=l\sin\beta$ and time of flight$=\dfrac{2u\sin(\alpha-\beta)}{g\cos\beta}$.

$$=\frac{2u}{g\cos\beta\sqrt{1+\cot^2(\alpha-\beta)}}$$

$$=\frac{2u}{g\cos\beta\sqrt{1+4\tan^2\beta}}=\frac{2u}{g\sqrt{1+3\sin^2\beta}}$$

Ex. 26. *Prove that the greatest range of a particle, projected with a given velocity on an inclined plane, is four times the greatest vertical altitude above the inclined plane.*

Sol. P be the point where the tangent is parallel to the inclined plane. If $PN=z$ be perpendicular from P on the inclined plane and PM the vertical altitude of P then evidently for all points on the path, P is the point where z is greatest and consequently PM is greatest.

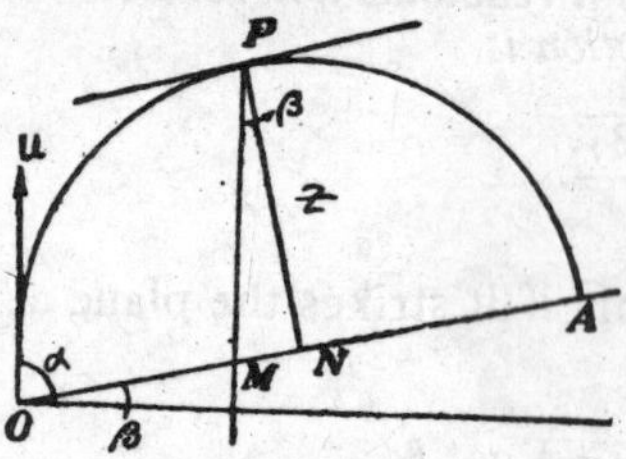

Now for the point P, velocity perpendicular to the inclined plane is zero. Now the velocity and acceleration perp. to the plane at O is $u\sin(\alpha-\beta)$ and $g\cos\beta$ and this velocity becomes zero at P.

$\therefore$
$$O=u^2\sin^2(\alpha-\beta)-2g\cos\beta.z$$

$$z=\frac{u^2\sin^2(\alpha-\beta)}{2g\cos\beta}$$

For max. range $\alpha=\frac{\pi}{4}+\frac{\beta}{2}$

or
$$\alpha-\beta=\frac{\pi}{4}-\frac{\beta}{2}$$

Hence
$$z=\frac{u^2}{2g\cos\beta}\sin^2\left(\frac{\pi}{4}-\frac{\beta}{2}\right)$$

$$=\frac{u^2}{4g\cos\beta}\left[1-\cos\left(\frac{\pi}{2}-\beta\right)\right]$$

$$=\frac{u^2}{4g\cos\beta}(1-\sin\beta)$$

or
$$PM=z\sec\beta=\frac{u^2}{4g\cos^2\beta}(1-\sin\beta)=\frac{u^2}{4g(1+\sin\beta)}$$

$$=\frac{1}{4}\text{ (max. range)}$$

Ex. 27. *A stone is thrown at an angle α with the horizontal from a point on the plane whose inclination to the horizon is β (the trajectory lying in the vertical plane containing the line of greatest slope* *through the point. Show that if the elevation above the horizon of the point of the path most distant from the inclined plane be θ, then*

$$2 \tan \theta = \tan \alpha + \tan \beta.$$

Sol. The point which is most distant from the inclined plane is the point the tangent at which is parallel to the plane. Hence the direction of motion there makes an angle β with horizon. Thus the point may be supposed to lie on an inclined plane of inclination θ, the direction of motion there making an angle β.

Hence from the result $\tan \theta = 2 \tan \beta - \tan \alpha$ where β is the angle of the plane, θ the direction of motion, we have here θ in place of β and β in place of θ so that

$$\tan \beta = 2 \tan \theta - \tan \alpha$$

$$\therefore \quad 2 \tan \theta = \tan \alpha + \tan \beta.$$

Ex. 28. *A particle is projected from O at an elevation α. Prove that there are two positions P on its path at which the direction of the velocity is perpendicular to OP. Prove that for real position to exist, α is not less than $\cos^{-1}(\frac{1}{3})$; also if in the two positions, OP makes angles θ_1, θ_2 with the horizontal, prove that*

$$\theta_1 + \theta_2 = \alpha.$$

Sol. This is the same as the case of striking an inclined plane of angle β at right angles, the condition for which is

$$\cot \beta = \tan \alpha - 2 \tan \beta$$

or

$$2 \tan^2 \beta - \tan \alpha \tan \beta + 1 = 0$$

For real roots $\tan^2 \alpha \geqslant 8$ or $\cos^2 \alpha \leqslant \frac{1}{9}$ or $\cos \alpha \leqslant \frac{1}{3}$

or

$$\alpha \geqslant \cos^{-1}(\tfrac{1}{3})$$

Also if θ_1, θ_2 are the two values of β,

then

$$\tan \theta_1 + \tan \theta_2 = \frac{\tan \alpha}{2} \text{ and } \tan \theta_1 \tan \theta_2 = \tfrac{1}{2}$$

$$\therefore \quad \tan (\theta_1 + \theta_2) = \frac{\frac{\tan \alpha}{2}}{1 - \frac{1}{2}} = \tan \alpha \quad \therefore \quad \theta_1 + \theta_2 = \alpha.$$

EXAMPLES III (D)

1. Prove that if a particle is projected from O at elevation α and if after time t the particle is at P then

$$\tan \beta = \tfrac{1}{2} (\tan \alpha + \tan \theta)$$

where β and θ are the inclinations to the horizontal of OP and of the direction of motion of the particle when at P.

2. A particle is projected at an elevation α and after t seconds it appears to have an elevation β as seen from the point of projection. Prove that the initial velocity was $\dfrac{gt \cos \beta}{2 \sin(\alpha-\beta)}$.

3. Show that for a given velocity of projection the maximum range down a plane of inclination α is greater than up the plane in the ratio

$$\frac{1+\sin \alpha}{1-\sin \alpha}.$$

4. Show that the greatest range up an inclined plane through the point of projection is equal to the distance through which a particle could fall freely during the time of flight.

5. A particle projected with velocity u from a point on an inclined plane whose inclination to the horizon is β, strikes it at right angles. Show that the range of the particle on the inclined plane is $\dfrac{2u^2}{g} \cdot \dfrac{\sin \beta}{1+3 \sin^2 \beta}$.

6. If from a point on the side of a hill two bodies are projected in the vertical plane through the line of greatest slope with the same velocity but in directions at right angles to each other, show the difference of their ranges is independent of their angles of projection.

3·5. Envelope of the paths.

For a particle projected with velocity u at an angle α, the path is

$$y=x \tan \alpha-\tfrac{1}{2}\frac{gx^2}{u^2 \cos^2 \alpha}.$$

Differentiating this with respect to α, we get

$$0=x \sec^2 \alpha-\frac{gx^2}{u^2} \tan \alpha \sec^2 \alpha$$

or

$$\tan \alpha=\frac{u^2}{gx}.$$

Substituting this value of $\tan \alpha$ in the original equation we get

$$y=x\,\frac{u^2}{gx}-\tfrac{1}{2}\,\frac{gx^2}{u^2}\left(1+\frac{u^4}{g^2x^2}\right)$$

i.e.,

$$y=\frac{u^2}{2g}-\frac{gx^2}{2u^2}$$

or

$$x^2=-\frac{2u^2}{g}\left(y-\frac{u^2}{2g}\right)$$

i.e., $x^2=-4h(y-h)$ where $h=\dfrac{u^2}{2g}$ = height to which the velocity of projection is due.

This is a *parabola* whose axis is vertical, focus is the point of projection and vertex is in the common directrix of the path.

Now the particles projected with a given velocity will not go beyond a certain range in a given direction and this *enveloping parabola gives the maximum distances covered in the different directions in the same vertical plane.* The particle can never reach any point outside this parabola and the maximum range along any line from the point of projection is its portion intercepted by this parabola.

Ex. 29. *If u be the velocity of projection and v_1 the velocity of striking the plane when projected so that the range up the plane is maximum and v_2 the velocity of striking plane so that the range down the plane is maximum, prove that u is a mean proportional between v_1 and v_2.*

Sol. The enveloping parabola is $x^2=-4h(y-h)$ where $h=\dfrac{u^2}{2g}$ and the inclined plane is $y=x\tan\beta$.

Hence the points of intersection are given by

$$y^2\cot^2\beta+4hy-4h^2=0.$$

If y_1, y_2 are the roots $\quad y_1+y_2=-4h\tan^2\beta$

$$y_1 \cdot y_2=-4h^2\tan^2\beta.$$

Now $v_1^2=2g(h-y_1)$ and $\quad v_2^2=2g(h-y_2)$,

since the directrix is at a height h

$$\therefore \quad v_1^2v_2^2=4g^2\,(h-y_1)(h-y_2)=4g^2\{h^2-h(y_1+y_2)+y_1y_2\}$$

$$=4g^2\{h^2+4h^2\tan^2\beta-4h^2\tan^2\beta\}$$

$$\therefore \quad v_1v_2=2gh=u^2.$$

Ex. 30. *A particle is projected under gravity with velocity $\sqrt{2ga}$ from a point at a height h above a level plane. Show that the angle of projection θ for the maximum range on the plane is given by $\tan^2\theta=\dfrac{a}{a+h}$ and the maximum range is $2\sqrt{a(a+h)}$.*

Sol. The enveloping parabola is $x^2=-4h(y-h)$

where $\quad h=\dfrac{u^2}{2g}=a \quad$ *i.e.,* $x^2=-4a(y-a)$

and the plane is $\quad y=-h.$

The maximum range is given by values of x from these two equations,

i.e., $\quad x^2=4a(h+a)$

i.e., the max. range is $2\sqrt{a(h+a)}$.

If θ be the angle of projection,

$$y=x\tan\theta-\tfrac{1}{2}g\frac{x^2}{u^2\cos^2\theta}$$

where $\quad y=-h$ and $x=2\sqrt{a(h+a)}$

Hence $\quad -h=2\tan\theta\sqrt{a(h+a)}-\tfrac{1}{2}g.\dfrac{4a(h+a)}{2ag}(1+\tan^2\theta)$

i.e., $\quad (h+a)\tan^2\theta-2\sqrt{a(h+a)}\tan\theta+a=0$

or $\quad \tan\theta=\sqrt{\dfrac{a}{a+h}}.$

Ex. 31. *A shot is fired with velocity $\sqrt{2gh}$ from the top of a mountain which is in the form of hemisphere of radius r. Show that the furthest points which can be reached by the shot are at a distance (measured in a straight line) $(r-\sqrt{r^2-4rh})$ from the point of projection.*

Sol. The enveloping parabola is $x^2=-4h(y-h)$

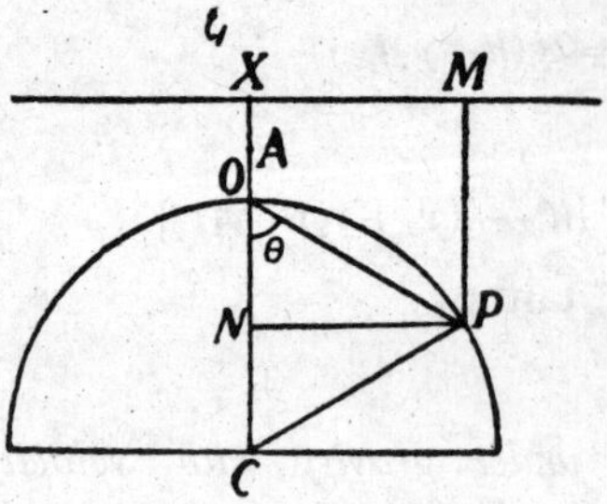

where $\quad h=\dfrac{u^2}{2g}=h.$

Its focus is the point of projection O, vertex is A where $OA=h$ and directrix is XM.

Let this enveloping parabola cut the sphere at P so that OP is the max. range. Draw PM perp. to directrix and PN perpendicular to CO, C the centre of the sphere.

Since O is the focus, $\quad PO=PM=XN=XO+ON$

$$=2h+PO\cos\theta.$$

Also $\quad PO=2r\cos\theta$

Hence $\quad PO\left(1-\dfrac{PO}{2r}\right)=2h$

or $\quad R\left(1-\dfrac{R}{2r}\right)=2h$ where $R=PO$

or $\quad R^2-2rR+4rh=0$

$$R=r\pm\sqrt{r^2-4rh}.$$

The plus sign corresponds to the second point where the parabola cuts the sphere again hence the minus sign is to be taken

$\therefore \quad R=r-\sqrt{r^2-4rh}$

From this resuit it follows that the max. value of h possible is $\frac{r}{4}$ when the two values of R become same and equal to r. In this case the parabola touches the sphere and the velocity of the projection is $\sqrt{2g \cdot h} = \sqrt{2g \cdot \frac{r}{4}} = \sqrt{\frac{1}{2} gr}$. The particle will then clear the sphere and the least value of the velocity of projection must be $\sqrt{\frac{1}{2}gr}$.

EXAMPLES III (E)

1. A rocket fired vertically upwards bursts at its highest point h feet above the ground. If each fragment starts with the same velocity u, prove that all fragments on reaching the ground lie within a circle of radius

$$\frac{u(u^2+2gh)^{\frac{1}{2}}}{g}.$$

2. Show that the area commanded by a gun placed on an inclined plane is an ellipse of eccentricity $\sin \alpha$ where α is the inclination of the plane and focus the point of projection.

[**Hint.** S is the position of the gun and $CASB$ the line of greatest slope through S. The trajectories of all bullets from the gun are enveloped by a paraboloid. APB is the section of this paraboloid by the vertical plane through ASB, placing its vertex and $CDFE$ is the directrix plane. This paraboloid can be obtained by rotating the parabola APB about SP. The dotted boundary is the section of this paraboloid by the given hill slope and is the area covered by the gun. The directrix plane cuts the hill slope in the line $B'CEB$. Let T be any point on the boundary of the covered area and TE, TF be perpendiculars to the line $B'CEB$ and to the directrix plane. $EF \perp B'B$ and $\angle FET = \alpha$. Use $ST = TF$ to get the required result.

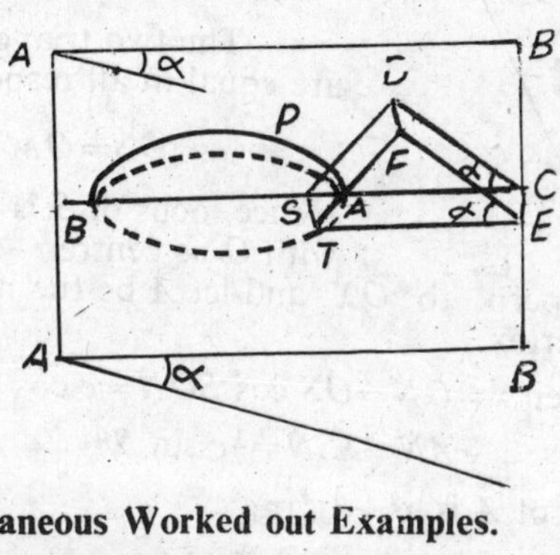

3·6. Miscellaneous Worked out Examples.

Ex. 32. *Particles slide from rest down the chords of a vertical plane curve, having its highest point as the pole and the horizontal line through it as the initial line; all these chords pass through the pole. If the particles be allowed to move freely after leaving the chords, prove that the locus of the foci of the subsequent paths is given by* $r = f\left(\frac{\theta}{2}\right) \cos \frac{\theta}{2}$ *where* $r = f(\theta)$ *is the equation to the curve.*

Sol. O is highest point and OP any chord. The velocity of the particle sliding from rest down OP is due to a fall from OX hence OX is the directrix of the parabolic path which begins from P, the line OPT being the tangent to this parabola.

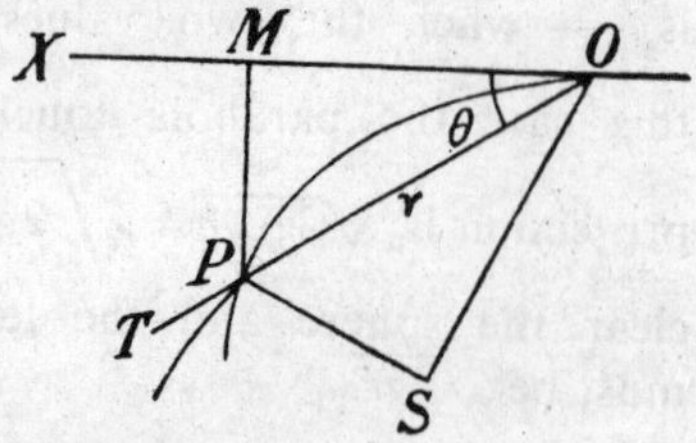

Draw $\angle OPS = \angle OPM$ and measure $PS = PM$, hence S is the focus of parabola.

Join PS. The two triangles OPS and OPM are equal in all respects.

$\therefore$ If S be the point (r', θ') then $\theta' = \angle SOX = 2\theta$,

and $r' = OS = OP \cos\theta = f(\theta)\cos\theta$

$$= f\left(\frac{\theta'}{2}\right)\cos\frac{\theta'}{2}$$

Hence the locus of S is $r = f\left(\dfrac{\theta}{2}\right)\cos\dfrac{\theta}{2}$.

Ex. 33. *Particles are allowed to fall from rest under gravity from a fixed point to a fixed vertical straight line along various rectilinear chords. If they be suffered to describe parabolic trajectories after leaving these chords, show that the locus of their foci is a circle and that of the vertices an ellipse.*

Sol. O is the fixed point and OP any rectilinear chord. As in example 32. OX is the directrix of the parabolic path which begins from P with OPT as tangent.

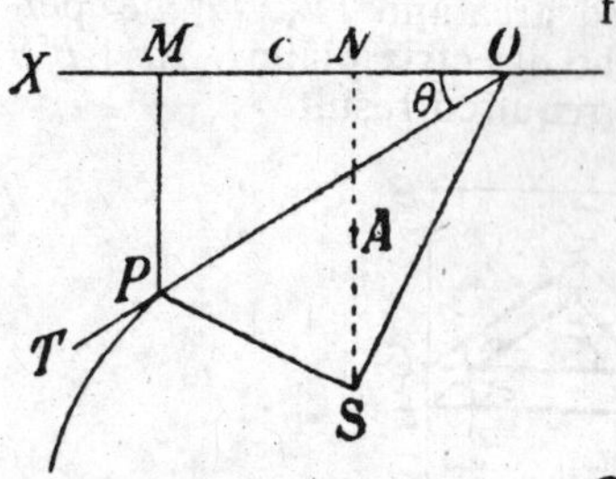

Make $\angle OPS = \angle OMP$ and $PS = PM$ so that S is the focus.

The two triangles OPS and OPM are equal in all respects.

$$OS = OM = \text{constant} = c.$$

Hence locus of S is a circle of radius c with O as centre.

Draw SN perp. to OX and let A be the middle point of SN so that A is the vertex.

If $A \to xy$, then $x = ON = OS\cos SON = c\cos 2\theta$

and $\qquad y = AN = \frac{1}{2}SN = \frac{1}{2}c\sin 2\theta$

$\therefore$ locus of A is $x^2 + 4y^2 = c^2$ ...an ellipse.

Ex. 34. *If v, v', v'' are the velocities at three points P, Q, R of the path of a projectile where the inclinations to the horizon are $\alpha, \alpha-\beta, \alpha-2\beta$ and if t, t' be the times of describing PQ, QR respectively, prove that*

$$v''t = vt', \quad \frac{1}{v} + \frac{1}{v''} = \frac{2\cos\beta}{v'}.$$

Sol. Resolving horizontally and vertically, we get

$$v \cos \alpha = v' \cos (\alpha - \beta)$$

$$= v'' \cos (\alpha - 2\beta) \quad \ldots(1)$$

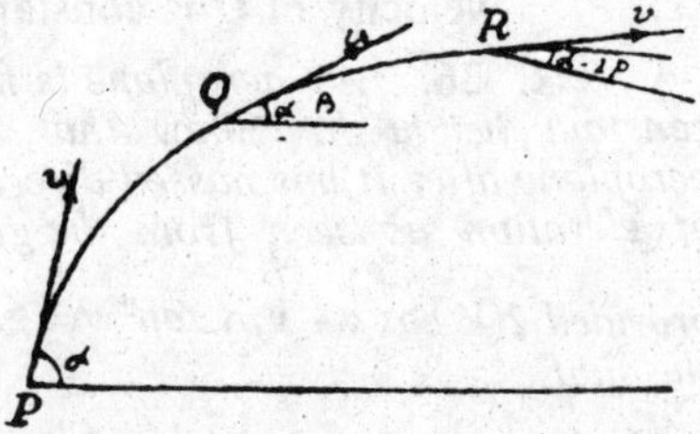

and $v' \sin (\alpha - \beta) = v \sin \alpha - gt \quad \ldots(2)$

$v'' \sin (\alpha - 2\beta) = v' \sin (\alpha - \beta) \quad gt' \; (3)$

$$\frac{1}{v} + \frac{1}{v''}$$

$$= \frac{1}{v'} \cdot \frac{\cos \alpha + \cos (\alpha - 2\beta)}{\cos (\alpha \quad \beta)} \text{ from (1)}$$

$$= \frac{1}{v'} \cdot \frac{2 \cos (\alpha - \beta) \cos \beta}{\cos (\alpha - \beta)} = \frac{2 \cos \beta}{v'}.$$

Also from (2) and (3)

$$gt = v \sin \alpha \quad v' \sin (\alpha - \beta)$$

$$gt' = v' \sin (\alpha - \beta) - v'' \sin (\alpha - 2\beta).$$

Multiplying these by v'' and v respectively and subtracting we get,

$$g(tv'' - t'v) = vv'' \sin \alpha - v'v'' \sin (\alpha - \beta) - vv' \sin (\alpha - \beta) + vv'' \sin (\alpha - 2\beta)$$

$$= vv'' \sin \alpha - \frac{v \cos \alpha}{\cos (\alpha - \beta)} \cdot v'' \sin (\alpha - \beta) - \frac{v'' \cos (\alpha - 2\beta)}{\cos (\alpha - \beta)} \times v \sin (\alpha - \beta) + vv'' \sin (\alpha - 2\beta)$$

$$= vv'' \left[\sin \alpha - \frac{\cos \alpha \sin (\alpha - \beta)}{\cos (\alpha - \beta)} - \frac{\cos (\alpha - 2\beta) \sin (\alpha - \beta)}{\cos (\alpha - \beta)} + \sin (\alpha - 2\beta)\right].$$

$$= 0$$

Hence $\qquad tv'' = t'v.$

Ex. 35. *Prove that, when a shot is projected from a gun at any angle of elevation, the shot as seen from the point of projection will appear to descend past a vertical target with uniform velocity.*

Sol. BQ is the vertical target, $OB = d$; P is the position of the shot at any instant; OP is the line of sight cutting BQ at Q. To find the velocity of Q,

Let $BQ = z$ and $\angle QOB = \phi$

Then $z = d \tan \phi = d. \dfrac{PM}{OM}$

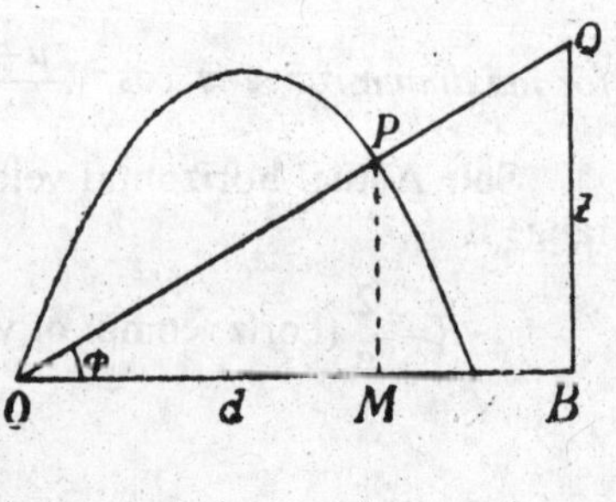

$$= d. \frac{y}{x} = d. \frac{u \sin \alpha \; t - \frac{1}{2} g t^2}{u \cos \alpha \; . \; t}$$

$$= d \; \frac{u \sin \alpha - \frac{1}{2} g t}{u \cos \alpha}$$

$$= d \left(\tan \alpha - \frac{1}{2} \frac{g}{u \cos \alpha} \; . \; t \right)$$

$$\therefore \quad z = -\frac{1}{2}\frac{gd}{u\cos\alpha} = \text{constant since } u\cos\alpha = \text{const.}$$

i.e., velocity of Q is constant.

Ex. 36. *An aeroplane is flying with constant velocity v and at a constant height h. Show that, if a gun is fired point blank at the aeroplane after it has passed directly over the gun and when its angle of elevation as seen from the gun is α, the shell will hit the aeroplane provided* $2(V\cos\alpha - v)\, v\tan^2\alpha = gh$ *when V is the initial velocity of the shot.*

Sol. A_0 is the position of the aeroplane when the shot was fired so that

$$O'A_0 = h\cot\alpha.$$

In time t, aeroplane has moved horizontally through vt and A is the point where the shot strikes it. A is the point

$$(h\cot\alpha + vt,\ h).$$

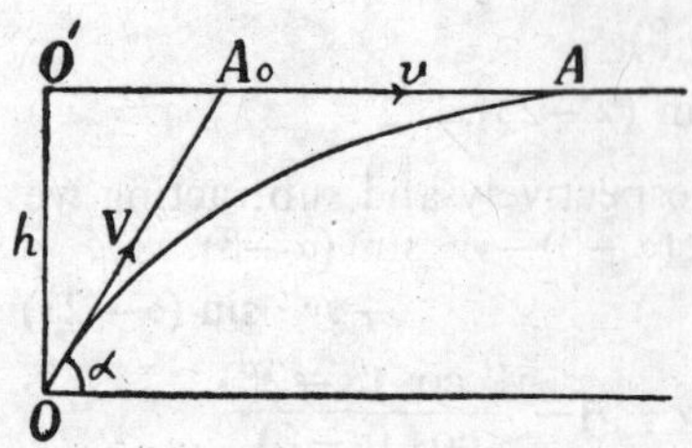

Hence $h\cot\alpha + vt$

$$= V\cos\alpha t$$

$$h = V\sin\alpha t - \tfrac{1}{2}gt^2$$

$$\therefore \quad t = \frac{h\cot\alpha}{V\cos\alpha - v} \quad \therefore \quad h = \frac{Vh\sin\alpha\cot\alpha}{V\cos\alpha - v} - \tfrac{1}{2}g\frac{(h\cot\alpha)^2}{(V\cos\alpha - v)^2}$$

i.e.,
$$h = \frac{2Vh\cos\alpha\,.\,(V\cos\alpha - v) - gh^2\cot^2\alpha}{2(V\cos\alpha - v)^2}$$

i.e.,
$$2(V\cos\alpha - v)^2 = 2V\cos\alpha\,(V\cos\alpha - v) - gh\cot^2\alpha$$

or
$$2(V\cos\alpha - v)\, v\tan^2\alpha = gh.$$

Ex. 37. *A battle ship is steaming ahead with velocity u. A gun is mounted on the ship so as to point straight backwards and is set at an angle of elevation α. If v be the velocity of projection relative to the gun, show that the range is* $\dfrac{2v}{g}\sin\alpha\,(v\cos\alpha - u)$ *and the angle for maximum range is* $\cos^{-1}\dfrac{u + \sqrt{u^2 + 8v^2}}{4v}$.

Sol. Actual horizontal velocity of projection is v, $\cos\alpha - v$, hence range is

$$R = \frac{2}{g}\,\text{(horiz. comp. of vel.) (initial vert. comp. of vel.)}$$

$$=\frac{2}{g}(v\cos\alpha-u)\,v\sin\alpha.$$

For maximum $\frac{dR}{d\alpha}=0$

$$\therefore\quad (v\cos\alpha-u)\cos\alpha-v\sin\alpha\sin\alpha=0$$

or $$v(\cos^2\alpha-\sin^2\alpha)=u\cos\alpha$$

or $$v(2\cos^2\alpha-1)-u\cos\alpha=0$$

or $$2v\cos^2\alpha-u\cos\alpha-v=0$$

$$\therefore\quad \cos\alpha=\frac{u\pm\sqrt{u^2+8v^2}}{4v}$$

the negative sign is to be discarded as α becomes obtuse, hence.

Ex. 38. *The radii of the front and hind wheels of a carriage are a and b and c is the distance between the axle-trees ; a particle of mud driven from the highest point of the hind wheel is observed to alight on the highest point of the front wheel. Show that the velocity of the carriage is* $\sqrt{\frac{g(c+b-a)(c+a-b)}{4(b-a)}}$.

Sol. If v be velocity of the carriage, the uelocity of the highest point is $2v$ horizontally. Thus the initial velocity of the mud particle is $2v$ horizontally, hence the velocity of the mud particle relative to carriage is v horizontally.

The horizontal distance moved is thus $=OM=\sqrt{c^2-(b-a)^2}$ and the vertical height fallen is $2(b-a)$.

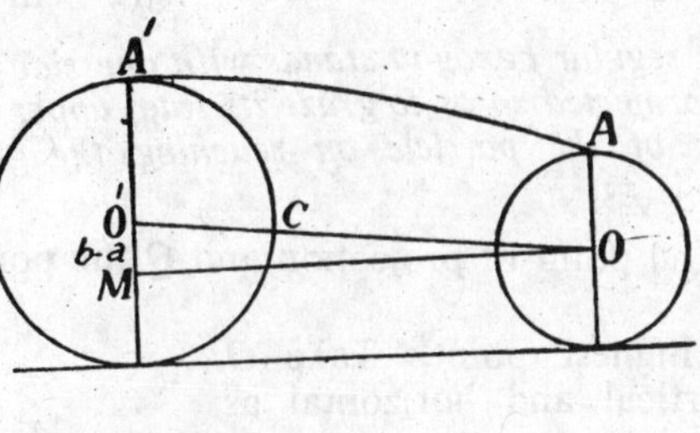

Hence $\sqrt{c^2-(b-a)^2}=vt$

and $$2(b-a)=\tfrac{1}{2}gt^2$$

$$\therefore\quad \sqrt{c^2-(b-a)^2}=v.\sqrt{\frac{4}{g}(b-a)}$$

$$v=\sqrt{\frac{g(c-b+a)(c+b-a)}{4(b-a)}}$$

Ex. 39. *A speck of mud is thrown off the rim of a wheel of radius a which is rolling in a straight line on horizontal ground with a speed of v ft. per second and just alights on the wheel again at a point on the level with its centre. When the speck of mud is thrown off, its angular distance from the top of the wheel is θ, measured in the sense in which the wheel is travelling. Show that*

$$2v^2 \cos\theta = ag\,(1-\sin\theta).$$

Sol.

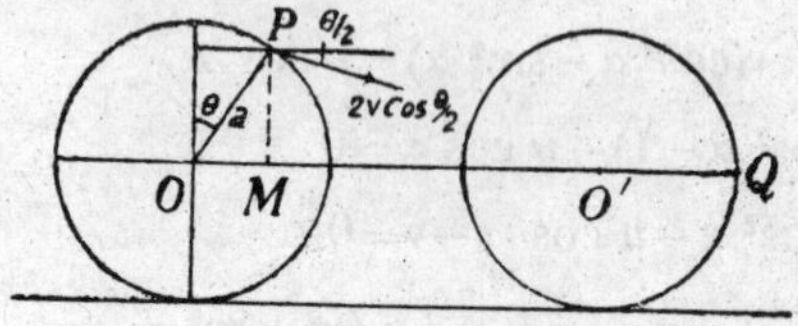

The speck of mud starts from a point P and falls at Q. Horizontal distance moved by the particle$=MQ=OO'+O'Q-OM$

$$=vt+a-a\sin\theta.$$

The velocity of the particle is $2v\cos\theta/2$ at an angle $\theta/2$ with the horizontal.

Hence horizontally $vt+a-a\sin\theta=2v\cos\theta/2.\cos\theta/2.t$

i.e., $\quad vt(2\cos^2\theta/2-1)=a(1-\sin\theta) \quad \therefore \quad vt=\dfrac{a(1-\sin\theta)}{\cos\theta}.$

Vertically $\quad a\cos\theta=2v\cos\theta/2.\sin\theta/2.t+\frac{1}{2}gt^2$

or $\quad a\cos\theta=vt\sin\theta+\frac{1}{2}gt^2$

$$=\frac{a(1-\sin\theta)\sin\theta}{\cos\theta}+\tfrac{1}{2}g\,.\,\frac{a^2(1-\sin\theta)^2}{v^2\cos^2\theta}$$

$\therefore\ 2v^2[\cos^2\theta-(1-\sin\theta)\sin\theta]\cos\theta=ag(1-\sin\theta)^2$

or $\quad 2v^2\cos\theta(1-\sin\theta)=ag(1-\sin\theta)^2$

or $\quad 2v^2\cos\theta=ag(1-\sin\theta)$

Ex. 40. *A regular hexagon stands with one side on the ground and a particle is projected so as to graze its four upper vertices. Show that the velocity of the particle on reaching the ground to its least velocity as $\sqrt{31}$ to $\sqrt{3}$.*

Sol. P is the point of projection and Q the point where it falls to the ground.

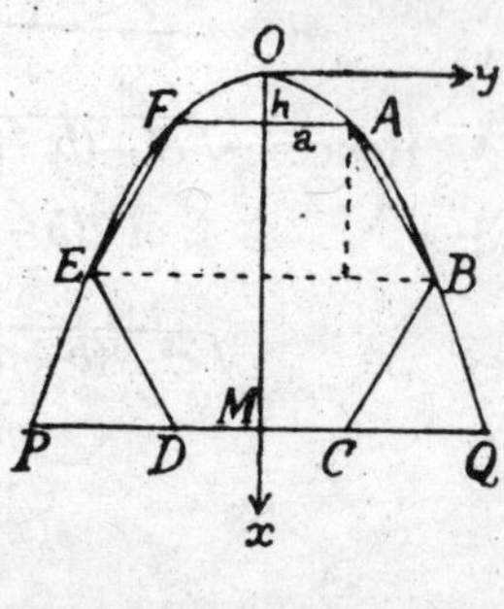

O is the highest point. Take O as origin and vertical and horizontal as axis of x and y.

Let parabola be $y^2=4bx$.

$A\to(h, a)$ where $h=$height of O above AF and $2a=$side of the hexagon.

$B\to(h+a\sqrt{3},\ 2a)$

Hence $\quad a^2=4bh$

and $\quad 4a^2=4b(h+a\sqrt{3})$

$\therefore \quad 4a^2=a^2+4ab\sqrt{3}$

$$\therefore \quad 4b=\sqrt{3}a \text{ and } h=\frac{a}{\sqrt{3}}$$

If u be the velocity at O which is entirely horizontal the equations are $y=ut$, $x=\frac{1}{2}gt^2$ $\therefore$ $y^2=\frac{2u^2}{g}x$ which is also the same parabola.

Hence $\frac{2u^2}{g}=4b=\sqrt{3}a$ $\therefore$ $u^2=\frac{\sqrt{3}}{2}ag$

If v be the velocity at Q, $v^2=u^2+2g.OM=u^2+2g(h+2a\sqrt{3})$

i.e.,
$$v^2=u^2+2g\left(\frac{a}{\sqrt{3}}+2a\sqrt{3}\right)$$

Hence
$$\frac{v^2}{u^2}=1+\frac{2ga\left(\frac{1}{\sqrt{3}}+2\sqrt{3}\right)}{\frac{\sqrt{3}}{2}ag}$$

$$=1+\frac{4}{\sqrt{3}}\left(\frac{1}{\sqrt{3}}+2\sqrt{3}\right)$$

$$=1+\frac{4}{3}+8=\frac{31}{3}$$

$$\therefore \quad \frac{v}{u}=\sqrt{\frac{31}{3}}.$$

Ex. 41. *AB is a horizontal straight line of length 2a; the middle point of AB is the centre of a sphere of radius b (<a). If a particle projected from A with velocity v just clears the sphere and passes through B, prove that*

$$v^2=\frac{g(2a^2-b^2)}{\sqrt{a^2-b^2}} \quad or \quad \frac{g(a^2+4b^2)}{2b}$$

according as $a<$ *or* $>b\sqrt{2}$.

Sol. Let the projectile just graze the sphere at the top O. With O as origin and downward vertical and horizontal through O as x and y-axes, the equation to the path is $y^2=4lx$ say where $4l$ is the latus rectum.

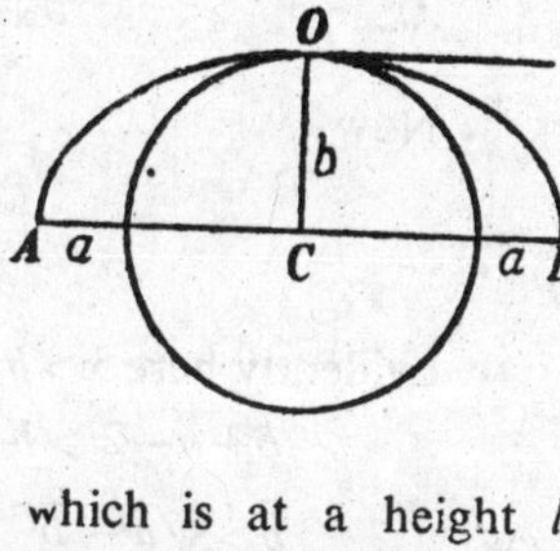

Since it passes through $B(b, a)$, $a^2=4lb$.

Now v is due to fall from directrix which is at a height l above O.

Hence $$v^2=2g(l+b)=2g\left(\frac{a^2}{4b}+b\right)=\frac{g(a^2+4b^2)}{2b}.$$

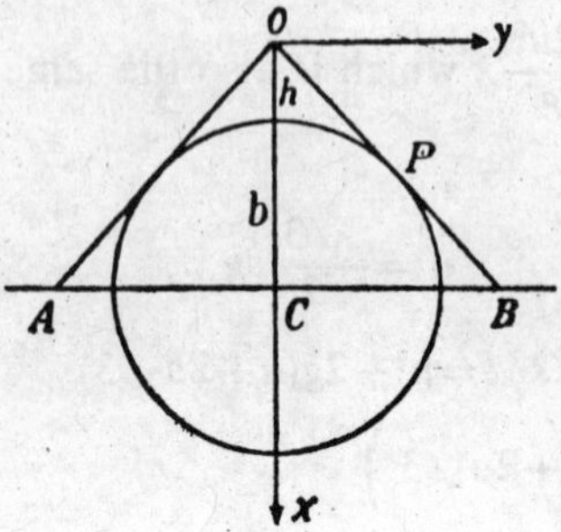

Let the particle graze the sphere at the same point P.

Here the vertex O of the parabola is above the top of the sphere say at a height h and with O as origin, axes vertically downward and horizontal, the parabola is

$$y^2=4lx$$

P be the point (xy) then the tangents to the parabola and the circle at P are same.

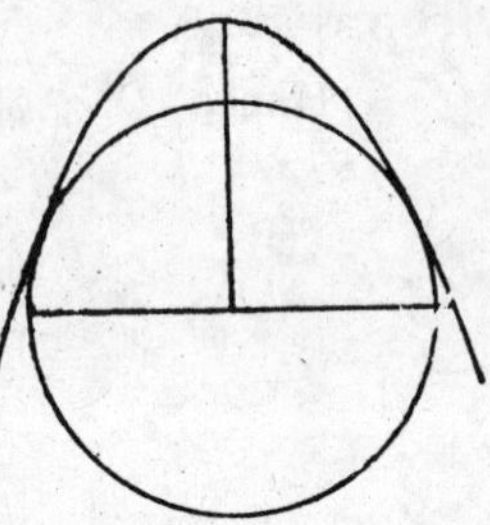

Now the circle is

$$y^2+(x-b-h)^2=b^2$$

for the parabola, $$\frac{dy}{dx}=\frac{2l}{y}$$

for the circle $$\frac{dy}{dx}=-\frac{x-b-h}{y}$$

Hence $$\frac{2l}{y}=-\frac{x-b-h}{y}$$

or $$x=b+h-2l$$

Substituting this in the equation of the circle

$$y^2+4l^2=b^2 \quad \text{or} \quad y^2=b^2-4l^2$$

For the point B, $$a^2=4l(b+h)$$

for the point P, $$b^2-4l^2=4l(b+h-2l)$$

$\therefore$ $$b^2-4l^2=a^2-8l^2 \quad \text{or} \quad 4l^2=a^2-b^2$$

$\therefore$ $$2l=\sqrt{a^2-b^2}$$

$\therefore$ $$b+h=\frac{a^2}{4l}=\frac{a^2}{2\sqrt{a^2-b^2}}.$$

Now $$v^2=2g(l+b+h)$$

$$=2g\left(\frac{\sqrt{a^2-b^2}}{2}+\frac{a^2}{2\sqrt{a^2-b^2}}\right)=g\,\frac{2a^2-b^2}{\sqrt{a^2-b^2}}$$

Evidently here $x>h$

$\therefore$ $$b+h-2l>h \quad \text{or} \quad b>2l$$

i.e., $$b>\sqrt{a^2-b^2} \quad \text{or} \quad b^2>a^2-b^2 \quad \text{or} \quad 2b^2>a^2$$

$$\text{or} \quad \sqrt{2}b>a.$$

EXAMPLES III (F)

1. Particles fall down the diameters of a vertical circle; prove that the locus of the foci of their subsequent parabolas is a circle.

2. If particles slide down the chords of a vertical circle to its lowest point and are then allowed to move freely, prove that the locus of the foci of their subsequent paths is a cardioide.

3. A number of bodies slide from rest down the chord of a vertical circle starting from its highest point and afterwards move freely; prove that the locus of the foci of their paths is a circle whose radius is half that of the given circle.

4. Prove that during the flight of a particle projected at an elevation α on a line through the point of projection of inclination β, the direction of motion turns through an angle whose cotangent is

$$\tfrac{1}{2}\cos\beta \sec\alpha \operatorname{cosec}(\alpha-\beta)-\tan\alpha$$

5. Prove that the locus of the foci of all the trajectories passing through two given points is a hyperbola.

6. Prove that in the parabolic path of a projectile the rate of change of the direction of motion is $\dfrac{u}{2y}$ where u is the horizontal velocity and y denotes the depth of the projectile below the directrix at any instant.

7. A gun is fired from a moving platform and the ranges of the shot are observed to be R and S when the platform is moving forwards and backwards respectively with velocity V. Prove that the elevation of the gun is

$$\tan^{-1}\left\{\frac{g}{4V^2}\cdot\frac{(R-S)^2}{R+S}\right\}.$$

8. Two points A and B are a distance a apart, their heights above the ground being h_1 and h_2. Prove that the least speed with which a stone can be thrown from the ground-level so as to pass through both the points is

$$\sqrt{g(h_1+h_2+a)}.$$

9. A shell fired with velocity V at elevation θ hits an airship at height H, which is moving horizontally away from the gun with velocity v. Show that if $(2V\cos\theta-v)(V^2\sin^2\theta-2gH)^{\frac{1}{2}}=vV\sin\theta$, the shell might also have hit the airship if the latter had remained stationary in the position it occupied when the gun was actually fired.

10. A stone is thrown at a flying bird. The direction of projection has an angle of elevation α and at the instant of projection the bird is directly overhead at a height h flying in a horizontal straight line with constant acceleration f and velocity U. Prove that no speed of projection will make the stone hit unless

$$U^2\tan\alpha>[(f^2h^2+2gh\,U^2)^{\frac{1}{2}}-fh].$$

11. A shot passes through two screens distant d from one another at heights h, k above the point of projection. If t be the time taken to pass between the screens, show that the muzzle velocity of the shot is that due to a fall through a height

$$\tfrac{1}{2}\left[k+h+\tfrac{1}{4}gt^2+\frac{d^2+(k-h)^2}{gt^2}\right].$$

12. The guns of a ship and a fort at a height a above sea-level have each a maximum range R on the level. Show that the ship will come under fire from the fort at a range $R+a$ but cannot return the fire till the range is reduced to $R-a$.

13. A gun on land is at a height h above the similar gun on a ship at sea. The muzzle velocity of their shells is v. Show that the gun on land can engage the gun at sea at a horizontal distance $\frac{v}{g}\left[(v^2+2gh)^{\frac{1}{2}}-(v^2-2gh)^{\frac{1}{2}}\right]$ greater than that at which the gun at sea can engage the gun on land.

14. A projectile aimed at a mark, which is in a horizontal plane through the point of projection, falls a feet short of it when the elevation is α and goes b feet too far when the elevation is β. Show that if the velocity of projection be the same in all cases the proper elevation is

$$\tfrac{1}{2}\sin^{-1}\frac{a\sin 2\beta+b\sin 2\alpha}{a+b}.$$

15. A particle is projected under gravity from A so as to pass through B. Show that for a given velocity of projection there are two paths. Show that if B has horizontal and vertical co-ordinates x, y referred to A and the velocity of projection is $\sqrt{2gh}$, the angle between the paths at B is a right angle if B lies on the ellipse $x^2+2y^2=2hy$.

16. A shot with a velocity v at an elevation α strikes a point P on a horizontal plane through the point of projection. If the point P is receding from the gun with velocity u, show that the elevation must be changed to θ where $\sin 2\theta=\sin 2\alpha+\frac{2u}{v}\sin\theta$.

17. A particle is projected so as to just graze the four upper corners of a regular hexagon, whose side is a and which is placed vertically with one side on the table. Show that the range on the table is $a\sqrt{7}$ and the square of the time of flight is $\frac{28a}{g\sqrt{3}}$.

18. A trapezium with parallel sides of length $2a$, $4a$ and the non-parallel sides each of length $2a$ is placed on the ground with its plane vertical and with the longest side in contact with the ground. A ball is projected from the corner lying on the ground so as just to graze the other three corners of the trapezium. Prove that the highest height reached by the particle is $\frac{4a}{\sqrt{3}}$ and the total time of flight is $\sqrt{\frac{32a}{g\sqrt{3}}}$.

19. A stone is thrown in such a manner that it would just hit a bird at the top of a tree and afterwards reach a height double that of a tree. If, at the moment of throwing the stone, the bird flies away horizontally, show that, notwithstanding this, the stone will hit the bird if its horizontal velocity be to that of the bird as $\sqrt{2}+1:2$.

20. A particle is to be projected so as just to pass through three equal rings, of diameter d, placed in parallel vertical planes at distances a apart, with their highest points in a horizontal straight line at a height h above the point of projection. Prove that the elevation must be $\tan^{-1}\frac{2\sqrt{hd}}{a}$.

21. Two inclined planes intersect in a horizontal line, their inclination to the horizon being α and β. If a particle be projected at right angles to the former from a point in it so as to strike the other at right angles, the velocity

of projection is $\sin\beta \left\{ \dfrac{2ag}{\sin\alpha \quad \sin\beta\cos(\alpha+\beta)} \right\}^{\frac{1}{2}}$, a being the distance of the point of projection from the intersection of the planes.

22. Two inclined planes, of equal height h and inclination α, are placed back to back. A ball projected along the surface of the plane at an angle β with the horizontal, flies over the top and falls at the foot of the other plane. Show that the velocity of projection is $\frac{1}{2}\sqrt{gh}\,\operatorname{cosec}\beta\,\sqrt{8+\operatorname{cosec}^2\alpha}$.

23. Two parallel straight lines are inclined to the horizon at an angle α. A particle is projected from a point midway between them so as to graze one of the lines and then strike the other at the right angles. Show that the angle of projection makes with either of the lines an angle $\tan^{-1}\{(\sqrt{2}-1)\cot\alpha\}$.

4

Work, Energy and Impulse

4·1. Work ; when is a force said to do work ?

A force is said to do work when its point of application moves in the direction in which the force acts. By the point of application is meant the particle on which the force is acting. When the point of application moves in the direction opposite to that in which the force acts the force is said to do negative work, or work is said to be done against the force. For example, when a body falls towards the earth, the weight of the body does work and when we raise the body away from the earth, we do work against the weight. *The work done by a constant force is measured by the product of the force and the distance through which the point of application is moved in the direction of the force.* Thus when a force moves a particle along its direction through a distance s, the work done by F is Fs units of work provided the same system of units is used to measure both F and s.

In the *F.P.S.* system the unit of work is called a **Foot Poundal** and is equal to the work done by a force of one poundal in moving its point of application through one foot in the direction of the force. In the *C.G.S.* system the unit of work is called an **Erg** and is equal to the work done by a force of one dyne in moving its point of application through one centimetre in the direction of the force.

One *foot pound* is the work done in raising a mass of one pound vertically through a height of one foot. *One foot pound*$=g$ *foot poundals*. Similarly one gramme-centimetre is the work done in raising one gramme vertically through one centimetre and one gramme-centimetre$=g$ ergs. In the M.K.S. system, the unit of work is called a joule.

If a force is variable and acts along a line, being some function of the distance x from a fixed point on the line, let its magnitude be F at a distance x from the fixed point O. Then when the particle on which it acts moves a small distance δx, then the work done is equal to $F.\delta x$, taking F to remain constant through this displacement.

Hence the total work done from the point $x=a$ to the point $x=b$ is

$$W=\int_a^b Fdx.$$

If the particle moves along a smooth plane curve, then only force which does work due to a small displacement δs along the curve is the tangential component of the force. If F be the tangential component of the force then the work done from a point $s=a$ to the point $s=b$ is $W=\int_a^b Fds.$

Work done in extending an elastic string.

When an elastic string of natural length l is extended beyond l, then at every point of its length a force of tension T acts, which is given by Hooke's Law

$$T=\lambda\frac{x}{l},$$

where x is the extension of the string, and λ a constant is called the modulus of elasticity. Hence the work done in extending the string through x is

$$W=\int_0^x \lambda\frac{x}{l}dx=\frac{1}{2}\frac{\lambda x^2}{l}$$

and the work done in extending it from an extension x_1 to extension x_2 is

$$W=\int_{x_1}^{x_2} \lambda\frac{x}{l}dx=\frac{1}{2}\frac{\lambda}{l}(x_2^2-x_1^2)$$

$$=\frac{\lambda}{l}\frac{x_1+x_2}{2}(x_2-x_1)=\frac{T_1+T_2}{2}.\text{ (Extension)}$$

where T_1 and T_2 are the tensions when the extensions are x_1 and x_2.

Power. The *power* of an agent (that which does work *e.g.*, a steam engine, a horse etc.) is measured by the rate at which the agent does work.

Horse Power. One horse power is the power of an agent which works at the rate of 500 foot-poundals per sec.

In M.K.S. system the absolute units of power are one joule/sec. In engineering practice, the units of power that are generally adopted are : One watt in the M.K.S. system,

Watt. One watt is the power of an agent which works at the rate of one joule per sec.

Hence one watt=one joule per sec.

$$=10^7 \text{ ergs per sec.}$$

It is easily seen, therefore,

$$1\ H.P.=746 \text{ watts approximately.}$$

When an engine is marked H horse-power it means that under favourable conditions it can work at the rate of 550 H foot-pounds per sec. Hence, if F be the force exerted by the engine when its velocity is v then we must have

$$Fv=550\ H$$

$$\therefore \quad H=\frac{Fv}{550} \quad \text{or} \quad F=\frac{550\ H}{v}.$$

Ex. 1. *A motor car weighing 10 quintals and travelling at 12 meters/sec. is brought to rest in 18 metres by the application of its brakes. Find the work done by the force of resistance due to brakes.*

Sol Let the resitance be assumed to be uniform. If r is retardation, then since u (initial velocity)=12 m/sec, v (final velocity)=om/sec and s (distance travelled)=18 metres, we get from

$$v^2-u^2=2\,(-r)\,.\,s$$

that $$0-144=-2r\,.\,18 \quad \text{or} \quad r=4 \text{ m/sec}^2$$

Now mass of motor car=1000 kg

$\therefore$ force of resistance=4000 Newtons.

Hence work done=4000×18 joules.

$$=7347 \text{ kg-meters (approx.)}$$

Ex. 2. *An engine working at the rate of 400 H.P. is pulling a train which with the engine weighs 150 tons, up an incline of 1 in 120 at a steady speed of 30 m.p.h. Find the average frictional resistance expressed in lb. wt. per ton.*

The brake-van weighing 20 tons, becomes detached. How far will it run before stopping ? If the engine continues to work at the same H.P., what will be the value of the speed when it again becomes steady.

Sol. Here G=comp. of weight down the plane

$$=150\times 2240\times \tfrac{1}{120}=2800 \text{ lbs.}$$

$\therefore$ If R be the resistance, total force down the plane is

$$(R+2800) \text{ lbs.}$$

$$=30 \text{ m.p.h.}=44 \text{ ft. per sec.}$$

$$\therefore \quad (R+2800)\,v=550\times H=550\times 400$$

$\therefore \qquad R=2200 \text{ lbs.}=\frac{2200}{150} \text{ lbs. wt. per ton.}$

$\qquad =14\frac{2}{3} \text{ lbs. wt. per ton.}$

When the brake-van is detached.

$$G=130\times 2240\times \tfrac{1}{120}=\tfrac{7280}{3} \text{ lbs.}, \; R=14\tfrac{2}{3}\times 130 \text{ lbs.}$$

Now $\qquad (R+G)V=550\times 400 \quad \therefore \quad V=\dfrac{550\times 400}{\frac{44}{3}\times 130+\frac{7280}{3}}$

$$=34\tfrac{18}{13} \text{ m.p.h.}$$

For the brake-van, u=initial velocity=44 ft. per sec.

$$R=\tfrac{44}{3}\times 20 \text{ lbs.}, \qquad G=20\times 2240\times \tfrac{1}{120} \text{ lbs.}$$

$\therefore \qquad R+G=\frac{2000}{3}$ lbs.

If f be the retardation,

$$f=\frac{R+G}{m}=\frac{2000}{3\times 20\times 2240} \; 32 \text{ ft. per sec}^2.$$

$$=\tfrac{10}{21} \text{ ft. per sec.}^2$$

It will stop after a distance s where

$$u^2=2fs \; \therefore \; s=\frac{44\times 44}{2\times \frac{10}{21}}=2032\tfrac{1}{5} \text{ ft.}$$

Ex. 3. *An engine exerting its full horse-power draws a train up a slope of 1 in n with constant speed v, frictional resistance to motion being* $\frac{1}{m}$ *of the weight of the train. Prove that the greatest speed with which the engine would draw the train along a level track against the same resistance is* $\left(1+\frac{m}{n}\right)v$.

Sol. Let W be the weight of the train.

Then $\qquad R$=frictional resistance$=\dfrac{W}{m}$

$\qquad G$=component of weight down the plane$=\dfrac{W}{n}$

$\therefore \qquad F$=total force down the plane$=W\left(\dfrac{1}{m}+\dfrac{1}{n}\right)$lbs.wt

$$H=\frac{Fv}{550}=\frac{Wv}{550}\left(\frac{1}{m}+\frac{1}{n}\right)$$

On the level

$$H=\frac{RV}{550}=\frac{WV}{550\,m}$$

$$\therefore \quad \frac{Wv}{550}\left(\frac{1}{m}+\frac{1}{n}\right)=\frac{WV}{550\,m}$$

$$\therefore \quad T=\left(1+\frac{m}{n}\right)v.$$

Ex. 4. *In starting a train the pull of the engine on the rails is at first constant and equals P ; after the speed attains a certain value u the engine works at a constant rate R=Pu. Prove that when the engine has attained a speed v > u, the time t and distance x from the start are given by*

$$t=\frac{1}{2}\frac{M}{Rg}(v^2+u^2),\ x=\frac{1}{3}\frac{M}{Rg}(v^3+\tfrac{1}{2}u^3)$$

where M is the mass of the engine and the train.

Sol. If f be the acceleration produced by the pull P then $Mf=Pg$

hence $$u=ft_1 \ \therefore\ t_1=\frac{u}{f}=\frac{Mu}{Pg}=\frac{Mu^2}{Rg} \text{ writing } P=\frac{R}{u}$$

and $$u^2=2fx_1 \ \therefore\ x_1=\frac{Mu^2}{2Pg}=\frac{Mu^3}{2\,Rg}.$$

If F be the force then $M\dfrac{dv}{dt}=Fg=\dfrac{R}{v}g$ where R is the rate of work,

or $$Mv\frac{dv}{dt}=Rg.$$

Integrating $\frac{1}{2}Mv^2=Rgt+\text{const.}=Rgt+\frac{1}{2}Mu^3$

$$\therefore \quad t=\frac{M}{2\,Kg}(v^2-u^2).$$

Hence the total time $=t_1+t=\dfrac{M}{2Rg}(v^2+u^2)$.

Also $$Mv\frac{dv}{dx}=\frac{R}{v}g \text{ or } Mv^2\frac{dv}{dx}=Rg$$

Integrating $\frac{1}{3}Mv^2=Rgx+\text{const.}=Rgx+\frac{1}{2}Mu^3$

$$\therefore \quad x=\frac{M}{3\,Rg}(v^3-u^3)$$

$$\therefore \quad \text{total distance}=x_1+x=\frac{M}{3\,Rg}(v^3+\tfrac{1}{2}u^3).$$

EXAMPLES IV (A)

1. A train whose total weight is 300 tons is drawn up an incline of 1 in 100. When it has reached a speed of 12 m.p.h., the H.P. exerted is 1000. Prove that the acceleration at this instant is 1·68 ft./sec^2.

2. A train of mass M lbs. is ascending a smooth incline of 1 in n and then the velocity of the train is v ft./sec., its acceleration is f ft./sec^2. Prove that the effective H.P. of the engine is $\dfrac{Mv\,(nf+g)}{550\,ng}$.

3. A constant force P lbs. wt. is applied vertically upwards to a body of mass M lbs. for a certain time. It then ceases to act and the body continues to move upwards to reach its greatest height in a total time t secs. from the commencement of the motion. Prove that the greatest H.P. developed by the force of P lbs. wt. is $\dfrac{(P-M)\,gt}{550}$.

4. A body of mass M lbs. starts with a speed of u ft./sec. The body moves under the action of a force which does work at rate of H horse-power. Prove that the equation which determines the speed v at a time t is

$$Mv\,\frac{dv}{dt}=550\,gH$$

5. Two trains of equal masses are drawn along smooth level lines by engines, one of which X exerts a constant pull while the other Y works at a constant rate. Both starts with the same initial velocity u and after a time t both again have the same velocity v. Show that Y describes a greater distance along the interval.

4·2. Energy. Energy of an agent is defined to be its capacity to do work. Energy may be of several kinds, mechanical energy, chemical energy, electrical energy, heat, light etc. Mechanical energy of a body is the energy possessed by it by virtue of its position or motion. Mechanical energy is again of two kinds, kinetic energy and potential energy.

Kinetic Energy. Kinetic energy of a body is its capacity to do work by virtue of its motion. It is measured by the amount of work which this body will do in coming to rest. A moving bullet has kinetic energy because when it strikes a wall of sand, it penetrates into it before coming to rest, thus doing work against the force of resistance offered by the sand. Hence it follows that before the bullet struck the wall it possessed K.E. *i.e.*, capacity to do work.

The kinetic energy of a particle of mass m moving with velocity v is $\frac{1}{2}mv^2$.

Let F be the resistance to its motion which produces a retardation f so that $F=mf$.

When the particle is moving in a straight line, $f=v\,\dfrac{dv}{dx}$ where v is its velocity at distance x from a fixed point O on the line.

Hence the work done in bringing the particle to rest

$$=-\int F dx=-m\int v\,\frac{dv}{dx}\,dx=-m\int_v^0 v\,dv=\tfrac{1}{2}mv^2.$$

When the particle is moving in a plane curve, the tangential force does work for displacement along the curve. Now the tangential component for acceleration is $f = v\,\dfrac{dv}{ds}$.

Hence the work done

$$= -\int F ds = -m \int v\,\frac{dv}{ds}\,ds$$

$$= -m \int_v^0 v\,dv = \tfrac{1}{2}\,mv^2.$$

Hence in either case $K.E. = \frac{1}{2}\,mv^2$.

Potential Energy. Potential energy of a body is its capacity to do work by virtue of its position. It is measured by the amount of work which the body will do in coming from its present position to some standard position usually called the zero position.

A bent spring has potential energy because it can always do work in recovering its natural length.

If a body of mass m is placed at a height h above the ground then when let fall, it will do work of amount mgh which is, therefore, potential energy of the body.

4·3. Principle of Energy. *A particle is made to move on a given smooth plane curve under the action of given forces in the plane, to find the motion.*

Let P be the position of the particle at time t where $AP = s$, A being a fixed point on the curve. Let v be the velocity at P, X, Y the components of the force acting on the particle parallel to axes of OX, OY; R the reaction along the normal at P.

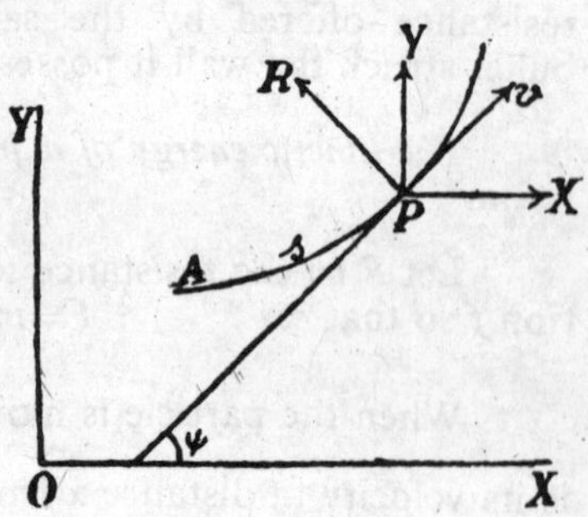

The tangential and normal equations of motion are

$$mv\,\frac{dv}{ds} = X\cos\psi + Y\sin\psi$$

and $$m\,\frac{v^2}{\rho} = -X\sin\psi + Y\cos\psi + R.$$

The first gives

$$mv\,\frac{dv}{ds} = X\,\frac{dx}{ds} + Y\,\frac{dy}{ds}$$

and therefore, integrating we get

$$\tfrac{1}{2}\,mv^2 = \int (X dx + Y dy) + C \text{ where } C \text{ is a constant.}$$

If $Xdx+Ydy$ be a complete differential of some function $\phi(x, y)$

so that $$X=\frac{\partial\phi}{\partial x},\ Y=\frac{\partial\phi}{\partial y}$$

then $$\tfrac{1}{2}mv^2=\int\left(\frac{\partial\phi}{\partial x}\,dx+\frac{\partial\phi}{\partial y}\,dy\right)+C=\phi(xy)+C. \quad \ldots(1)$$

If the particle be projected from a point (x_0, y_0) with velocity V

then $$\tfrac{1}{2}mV^2=\phi(x_0, y_0)+C.$$

Hence $\tfrac{1}{2}mv^2-\tfrac{1}{2}mV^2=\phi(xy)-\phi(x_0, y_0)$ ...(2)

This result is quite independent of the path of the particle and depends only on the initial and final positions of the particle.

Now $(Xdx+Ydy)$ is the work done by the forces (X, Y) during a small displacement ds along the curve so that $\int[Xdx+Ydy]=\phi(xy)$ is the total work done on the particle during its motion from its point of projection to any point P. The quantity $\varphi\,(xy)$ is called the *Work-Function* of the system of forces and forces are called the *conservative forces.* Thus conservative forces are such that the work done by them depends only upon the initial and final values of the work function, *i.e.*, upon the initial and final values of the forces and not upon the path followed by the particle. These help the motion when it is in their direction and oppose it when the motion is in the opposite direction and maintain their character irrespective of the direction of motion.

Now the left hand side of (2) represents the change in the kinetic energy of the particle and the right hand side represents the work done as the particle moves from its initial position to its final position. Thus we get the result that *the change in kinetic energy is equal to the work done by the external forces, if the force system be conservative.* This is called the **Principle of Energy**.

Now the potential energy of the particle when at P

= work done by the forces as the particle moves from P to some standard position, say $(x_1\, y_1)$.

$$\int_{(x\,y)}^{(x_1y_1)}(Xdx+Ydy)=\Big[\phi(xy)\Big]_{(x\,y)}^{(x_1y_1)}=\phi(x_1y_1)-\phi(xy).$$

Hence (Kinetic Energy+Potential Energy) of the particle when at P

$$=\phi(xy)+C+\phi(x_1y_1)-\phi(xy)$$
$$=C+\phi(x_1y_1)=\text{constant}.$$

Hence *when a particle moves under the action of a Conservative System of Forces, the sum of its Kinetic and Potential Energies is constant throughout the motion.* This is called the principle of **Conservation of Energy**.

Ex. 5. *If an elastic string whose natural length is that of a uniform rod, be attached to the rod at both ends suspended by the middle point, show that the rod will sink until the strings are inclined to the horizon at an angle θ given by the equation*

$$cot^3 \frac{\theta}{2} - cot \frac{\theta}{2} = 2n,$$

given that the modulus of elasticity of the string is n times the weight of the rod.

Sol. Let AB be the rod, length $2a$; O the middle point of the string.

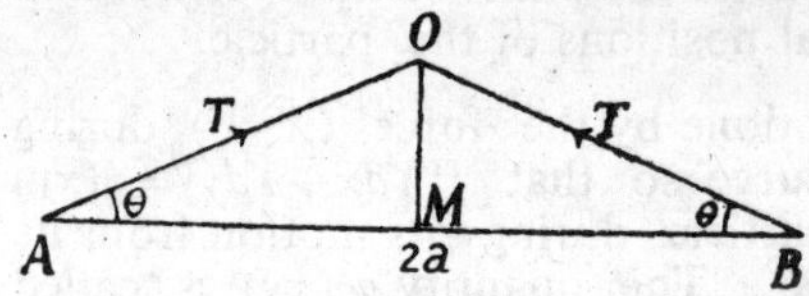

Height descended by the rod $= OM = a \tan\theta$

$\therefore$ work done by its weight $= mg\,a \tan\theta$.

Work done against the tension of the string

$$= 2.\frac{O+T}{2} \text{extension} = \lambda.\ \frac{OA-a}{a}.\ (OA-a)$$

$$= nmg.\frac{a\sec\theta - a}{a}\ (a\sec\theta - a)$$

$$= nmg\ a\ (\sec\theta - 1)^2.$$

Since the system starts from rest and comes to rest, kinetic energy is zero both initially and finally hence the work done is also zero *i.e.*, work done by the weight of the rod = that by tension,

i.e., $\quad mg\,.\,a\tan\theta = nmga\ (\sec\theta - 1)^2$

or $\quad \tan\theta = n(\sec\theta - 1)^2$

or $\quad \tan\theta = n\ \dfrac{(1-\cos\theta)^2}{\cos^2\theta}$

or $\quad \sin\theta\cos\theta = n(1-\cos\theta)^2$

or $\quad 2\sin\theta/2\cos\theta/2\ (\cos^2\theta/2 - \sin^2\theta/2) = 4n\sin^4\theta/2,$

or $\quad \cot^3\dfrac{\theta}{2} - \cot\dfrac{\theta}{2} = 2n.$

Ex. 6. *A heavy uniform chain of length 2l, hangs over a small smooth fixed pulley, the length l+c being at one side and l−c at the other. If the end of the shorter portion be held and then let go, show that the chain will slip off the pulley in time*

$$\sqrt{\frac{l}{g}}.\ log\ \frac{l+\sqrt{l^2-c^2}}{c}.$$

Sol. When the lengths on the two sides are $l+c+x$ and $l-c-x$, the depth of the centre of gravity of the string below the pulley is ζ where

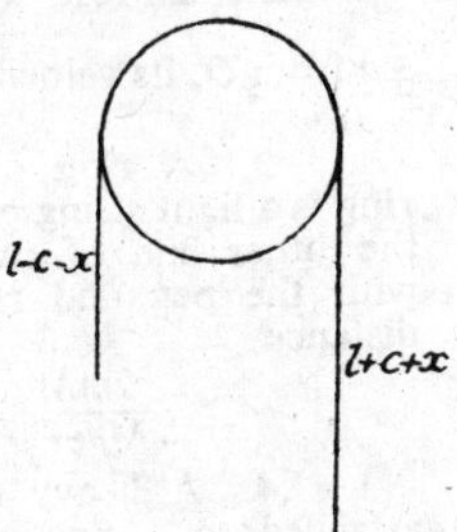

$$m\zeta=\frac{m}{2l}(l+c+x).\frac{l+c+x}{2}$$

$$+\frac{m}{2l}(l-c-x).\frac{l-c-x}{2}$$

$$=\frac{m}{4l}[(l+c+x)^2+(l-c-x)^2]$$

Original depth of $C.G.$ is

$$\zeta_0=\frac{1}{4l}[(l+c)^2+(l-c)^2].$$

Hence the work done by the weight of the chain $=m(\zeta-\zeta_0)g$

$$=\frac{mg}{4l}[(l+c+x)^2+(l-c-x)^2-(l+c)^2-(l-c)^2]$$

$$=\frac{mg}{2l}[(x+c)^2-c^2].$$

Hence the Principle of the Energy gives

$$\tfrac{1}{2}m\dot{x}^2=\frac{mg}{2l}[(x+c)^2-c^2]$$

or
$$\frac{dx}{dt}=\sqrt{\frac{g}{l}}\sqrt{(x+c)^2-c^2}$$

$$\therefore \quad \sqrt{\frac{g}{l}}\,.\,t=\int_0^{l-c}\frac{dx}{\sqrt{(x+c)^2-c^2}}$$

$$=\log\left[\frac{(x+c)+\sqrt{(x+c)^2-c^2}}{c}\right]_0^{l-c}$$

$$=\log\frac{l+\sqrt{l^2-c^2}}{c}$$

Hence
$$t=\sqrt{\frac{l}{g}}\log\frac{l+\sqrt{l^2-c^2}}{c}.$$

EXAMPLES IV (B)

1. An elastic cord, modulus λ, natural length l, hangs vertically, one end being attached to a fixed point and the other to a body of mass M. The mass is raised till the string has its natural length and is then released from rest. Show that its greatest fall in the subsequent motion is $\frac{2Mgl}{\lambda}$.

2. A light string $ABCDE$ whose middle point is C passes through smooth rings B, D which are fixed in a horizontal plane at a distance $2a$ apart. To each of the points A, C, E, is attached a mass m. Initially C is held at O the middle point of BD and is then set free. Show that C will come to instantaneous rest when $OC=\frac{4}{3}a$. Show also that when C has fallen through $\frac{3}{4}a$ from O, its velocity is $\sqrt{\frac{45}{86}ag}$.

3. A ring of mass m slides on a smooth vertical rod; attached to the ring is a light string passing over a smooth peg distant a from the rod, and at the other end of this string is a mass $M(>m)$. The ring is held on a level with the peg and released. Show that it first comes to rest after falling a distance

$$\frac{2mMa}{M^2-m^2}.$$

4. A heavy ring of mass m slides on a fixed smooth vertical rod and is attached to a fine string which passes over a smooth peg distant a from the rod, and then after passing through a smooth ring of mass $3m$, is tied to the peg. Show that, if m is dropped from the point in the rod at the same horizontal level as the peg, then it will oscillate through a distance $\frac{12}{5}a$.

5. ACB is a horizontal straight line of length $a+b$. A heavy particle at C is attached by an inextensible string of length b to A, and by an unstretched elastic string of length a to B. If, when the particle starts from rest the string AC just oscillates through a right angle and the modulus of elasticity of the elastic string is equal to twice the weight of the particle prove

$$4a^2=b(a+2b)^2.$$

6. A uniform string of mass M and length $2a$, is placed symmetrically over a smooth peg and has particles of masses m and m' attached to its ends; show that when the string runs off the peg its velocity is

$$\sqrt{\frac{M+2(m-m')}{M+m+m'}ag}.$$

7. ACB is a horizontal straight line of length $a+b$. Heavy particle at C is attached by an inextensible string of length b to A, and by an unstretched elastic string of length a to B. If, when the particle starts from rest, the string AC just oscillates through a right angle and the modulus of elasticity of the elastic string is equal to twice the weight of the particle, prove that $4a=b(a+2b)^2$.

4·4. Impulse. If a force F acts on a particle of mass m always in the same direction, the equation of motion is $m\frac{dv}{dt}=F$, where v is the velocity of the particle at time t. Let T be the interval during which the force acts and v, v' be the velocities at the beginning and end of that interval. Then integrating the above equation, we get

$$m(v'-v)=\int_0^T Fdt.$$

Now suppose the force F to increase without limit while the interval T to decrease without limit. Then the above integral may have finite limit. Let this limit be P. Then equation becomes $m(v'-v)=P$.

The velocity in the interval has changed from v to v'. Supposing the velocity to have remained finite, let V be its greatest value during the interval. Then the space described by the particle is less than VT. Since this vanishes in the limit the particle has not moved during the action of F. *It has not had time to move but its velocity has changed.* Thus in the case of infinite forces which act for an indefinitely short time, the change of position is zero and the change of velocity is the measure of the force. A force so measured is called an **impulse**. *It may be defined as the limit of a force which is infinitely great and acts only during an infinitely short time.* The blow of a hammer is a force of this kind. *An impulse force is measured by the whole momentum generated by the impulse.* Thus if I be the impulse then $I=m(v-u)$ where u, v are the velocity just before and just after the impulse.

Work done by the impulse = change in Kinetic energy

$$=\tfrac{1}{2}m(v^2-u^2)=\tfrac{1}{2}m(v+u)(v-u)=I\frac{v+u}{2}$$

= product of the impulse and mean of the velocities before and after the blow.

4·5. Conservation of Linear Momentum. If a body of mass m moves with velocity v under the action of a force F in the line of motion then $\frac{d}{dt}(mv)=F$.

The follows from Newton's second law of motion.

If $F=0$, then $\frac{d}{dt}(mv)=0$ *i.e.*, $mv=$constant ;

the body moves with a constant momentum.

If there be two bodies of masses m, m' moving with velocities u and u' in the same straight line under forces F_1, F_2 respectively, then since the mutual action between the bodies are equal and opposite, we have

$$\left.\begin{aligned}\frac{d}{dt}(mu) &= F_1+R\\ \frac{d}{dt}(m'u') &\ \ F_2-R\end{aligned}\right\}\ R \text{ being the mutual action and reaction.}$$

$$\therefore\quad \frac{d}{dt}(mu+m'u')=F_1+F_2.$$

If no external force acts on the system,

$$\frac{d}{dt}(mu+m'u')=0$$

i.e., $$mu+m'u'=\text{Constant}.$$

Thus, if the sum of the external forces acting on a system of particles be zero in any direction, then the total momentum of the system in that direction remains constant during the motion. This is called the principle of **Conservation of Linear Momentum.**

Worked out Examples

Ex. *Two particles of masses m and m′ are moving in parallel straight lines with given velocities v and v′; the particles are connected by a string of such a length that the instant when it becomes tight it is inclined at an angle α to the two lines; assuming that v>v′, show that the impulsive tension of the string at the instant it tightens is*

$$\frac{mm'}{m+m'}(v-v')\cos\alpha$$

Sol. If V be the common velocity just when the string becomes tight then momentum along the line joining them is constant

$$\therefore \quad (mv+m'v')\cos\alpha=(m+m')V\cos\alpha$$

T = impulsive tension on each
= change of momentum of one along the line joining then

$$=m(V-v)\cos\alpha$$

$$=m\left(\frac{mv+m'v'}{m+m'}-v\right)\cos\alpha \text{ - etc.}$$

Ex. *Two masses m_1, m_2 are connected by a fine inelastic string which passes over a smooth fixed pulley. The larger mass m_2 rests on a horizontal table; m_1 is then raised h feet vertically above its position of rest and let fall. Show that the interval between m_2 leaving the table and returning to it again is*

$$\frac{2m_1}{m_2-m_1}\sqrt{\frac{2h}{g}}$$

Sol. The velocity of m_1 when it has fallen h feet is $\sqrt{2gh}$. At that instant the string becomes tight and m_2 is jerked off the table. The velocity imparted to m_2 is given by

$$(m_1+m_2)v=m_1\sqrt{2gh}$$

The common acceleration f of m_1 and m_2 during subsequent motion is given by

$$m_1 f=m_1 g-T$$

and

$$m_2 f=T-m_2 g$$

so that

$$f=\frac{m_2-m_1}{m_2+m_1}g.$$

The mass m_2 starts with velocity v and acc. f.

Its velocity vanishes after time t where

$$0=\frac{m_1}{m_1+m_2}\sqrt{2gh}-\frac{m_2-m_1}{m_2+m_1}gt$$

$$\therefore \qquad t=\frac{m_1}{m_2-m_1}\sqrt{\frac{2h}{g}}.$$

At this time, since $m_2>m_1$, m_2 will descend to the table. It will take the same time the required time$=2t$.

Ex. 9. *Two buckets, each weighing 10 ozs. are connected by a cord over a frictionless pulley and an animal weighing 3 ozs. is placed in one of the buckets. The system is let go from rest and at the end of $1\frac{1}{2}$ secs., the animal springs out of the bucket reaching a height of 8 inches above the starting point of the jump. Show that the velocity of the buckets while the animal is clear of them will be 5·11 ft./sec.*

Sol. The bucket with animal moves down with common acc. f where

$$\frac{19}{16}f=\frac{19}{16}g-T$$

and

$$f=T-g$$

$$\therefore \qquad f=\frac{3}{35}g$$

Downward velocity of the animal in the bucket after $1\frac{1}{2}$ sec. is

$$v=\frac{3}{35}\times 32\times\frac{3}{2}=\frac{144}{35} \text{ ft /sec.}$$

The velocity with which the animal springs up, so that it may travel up against gravity a distance 8 inches is

$$0=v_1^2-\frac{2\times 32\times 8}{12}$$

$$\therefore \qquad v_1=8\sqrt{\frac{2}{3}} \text{ ft./sec.}$$

The jumping of the animal will disturb the velocity of the buckets and if the new velocity of the buckets be v_2, we have by the law of conservation of momentum,

impulse on the animal$=$that on the buckets

$$\frac{3}{16}\left[8\sqrt{\frac{2}{3}}-\left(-\frac{144}{35}\right)\right]=\frac{32}{16}\left(v_2-\frac{144}{35}\right)$$

giving

$$v_2=5\cdot 11 \text{ ft./sec.}$$

Ex. *Two men each of mass M, stand on two inelastic platforms, each of mass m, hanging over a small smooth pulley. One of the men leaping from the ground would be able to raise his centre of gravity through a distance h; show that if he leaps with the same* *energy from his platform his C.G. would rise a distance*

$$\frac{(M+2m)h}{2(M+m)}$$

and that he would fall on the platform in its original position and the whole system be again at rest.

Sol. Let V be the velocity of the man when he leaps from the ground

$$\therefore \qquad V^2=2gh.$$

Let u, v be the velocities of the platform and the man. Since the momentum of the man is equal and opposite to that of the system consisting of the other man and the two platforms, we have

$$Mv=(M+2m)u$$

Since he leaps with the same energy in the two cases

$$\tfrac{1}{2}MV^2=\tfrac{1}{2}Mv^2+\tfrac{1}{2}(M+2m)u^2$$

These give $\qquad v^2=\dfrac{M+2m}{M+m}gh$

$$\therefore \quad \text{height risen}=\frac{v^2}{2g}=\frac{(M+2m)h}{2(M+m)}$$

$$\text{Time of return}=\frac{2v}{g}$$

After the man has left the platform, the platform starts with velocity u downwards but acceleration is $\dfrac{M}{M+2m}g$ upwards hence the platform comes back after time t

if $\qquad 0=ut-\tfrac{1}{2}\dfrac{M}{M+2m}gt^2$

$$\therefore \qquad t=\frac{2(M+2m)}{M}\cdot\frac{u}{g}=\frac{2v}{g}$$

Hence they come back to the original position after the same time and since their momenta are equal and opposite, the system comes to rest.

Ex. 11. *Prove that the mean kinetic energy of a particle of mass m moving under a constant force, in any interval of time, is* $\frac{1}{6}m(u_1^2+u_1u_2+u_2^2)$ *where* u_1 *and* u_2 *are initial and final velocities.*

Prove that it is greater than the kinetic energy of the middle of the interval but less than that when the particle is half-way.

Sol. Since the force is constant, acceleration is constant, say f.

If T be the interval then

mean $K.E.=\frac{m}{2T}\int_0^T v^2 dt$ where v u_1+ft

$$=\tfrac{1}{2}m\left[u_1^2+\frac{f^2T^2}{3}+u_1fT\right],$$

but $u_2=u_1+fT$ $\therefore$ $fT=u_2-u_1$

$\therefore$ mean $K.E.=\frac{1}{6}m(u_1^2+u_2^2+u_1u_2)$

$K.E$ at the middle of the interval

$$=\tfrac{1}{2}m\,(u_1+\tfrac{1}{2}fT)^2=\frac{m}{8}(u_1+u_2)^2$$

$K.E.$ half-way $=\frac{1}{2}m(u_1^2+fs)$

$$=\tfrac{1}{2}m\left(u_1^2+\frac{u_2^2-u_1^2}{2}\right)=\tfrac{1}{4}m(u_1^2+u_2^2).$$

Ex. 12. *A bullet of mass m moving with velocity u strikes a block of mass M which is free to move in the direction of motion of the bullet and is embedded in it. Show that a portion* $\frac{M}{M+m}$ *of the kinetic energy is lost.*

If the bullet be embedded a feet, show that the resistance of the block to the bullet supposed uniform is $\frac{Mm}{M+m}\cdot\frac{v^2}{2ga}$ *lbs. wt. and that the time of penetration is* $\frac{2a}{v}$ *secs. during which time the block will move* $ma/(M+m)$ *ft.*

Sol. Let V be the velocity of the combined body when embedded then by the principle of conservation of momentum we get

$$(M+m)V=mv$$

Loss of $K.E.=\frac{1}{2}mv^2-\frac{1}{2}(M+m)V^2$

$$=\tfrac{1}{2}mv^2-\tfrac{1}{2}(M+m)\frac{m^2v^2}{(M+m)^2}=\tfrac{1}{2}\frac{Mmv^2}{M+m}$$

$$=\frac{M}{M+m}\text{ (initial energy).}$$

If R be the resistance

$$Ra=\text{loss of }K.E.=\tfrac{1}{2}\frac{Mmv^2}{M+m}$$

$$\therefore\quad R=\tfrac{1}{2}\frac{Mmv^2}{(M+m)a}\text{ poundals}=\tfrac{1}{2}\frac{Mmv^2}{ag(M+m)}$$

lbs. wt.

Rt=impulse on the bullet=change of momentum

$$=m\,(v-V)=\frac{Mmv}{M+m}$$

$$\therefore \qquad t=\frac{2a}{v}, \qquad f=\text{acceleration}=\frac{V}{t}$$

$$\therefore \qquad s=\tfrac{1}{2}\,ft^2=\tfrac{1}{2}vt=\frac{ma}{M+m}$$

Ex. 13. *An elastic ball of mass m is dropped from a height h above the ground and at the same time a second ball of mass m_1 is projected vertically upwards to meet the former. Show that in order that immediately after collision, the balls may be at rest, the second ball must be projected with a velocity*

$$\left(\frac{m+m_1}{m_1}\,gh\right)^{\frac{1}{2}}$$

Sol. Let them meet at a distance s up the ground so that for the first ball

$$v_1^2=2g(h-s)$$

and for the second ball

$$v_2^2=u^2-2gs$$

if v_1, v_2 be their velocity with which they collide and u the initial velocity for the second ball.

Also for first $h-s=\frac{1}{2}gt^2$

for second $s=ut-\frac{1}{2}gt^2$

$$\therefore \qquad h=ut \text{ etc.}$$

Also $mv_1-m_1v_2=0$

Now substituting $t=\dfrac{h}{u}$ in the third and fourth equations we can find $h-s$ and s. These give

$$\frac{mgh}{u}=m_1\left(u-\frac{gh}{u}\right)$$

$$\therefore \qquad u^2=\frac{m+m_1}{m_1}\,gh.$$

Ex. 14. *Assuming that in a cannon the force on the ball depends only on the volume of gas generated by the gun-powder, show that the ratio of the final velocity of the ball when the gun is free to recoil to its velocity when the gun is fixed, is $\sqrt{\dfrac{M}{M+m}}$ where M and m are the masses of the cannon and the ball respectively.*

Sol. Let V and v be the velocities of the cannon and the ball when the cannon is free to recoil and E the kinetic energy released by the explosion, then $E=\frac{1}{2}MV^2+\frac{1}{2}mv^2$

and $$MV=mv$$

$$\therefore \quad E=\tfrac{1}{2}\left[\frac{m^2v^2}{M}+mv^2\right]=\tfrac{1}{2}mv^2.\ \frac{m+M}{M}.$$

If v' be the velocity of the ball when the cannon is fixed

$$E=\tfrac{1}{2}mv'^2$$

$$\therefore \quad mv^2\,\frac{M+m}{M}=mv'^2 \ \ldots\ \frac{v}{v'}=\sqrt{\frac{M}{M+m}}.$$

Ex. 15. *A gun is mounted on a gun-carriage, movable on a smooth horizontal plane and the gun is elevated at an angle* α *to the horizon ; a shot is fired and leaves the gun in a direction inclined at an angle* θ *to the horizon ; if the mass of the gun and its carriage be* n *times that of the shot, show that*

$$\tan\theta=\left(1+\frac{1}{n}\right)\tan\alpha.$$

Sol. Let u be the velocity of the shot relative to the gun and v the backward horizontal velocity of the gun and carriage. Let M be the mass of the shot so that nM is the mass of the gun and carriage. Since the impulsive force exerted by the explosion on the shot is equal and opposite to that exerted on the gun and carriage, the horizontal components of these forces are also equal and hence the horizontal momentum of the shot is equal and opposite to the horizontal momentum of the gun and carriage, so that

$$nMv=M(u\cos\alpha-v).$$

If the resultant velocity be V at an angle θ to the horizon,

then $$V\cos\theta=u\cos\alpha-v$$

and $$V\sin\theta=u\sin\alpha$$

$$\therefore \quad \tan\theta=\frac{u\sin\alpha}{u\cos\alpha-v}=\left(1+\frac{1}{n}\right)\tan\alpha.$$

Ex 16. *A shell of mass M is moving with velocity V. An internal explosion generates an amount of energy E and breaks the shell into two portions whose masses are in the ratio* $m_1 : m_2$. *The fragments continue to move in the original line of motion of the shell. Show that their velocities are*

$$V+\sqrt{\frac{2m_2E}{m_1M}}\quad \textit{and}\quad V-\sqrt{\frac{2m_1E}{m_2M}}.$$

Sol. Let v_1, v_2 be the velocities of the two portions whose masses be M_1, M_2.

Then $\dfrac{M_1}{m_1}=\dfrac{M_2}{m_2}=\dfrac{M}{m_1+m_2}$ $\therefore$ $M_1=\dfrac{m_1M}{m_1+m_2}$, $M_2=\dfrac{m_2M}{m_1+m_2}$

By the conservation of momentum

$$M_1v_1+M_2v_2=MV$$

and by the equation of energy

$$\tfrac{1}{2}M_1v_1^2+\tfrac{1}{2}M_2v_2^2-\tfrac{1}{2}MV^2=E.$$

Put $v_1=V+\dfrac{x}{M_1}$ and $x_2=V-\dfrac{x}{M_2}$ so that the first equation is satisfied since $M=M_1+M_2$

From the second equation, we get

$$M_1\left(V+\frac{x}{M_1}\right)^2+M_2\left(V-\frac{x}{M_2}\right)^2-MV^2=2E.$$

or $$M_1V^2+2xV+M_2V^2-2xV+\frac{x^2}{M_1}+\frac{x^2}{M_2}-MV^2=2E$$

or $$x^2\left(\frac{1}{M_1}+\frac{1}{M_2}\right)=2E \text{ or } x=\sqrt{\frac{2EM_1M_2}{M_1+M_2}}$$

$\therefore$ $$v_1=V+\sqrt{\frac{2EM_2}{M_1(M_1+M_2)}}=V+\sqrt{\frac{2m_2E}{m_1M}}$$

and $$v_2=V-\sqrt{\frac{2m_1E}{m_2M}}$$

Ex. 17. *A cannon ball of mass m is shot from a gun of mass M which is so mounted as to be free to recoil in a horizontal direction, so that muzzle velocity, when fired horizontally relative to the ground is V. Prove that the greatest range on a horizontal plane is* $\dfrac{V^2}{g}\left(1+\dfrac{m}{M}\right)$ *and that this is attained by giving the gun the elevation 45°.*

Sol. Let u be the velocity of the ball along the muzzle relative to the gun and v the backward horizontal velocity of the gun. Let the inclination of the gun be α and let the ball leave the muzzle with velocity v' at an angle θ.

Momentum equation is $Mv=m(u\cos\alpha-v)$ ...(1)

Resolving horizontally and vertically for the velocity of the ball when it leaves the muzzle

$$v' \cos\theta = u\cos\alpha - v$$
$$v' \sin\theta = u\sin\alpha$$

$\therefore \quad R = \text{range} = \dfrac{2v'\sin\theta}{g}\cdot v'\cos\theta$

$$= \frac{2u\sin\alpha}{g}(u\cos\alpha - v) = \frac{2u\sin\alpha}{g}\left(u\cos\alpha - \frac{mu\cos\alpha}{M+m}\right) \text{ from (1)}$$

$$= \frac{u^2\sin 2\alpha}{g}\cdot\frac{M}{M+m}$$

This is maximum when $\alpha = \pi/4$ and its maximum value is

$$R = \frac{u^2}{g}\cdot\frac{M}{M+m}.$$

But when $\alpha = 0$ *i.e.*, fired horizontally, V is the velocity of the ball relative to the ground

i.e. $\quad u - v = V \quad$ when $\alpha = 0$

Putting $\alpha = 0$ in (1) we get $(M+m)v = mu$

$\therefore \quad (M+m)(u-V) = mu$ or $Mu = (M+m)V$

hence the maximum range

$$= \frac{M}{(M+m)g}\cdot\frac{(M+m)^2V^2}{M^2} = \frac{(M+m)V^2}{Mg}.$$

Ex. 18. *A shell of mass m_1+m_2 is fired with a given velocity in a given direction At the highest point of its path the shell explodes into two portions of masses m_1 and m_2. The explosion produces an additional kinetic energy E and the portions separate in a horizontal direction. Show that the portions strike the ground at a distance* $\dfrac{v}{g}\sqrt{2E\left(\dfrac{1}{m_1}+\dfrac{1}{m_2}\right)}$ *where v is the vertical component of the velocity of projection.*

Sol. Time to reach highest point $= \dfrac{v}{g}$

If u be the horizontal component of the velocity of projection and u_1, u_2 those for the fragments then by the law of conservation of momentum we have

$$(m_1+m_2)u = m_1u_1 \quad m_2u_2$$

By the law of conservation of energy

$$\tfrac{1}{2}(m_1+m_2)u^2 + E = \tfrac{1}{2}m_1u_1^2 + \tfrac{1}{2}m_2u_2^2$$

These give $\quad u_2 - u_1 = \left\{2E\left(\dfrac{1}{m_1}+\dfrac{1}{m_2}\right)\right\}^{\frac{1}{2}}$

Time to reach the ground for each is $\frac{v}{g}$

Hence the distance between their positions of striking

$$=\frac{v}{g}\left(u_2-u_1\right)=\frac{v}{g}\left\{2E\left(\frac{1}{m_1}+\frac{1}{m_2}\right)\right\}^{\frac{1}{2}}$$

Ex. 1 *An inelastic pile of mass m lbs. is driven vertically into the ground a distance of a feet at each blow of a hammer of mass M lbs., which falls vertically through h feet. Show that the weight which would have to be placed on the top of the pile to drive it slowly into the ground would be*

$$M+\frac{M^2h}{(M+m)a} \text{ lbs.}$$

and the time during which the pile is in motion is

$$\frac{(M+m)a}{M}\sqrt{\frac{2}{gh}}.$$

Sol. Let u be the velocity of the hammer on hitting the pile so that

$$u^2=2gh \qquad \ldots(1)$$

Let v be the velocity of the hammer and pile immediately after the impact. By the principle of conservation of momentum

$$(M+m)v=Mu \qquad \ldots(2)$$

If R be the resistance of the ground in poundals, the force to resist the driving of the pile is

$$R-(M+m)g$$

The principle of conservation of energy gives

$$\tfrac{1}{2}(M+m)v^2=[R-(M+m)g]a$$

$$\therefore \qquad R=(M+m)g+(M+m)\frac{v^2}{2a}$$

$$=(M+m)g+\frac{M^2gh}{(M+m)a} \qquad \text{from (1)+(2)}$$

A weight of slightly more than $\frac{M^2h}{(M+m)a}$ lbs. placed on the pile would thus slowly overcome the resistance and just drive the pile down.

The principle of momentum gives

$$[R-(M+m)g]t=\text{Change in momentum}$$

$$=(M+m)v=Mu$$

or $$t\times\frac{M^2}{M+m}\,\frac{u^2}{2a}=Mu$$

$$\therefore \quad t=\frac{M+m}{M}\,\frac{2a}{u}=\frac{M+m}{M}\,a\sqrt{\frac{2}{gh}}$$

$$\text{K.E. lost}=\tfrac{1}{2}Mu^2-\tfrac{1}{2}(M+m)v^2$$

$$=\frac{m}{M+m}.\tfrac{1}{2}Mu^2$$

$$=\frac{m}{M+m}\text{ (energy of the hammer on striking the pile).}$$

EXAMPLES IV (C)

1. A bullet moving horizontally with velocity u strikes and penetrates a fixed vertical target of thickness a and emerges with horizontal velocity v. Prove that the same bullet would have just failed to penetrate the target, had the thickness been $\frac{au^2}{u^2-v^2}$, the average resistance to penetration being the same in the two cases.

2. A shot of mass m penetrates a thickness s of a fixed plane of mass M; prove that if M is free to move thickness penetrated is $\frac{Ms}{M+m}$.

3. If a shot of mass m lbs. striking a fixed metal plate with velocity u penetrates it through a distance a, show that it will completely pierce through a plate, free to move, mass M lbs. and thickness b if $b<\frac{Ma}{m+M}$ the resistance being supposed uniform.

4. A set of n equal trucks with S meters clear distance between them are inelastic and are set in motion by starting the end one with velocity V towards the next. Prove that the time taken for the last truck to start is $\frac{n(n-1)S}{2V}$ and the final velocity is $\frac{V}{n}$. Show also that the proportion of the initial energy wasted when all the trucks are in motion is $\frac{n-1}{n}$.

5. Two particles A and B of equal mass lie close together on a horizontal table and are connected by a light inextensible string of length l. A is projected vertically upwards with velocity $\sqrt{10gl}$. Prove that it reaches the table again with velocity $2\sqrt{gl}$.

6. At the highest point of its path a shell explodes into two equal parts of which one falls vertically from rest. Prove that the other will describe a parabola of which the latus rectum will be four times the latus rectum of the original parabola.

7. A shell lying in a straight smooth horizontal tube, suddenly explodes and breaks into two portions of masses m and m'. If x is the distance apart in the tube of the mass after a time t, show that the work done by the explosion is $\frac{1}{2}\frac{mm'}{m+m'}\cdot\frac{x^2}{t^2}$.

8. A gun of mass M fires a shell of mass m horizontally and the energy of the explosion is such as would be sufficient to project the shell vertically to a height h. Show that the velocity of recoil is

$$\left\{\frac{2m^2gh}{M(M+m)}\right\}^{\frac{1}{2}}$$

9. A man of weight W hangs at the end of light extensible rope whose modulus is nW, the other end being fastened to a fixed point. He proceeds to climb up rope. Prove that when he reaches the fixed point he has done $\frac{2n+1}{2n+2}$ times the work he would have done in climbing the same distance up an inextensible rope.

5

Impact

5·1. Impact. When two bodies collide with or strike against each other, they are said to impinge on one another. This phenomenon of impinging or colliding is called an *impact*.

An impact between two bodies is said to be *direct* if the direction of motion of each, just before impact, is along the common normal at the point of contact. The direction of this common normal at the point of contact is called the *line of impact*.

When two bodies A and B impinge on each other, then by Newton's 3rd law, the action of A on B is equal and opposite to that of B on A; also the time of contact for both A and B is the same. Hence the impulse of the action of A on B is equal and opposite to that of B on A. That is, the change of momentum of A is equal and opposite to that of B; or in other words, if the momentum is measured in the same direction, the sum of the two changes of momentum is zero. Thus if the masses of A and B are respectively m and m' their initial velocities are u, u' and their velocities (after impact) are v and v' such that the velocities are measured positive in the same direction. Then

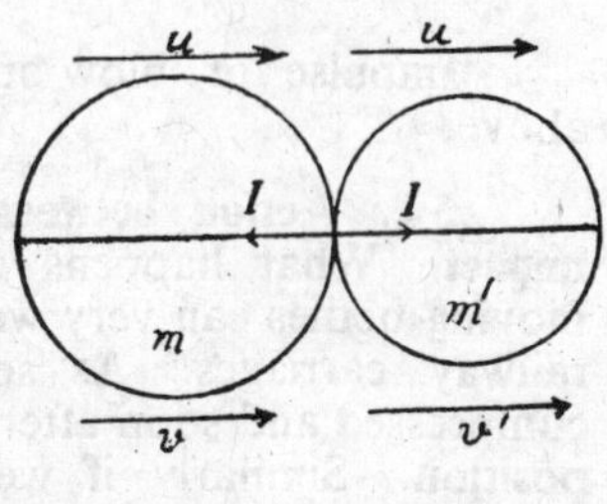

$$mv+m'v'=mu+m'u' \quad \text{...(1)}$$

Newton's experimental law of direct impulse. Newton found by experiments that when two bodies impinge directly, their relative velocity after impact is in a constant ratio with their relative velocity before impact and is in opposite direction, this constant ratio depending only on the material of the bodies and not on their masses or velocities. This constant ratio is called the *modulus* (or *co-efficient*) *of restitution* and is denoted by the letter e.

Thus $$v-v'=-e(u-u') \qquad ...(2)$$

The quantity e has different values for different pairs of substances but it always lies between 0 and 1. Bodies for which $e=0$ are called *inelastic bodies*, whilst those for which $e=1$ are *perfectly elastic bodies.*

Multiplying (2) by m' and adding to (1) we get

$$v=\frac{(m-em')u+m'(1+e)u'}{m+m'}$$

Similarly $$v'=\frac{m(1+e)u+(m'-em)u'}{m+m'}$$

Cor. If we put $m=m'$, and $e=1$, we get

$$v=u',\ v'=u.$$

Thus if two equal and perfectly elastic balls moving in the same direction impinge directly they interchange their velocities.

Impulse of the blow. The impulse of the blow on the ball of mass m

$$=\text{change of momentum of } m$$

$$=m(u-v)=\frac{mm'(1+e)(u-u')}{m+m'}$$

Impulse of blow on the other ball is equal and opposite to the above.

5·2. Action between two elastic bodies during the period of impact. What happens during the period of impact of any two moving bodies can very well be exemplified by the collision of two railway carriages. As soon as they collide, both the buffers are compressed and soon after they expand and gradually regain their position. Similarly if we drop a tennis ball on the ground covered with fine coloured powder, we observe that the surface of the ball in contact with the ground is a circular spot and not a point. This shows that during the earlier portion of the period of impact the ball is compressed and deformed. But then it is observed to rebound and come back in its original shape. This shows that during the latter portion ot impact the compressed ball expands and recovers its original shape.

We can thus divide the period of impact into two periods ; firstly the *period of compression*, during which each solid is approaching and compressing the other, and secondly *the period of restitution*, during which each solid tries to regain its original shape and in doing so is pushing the other away from it. The first period ends when the compression is the greatest ; at this moment the two centres of gravity have the same instantaneous velocity. If, however, solids are perfectly inelastic, there will be no period of restitution, the solids will not attempt to regain their original shape but will continue to move in their compressed state with the common velocity.

Just as the period of impact is divided into two periods, the total impulse of the blow can also be divided into impulses—the *impulse of compression* and *the impulse of restitution*. Thus if I and I' be the two impulses, then the total impulse$=I+I'$.

But for two balls, we have the total impulse

$$=\frac{mm'(1+e)(u-u')}{m+m'}$$

If the balls are inelastic, $I'=0$. So by putting $e=0$ in the above we have

$$I=\frac{mm'(u-u')}{m+m'}$$

Hence by subtraction

$$I'=\frac{mm'\,e(u-u')}{m+m'}$$

therefore $\qquad I'=eI \qquad \ldots(1)$

Further if U be the common velocity at the instant of greatest compression, we have the following equations :—

$$\text{on } m,\ -I=m\,(U-u),\ -I'=m\,(v-U)$$

$$\text{on } m',\ \ I=m'\,(U-u'),\ I'=m'\,(v'-U).$$

These give $mu+m'u'=(m+m')U=mv+m'v'$.

Also eliminating U, we get

$$I=\frac{mm'}{m+m'}(u-u') \text{ and } I''=\frac{mm'}{m+m'}(v'-v)$$

giving $\qquad \dfrac{I'}{I}=-\dfrac{v-v'}{u-u'}. \qquad \ldots(2)$

If we take Newton's law as

$$v-v'=-e\,(u-u') \qquad \ldots(3)$$

then from (2) we get $I'=eI$, same as (1)

or if we take $\qquad I'=eI$ then we get (3).

Total impulse $=I+I'=(1+e)I$

$$=\frac{mm'}{m+m'}(1+e)(u-u')$$

same as found before.

5·3. Oblique impact. An impact which is not direct is oblique, that is when the direction of motion of one, or both, before impact, is not along the common normal at the point of contact.

Newton's experimental law on oblique impact. When two bodies impinge obliquely their relative velocity resolved along their common normal after impact is in a constant ratio with their relative velocity before impact resolved in the same direction, and is of opposite sign. The constant ratio is *e*.

Oblique impact of two perfectly smooth spheres. Let two smooth spheres of masses m, m' impinge obliquely and let their line of centres be AB, A, B being the centres. Let their velocities before impact be u, u' and their velocities after impact be v, v' respectively. Let also the angles which the directions of u, u', v, v' make with AB be α, β, θ, ϕ.

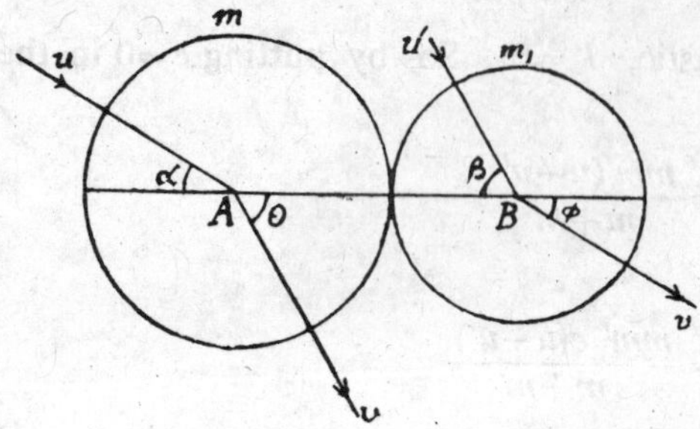

Since there is no friction between the spheres ; on each sphere, only force is along the line of centres ; there being no force perp. to the line of centres, therefore the components of the velocity of each perp. to AB are unaltered. Hence

$$v \sin \theta = u \sin \alpha \qquad \text{...(1)}$$

$$v' \sin \phi = u' \sin \beta \qquad \text{... (2)}$$

Also by Newton's law

$$v \cos \theta - v' \cos \phi = -e(u \cos \alpha - u' \cos \beta) \qquad \text{...(3)}$$

Again, the only force acting on the spheres during impact is the blow along the line of centres so that on the two spheres taken together there is no impulsive force along the line of centres, therefore, the total momentum in that direction is unaltered by impact. Hence

$$mv \cos \theta + m'v' \cos \phi = mu \cos \alpha + m'u' \cos \beta \qquad \text{...(4)}$$

These four equations are sufficient to determine the four unknowns, v, v', θ and ϕ.

Cor. 1. If $u'=0$, we have from (2), $\phi=0$.

Hence *if a moving smooth sphere impinges obliquely on another smooth sphere at rest, the latter begins to move along the line of centres.*

Cor. 2. If $m=m'$, and $e=1$, we get

$$v \cos \theta = u' \cos \beta, \; v' \cos \phi = u \cos \alpha.$$

Hence, *if two perfectly elastic and smooth spheres of equal masses impinge obliquely, then they interchange their velocities along the line of centres.*

Impulse of the blow. The impulse of the blow on the sphere of mass m'

=change produced in its momentum along the line of centres

$= m' (v' \cos \phi - u' \cos \beta)$

$= \dfrac{mm' (1+e)(u \cos \alpha - u' \cos \beta)}{m+m'}$ on substituting from the above equations.

Impulse of the blow on the sphere m is equal and opposite to the above.

Ex. 1. *A series of n elastic spheres whose masses are 1, e, e^2, etc., are at rest, separated by intervals, with their centres on a straight line. The first is made to impinge directly on the second with velocity u. Show that finally the first (n—1) spheres will be moving with the same velocity (1—e) u and the last with the velocity u.*

Prove that the final K.E. of the system is $\frac{1}{2}(1-e+e^n)u^2$.

Sol. Let v_1, v'_1 be the velocities of the first and second spheres immediately after the first sphere strikes the second.

Then $v_1 = \dfrac{mu - em' u'}{m+m'} = \dfrac{(1-e^2)\, u}{1+e} = (1-e)\, u \;\because\; m=1, m'=e.$

$$v_1' = \frac{1+e}{1+e}\, u = u.$$

Similarly
$$v_2 = (1-e)\, v_1' = (1-e)u, \; v_2' = v_1' = u$$
$$v_3 = (1-e)\, v'_2 = (1-e)\, u, \; v_3' = v_2' = u,$$

and so on.

$$K.E. = \tfrac{1}{2}.\, 1.\, (1-e)^2 u^2 + \tfrac{1}{2} e\, (1-e)^2 u^2 + \ldots\ldots + \tfrac{1}{2} e^{n-2} (1-e)^2 u^2 + \tfrac{1}{2} e^{n-1}.\, u^2$$

$$= \tfrac{1}{2}(1-e)^2 u^2 (1+e+e^2+\ldots\ldots+e^{n-2}) + \tfrac{1}{2} e^{n-1} u^2$$

$$= \tfrac{1}{2}(1-e)^2 u^2 . \frac{1-e^{n-1}}{1-e} + \tfrac{1}{2} e^{n-1} u^2$$

$$= \tfrac{1}{2}(1-e+e^n)\, u^2.$$

Ex. 2. *Three balls of masses m, m′ and M respectively are in a straight line, the two last being at rest ; the first moving in the straight line with velocity u, strikes the second which afterwards strikes the third. Find the velocity of the third ball after impact.*

Prove that if the masses of the first and third balls be given, the velocity of the third ball after impact is greatest when the mass of the second ball is a mean proportional between the masses of the other two and the velocity is then equal to

$$\frac{(1+e)^2\, mu}{(\sqrt{m}+\sqrt{M})^2}.$$

Sol. For the impact between m, m', the equation of momentnm and Newton's law give, if v, v' be their velocities just after impact,

$$mv+m'v'=mu$$

and
$$v-v'=-eu$$

$$\therefore \qquad v'=\frac{um\,(1+e)}{m+m'}$$

For the impact between m' and M, if V be the velocity of M after impact, we have similarly, $V=\dfrac{v'\,m'\,(1+e)}{m'+M}$

i.e.,
$$V=\frac{umm'\,(1+e)^2}{(m+m')(m'+M)}$$

i.e.,
$$V=\frac{um\,(1+e)^2}{\left(1+\dfrac{m}{m'}\right)(m'+M)}.$$

Given m, M, this V is greatest when $\left(1+\dfrac{m}{m'}\right)(m'+M)$ is least

i.e. when
$$m'+m+M+\frac{mM}{m'} \text{ is least}$$

i.e., when $1-\dfrac{mM}{m'^2}=0$ *i.e.,* $m'=\sqrt{mM}.$

Then
$$V=\frac{mu(1+e)^2}{(\sqrt{m}+\sqrt{M})^2}.$$

Ex. 3. *A smooth sphere of mass m travelling with velocity u, impinges obliquely on a smooth sphere of mass M at rest, its original line of motion making an angle θ with the line of centres at the moment of impact. Show that the spheres of mass m will be reflected through a right angle if* $\tan^2\theta=\dfrac{eM-m}{M+m}$.

Sol. Let v_1, v_2 be the velocities of m, M after impact, v_1 making $\angle\phi$ with the line of centres and v_2 along the line of centres.

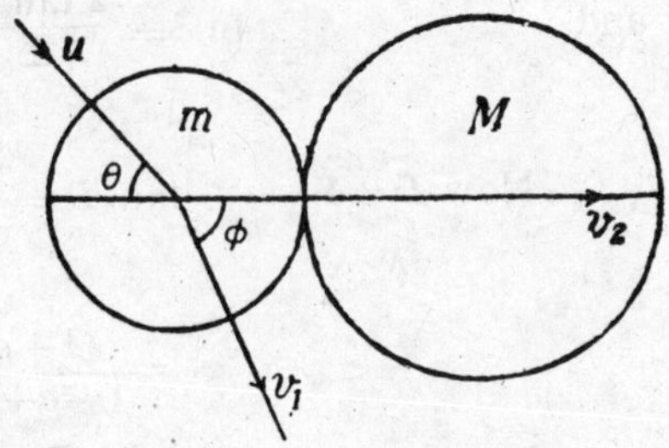

For m, the equation perpendicular to the line of centres gives $v_1 \sin\phi = u \sin\theta$.

The equation of momentum gives $mv \cos\phi + Mv_2 = mu \cos\theta$.

Newton's law gives

$v_1 \cos\phi - v_2 = -eu \cos\theta$.

If $\quad \phi - \theta = 90°$

i.e., $\quad \phi = \theta + 90°$

then $\quad v_1 \cos\theta = u \sin\theta$

$$-mv_1 \sin\theta + Mv_2 = mu \cos\theta$$

and $\quad v_1 \sin\theta + v_2 = eu \cos\theta$

Hence $mu \cos\theta = -mv_1 \sin\theta + M(eu \cos\theta - v_1 \sin\theta)$

$$= M\, eu \cos\theta - (M+m)\, v_1 \sin\theta$$

$$= Meu \cos\theta - (M+m)u \frac{\sin^2\theta}{\cos\theta}$$

or $\quad (M+m)\sin^2\theta = (eM - m)\cos^2\theta$

$$\therefore \quad \tan^2\theta = \frac{eM-m}{M+m}.$$

Ex. 4. *A smooth ball impinges on another smooth equal ball at rest in a direction that makes an angle θ with the line of centres at the moment of the impact. Prove that if D be the angle through which the direction of motion of the impinging ball is deviated then*

$$\tan D = \frac{(1+e)\tan\theta}{1-e+2\tan^2\theta}.$$

If the direction of the impinging ball bisects the angle between the future directions of the ball then $\tan\theta = \sqrt{e}$.

Prove also that the maximum deviation is $\tan^{-1} \dfrac{1+e}{2\sqrt{2(1-e)}}$.

Sol. As in Ex. 3, the equatians are

$$v_1 \sin\phi = u \sin\theta$$

$$v_1 \cos\phi + v_2 = u \cos\theta$$

and $\quad v_1 \cos\phi - v_2 = -eu \cos\theta$

$$\therefore \quad u\cos\theta = 2v_1 \cos\phi + eu \cos\theta$$

Hence $\quad \cos\phi = \dfrac{(1-e)\, u\cos\theta}{2v_1}$

and $$\sin \phi = \frac{u \sin \theta}{v_1}$$

and $$\tan \phi = \frac{2 \tan \theta}{1-e}.$$

Now $D = \phi - \theta$ $\therefore$ $$\tan D = \frac{\tan \phi - \tan \theta}{1 + \tan \phi \tan \theta} = \frac{\frac{2 \tan \theta}{1-e} - \tan \theta}{1 + \frac{2 \tan^2 \theta}{1-e}}$$

$$= \frac{(1+e) \tan \theta}{1 - e + 2 \tan^2 \theta}.$$

If $\phi = 2\theta$ then $\tan 2\theta = \frac{2 \tan \theta}{1-e}$ or $\frac{2 \tan \theta}{1 - \tan^2 \theta} = \frac{2 \tan \theta}{1-e}$

or $$\tan^2 \theta = e \quad \text{or} \quad \tan \theta = \sqrt{e}.$$

Now $$\tan D = \frac{1+e}{(1-e) \cot \theta + 2 \tan \theta}$$

$\therefore$ D is maximum when $(1-e) \cot \theta + 2 \tan \theta$ is minimum

i.e., when $$-(1-e) \operatorname{cosec}^2 \theta + 2 \sec^2 \theta = 0$$

or $$\tan^2 \theta = \frac{1-e}{2} \quad \text{or} \quad \tan \theta = \sqrt{\frac{1-e}{2}}.$$

Then $$\tan D = \frac{1+e}{2\sqrt{2(1-e)}}.$$

Ex. 5. *Two smooth spheres of mass m, m′ moving with velocities u, u′ at right angles to one another, collide. Prove that, if their directions of motion after impact are also at right angles*

$$e = \frac{m^2 u \cos \alpha + m'^2 u' \sin \alpha}{mm' (u \cos \alpha + u' \sin \alpha)}.$$

If the directions of motion before impact make angles α, α′ with the line of centres and if their directions of motion after impact be perpendicular to their directions before impact

$$e = \frac{m \sin^2 \alpha' + m' \sin^2 \alpha}{m \cos^2 \alpha' + m' \cos^2 \alpha}.$$

Sol.

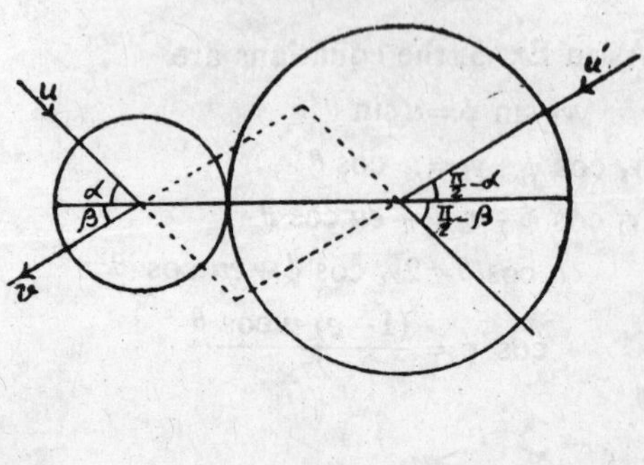

Let u, u' be the velocities of the balls m, m' before impact at angles α, and $\left(\frac{\pi}{2}-\alpha\right)$ with the line of centres and v, v' their velocities after impact at angles β and $\left(\frac{\pi}{2}-\beta\right)$, their senses being indicated by the arrow-heads.

Equations perp. to the line of centres, give for each ball,

$$u \sin \alpha = v \sin \beta \qquad \ldots(1)$$

$$u' \cos \alpha = v' \cos \beta \qquad \ldots(2)$$

Momentum equation gives

$$m'v' \sin \beta - mv \cos \beta = mu \cos \alpha - m'u' \sin \alpha \qquad \ldots(3)$$

Newton's law gives

$$v' \sin \beta + v \cos \beta = e\,(u' \sin \alpha + u \cos \alpha) \qquad \ldots(4)$$

$\therefore$ from (1), (2) and (3), we get

$$m'u' \cos \alpha \tan \beta - mu \sin \alpha \cos \beta = mu \cos \alpha - m'u' \sin \alpha$$

or $$m'u' (\cos \alpha \tan \beta + \sin \alpha) = mu (\cos \alpha + \sin \alpha \cot \beta)$$

or $$m'u' \frac{\sin(\alpha+\beta)}{\cos \beta} = mu \frac{\sin(\alpha+\beta)}{\sin \beta} \quad \therefore \quad \frac{m'u'}{\cos \beta} = \frac{mu}{\sin \beta}$$

i.e., $$\tan \beta = \frac{mu}{m'u'}.$$

Then from (4),

$$e = \frac{v' \sin \beta + v \cos \beta}{u' \sin \alpha + u \cos \alpha} = \frac{u' \cos \alpha \tan \beta + u \sin \alpha \cot \beta}{u' \sin \alpha + u \cos \alpha}$$

$$= \frac{u' \cos \alpha \,.\, \dfrac{mu}{m'u'} + u \sin \alpha \,.\, \dfrac{m'u'}{mu}}{u' \sin \alpha + u \cos \alpha} = \frac{m^2 u \cos \alpha + m'^2 u' \sin \alpha}{mm' (u \cos \alpha + u' \sin \alpha)}$$

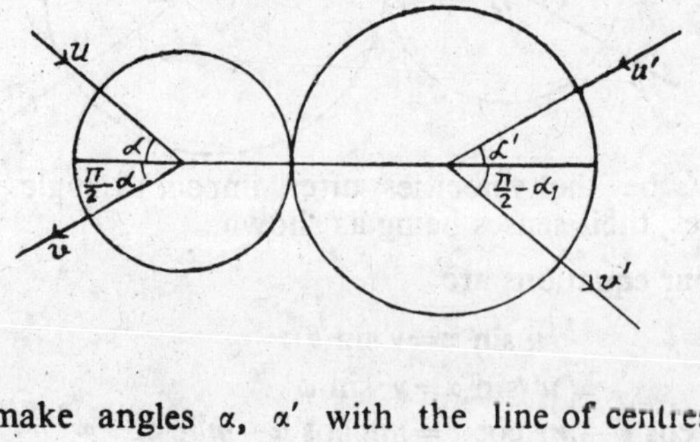

u, u' make angles α, α' with the line of centres, v, v' make angles $\left(\frac{\pi}{2}-\alpha\right)$ and $\left(\frac{\pi}{2}-\alpha'\right)$.

The four equations are

$$u \sin \alpha = v \cos \alpha \qquad \text{...(1)}$$

$$u' \sin \alpha' = v' \cos \alpha' \qquad \text{...(2)}$$

$$mu \cos \alpha - m'u' \cos \alpha' = m'v' \sin \alpha' - mv \sin \alpha \qquad \text{...(3)}$$

$$v \sin \alpha + v' \sin \alpha' = e\,(u \cos \alpha + u' \cos \alpha') \qquad \text{...(4)}$$

From (1), (2) and (3),

$$mu\left(\cos \alpha + \frac{\sin^2 \alpha}{\cos \alpha}\right) = m'u'\left(\cos \alpha' + \frac{\sin^2 \alpha'}{\cos \alpha'}\right)$$

$$\therefore \quad \frac{mu}{\cos \alpha} = \frac{m'u'}{\cos \alpha'}.$$

From (4), $$e = \frac{v \sin \alpha + v' \sin \alpha'}{u \cos \alpha + u' \cos \alpha'} = \frac{\dfrac{u \sin^2 \alpha}{\cos \alpha} + \dfrac{u' \sin^2 \alpha'}{\cos \alpha'}}{u \cos \alpha + u' \cos \alpha'}$$

$$= \frac{\dfrac{u \sin^2 \alpha}{\cos \alpha} + \dfrac{mu \sin^2 \alpha'}{m' \cos \alpha}}{u \cos \alpha + \dfrac{mu \cos^2 \alpha'}{m' \cos \alpha}} = \frac{m' \sin^2 \alpha + m \sin^2 \alpha'}{m' \cos^2 \alpha + m \cos^2 \alpha'}.$$

Ex. 6. *Two balls of elasticity e, moving in parallel directions with equal momenta impinge. Prove that if their directions of motion be opposite, they will move after impact in parallel directions with equal momenta.*

Sol. Let u, u' be the velocities of m, m' before impact, their directions being parallel and opposite, each making $\angle \alpha$ with the line of

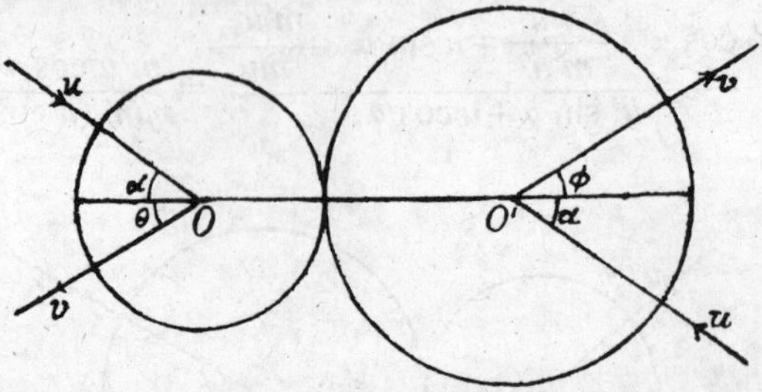

centres v, v' be the velocities after impact at angles θ, ϕ with the line of centres, their senses being as shown.

The four equations are

$$u \sin \alpha = v \sin \theta \qquad \text{...(1)}$$

$$u' \sin \alpha = v' \sin \phi \qquad \text{...(2)}$$

$$m'v' \cos \phi - mv \cos \theta = mu \cos \alpha - m'u' \cos \alpha \qquad \text{...(3)}$$

$$v' \cos\phi + v\cos\theta = e(u\cos\alpha + u'\cos\alpha) \qquad ...(4)$$

Given $mu = m'u'$ $\quad\therefore\quad$ (3) gives $m'v'\cos\phi = mv\cos\theta$

i.e., $$\frac{v\cos\theta}{v'\cos\phi} = \frac{m'}{m} = \frac{u}{u'} = \frac{v\sin\theta}{v'\sin\phi} \text{ from (1) and (2)}$$

i.e., $$\tan\phi = \tan\theta \quad \text{or} \quad \phi = \theta$$

i.e., the balls move in parallel directions after impact.

Also since $\phi = \theta$, we have $m'v' = mv$

i.e., their momenta after impact are equal.

From (4)

$$\frac{mv}{m'}\cos\theta + v\cos\theta = e\left(u\cos\alpha + \frac{mu}{m'}\cos\alpha\right)$$

or $$\cos\theta = eu\cos\alpha$$

$\therefore$ From (1) $$\tan\theta = \frac{\tan\alpha}{e}$$

But $e < 1$ $\therefore$ $\theta > \alpha$, unless $e = 1$ when $\theta = \alpha$.

The distance between the directions of the balls before impact $= OO'\sin\alpha$ and after impact $= OO'\sin\theta$; O, O' being the centres. But $OO'\sin\theta > OO'\sin\alpha$ $\therefore$ parallel lines after impact are at a greater distance than those before impact.

If the directions turn through a right angle,

$$\theta = \frac{\pi}{2} - \alpha \qquad \therefore \quad \cot\alpha = \frac{\tan\alpha}{e}.$$

or $$\tan\alpha = \sqrt{e}$$

Ex. 7. *Three equal spheres are in a straight line on a table and one of them moves towards the other two which are at rest and not in contact ; if $e = \frac{1}{2}$, find how many impacts will take place and show that the ultimate speeds of the spheres are in ratios 13 : 15 : 36.*

Sol. Let the spheres be A, B, C and let A move with velocity u to strike B which is at rest.

First impact between A and B :—Since the masses are the same we have the following equations. Let v be the velocity of A and v' that of B after impact.

$$v - v' = -e(u - 0) = -\tfrac{1}{2}u$$

and $$v + v' = u$$

$\therefore$ $$v = \frac{u}{4},\ v' = \frac{3u}{4}.$$

Then B with velocity $\frac{3u}{4}$ strikes C at rest. Let v_1, v_1' be the velocities of B and C after impact.

Then $$v_1 - v_1' = -\frac{1}{2} \cdot \frac{3u}{4}$$

and $$v_1 + v_1' = \frac{3u}{4}$$

$\therefore$ $$v_1 = \frac{3}{16}u, \quad v_1' = \frac{9}{16}u.$$

Then A with velocity $\frac{u}{4}$ strikes B which is moving with $\frac{3}{16}u$ since $\frac{1}{4}u > \frac{3}{16}u$. Let v_2, v_2' be the velocities of A and B after this impact. Then

$$v_2 - v_2' = -\frac{1}{2}\left(\frac{u}{4} - \frac{3}{16}u\right) = -\frac{1}{32}u$$

and $$v_2 + v_2' = \frac{u}{4} + \frac{3}{16}u = \frac{7}{16}u$$

$\therefore$ $$v_2 = \frac{13}{64}u, \quad v_2' = \frac{15}{16}u.$$

Now velocity of C *i.e.*, $\frac{9}{16}u$ is greater than that of B *i.e.*, $\frac{15}{64}u$, also the velocity of B *i.e.*, $\frac{15}{16}u$ is greater than that of A *i.e.*, $\frac{13}{64}u$, therefore there will be no more impacts. Hence there will be three impacts and the ultimate velocities of A, B, C, are

$$\frac{13}{64}u, \quad \frac{15}{64}u, \quad \frac{9}{16}u.$$

i.e., $$\text{ratios} = \frac{13}{64} : \frac{15}{64} : \frac{9}{16} = 13 : 15 : 36.$$

Ex. 8. *Three small spheres A, B, C whose masses are 8m, m, 7m are at rest in line and AB=BC=a. The middle sphere B is projected towards C with velocity u. Assuming all the spheres to be perfectly elastic, find the times and positions of impacts.*

Sol. First impact between B and C :

If u, v be the velocities of B and C after impact

$$mU = mu + 7mv \text{ and } u - v = -U$$

$\therefore$ $$u = -\frac{3}{4}U \text{ and } v = \frac{U}{4},$$

i.e., B goes back towards A with velocity $\frac{3}{4}U$.

Impact between B and A :

If u_1, v_1 be the velocities of B and A after impact

$$m.\frac{3}{4}U=mu_1+8mv_1 \quad \text{and} \quad v_1-u_1=\frac{3}{4}U$$

$$\therefore \quad v_1=\frac{U}{6},\ u_1=-\frac{7}{12}U$$

i.e., B goes towards C with velocity $\frac{7}{12}U$.

Second impact between B and C :

If u_2, v_2 be the velocities of B and C then

$$m.\frac{7}{12}U+7m.\frac{1}{4}U=u_2+7v_2$$

$$u_2-v_2=-\left(\frac{7}{12}U-\frac{U}{4}\right)$$

giving $$u_2=0, \quad v_2=\frac{1}{3}U$$

There are no further collisions.

Initial positions are $A=-a, \quad B=0, \quad C=a$

Positions of first impact are

$$A=-a, \quad B, C=a$$

$$t_1=\text{Time to first impact}=\frac{a}{U}$$

$$t_2=\text{Time from first to second impact}=\frac{2a}{\frac{3}{4}U}=\frac{8a}{3U}$$

and the position of C at the second impact

$$=a+\frac{1}{4}U.\frac{8a}{3u}=\frac{5}{3}a$$

thus the positions at second impact are

$$A, B=-a, \ C=\frac{5}{3}a.$$

t_3=Time from second impact to third is obtained as follows :

Velocity of B relative to $C=\frac{7}{12}U-\frac{1}{4}U$

$$=\frac{U}{3} \text{ and distance}=a+\frac{5}{3}a=\frac{8}{3}a$$

Hence $$t_3=\frac{8a}{3}\div\frac{U}{3}=\frac{8a}{U}$$

Thus times of impact are

$$t_1=\frac{a}{U},\quad t_1+t_2=\frac{a}{U}+\frac{8}{3}\frac{a}{U}=\frac{11}{3}\cdot\frac{a}{U}$$

$$t_1+t_2+t_3=\frac{11}{3}\frac{a}{u}+\frac{8a}{U}=\frac{35}{3}\frac{a}{U}.$$

Ex. 9. *A ball impinges on another equal ball moving with the same speed in a direction perpendicular to its own, the line joining the centres of the ball at the instant of impact being perpendicular to the direction of motion of the second ball; if e be the co-efficient of elasticity, show that the directions of motion of second ball is turned through an angle* $\tan^{-1}\frac{1+e}{2}$.

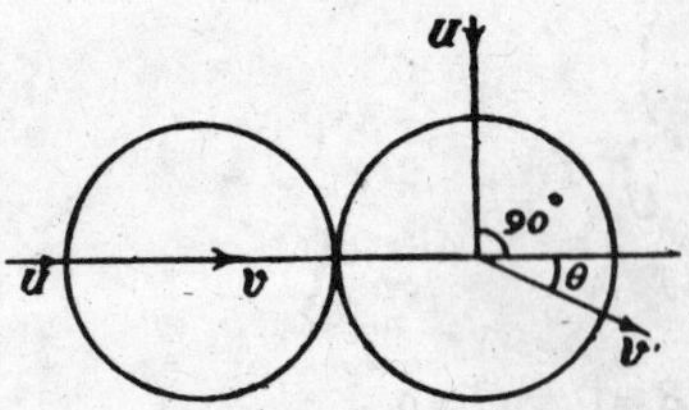

Sol. Resolving perp. to the line of centres for the second ball

$$v'\sin\theta=u$$

Momentum equation along the line of centres is

$$mu=mv+mv'\cos\theta$$

or $$u=v+v'\cos\theta.$$

Newton's law along the line of centre is

$$v-v'\cos\theta=-eu$$

$\therefore$ from second, $u=v'\cos\theta-eu+v'\cos\theta$

$$\therefore\quad v'=\frac{(1+e)u}{2\cos\theta}$$

Hence from first $$\frac{(1+e)u}{2\cos\theta}\sin\theta=u$$

$$\therefore\quad \tan\theta=\frac{2}{1+e}$$

Required angle $=90+\theta=\phi$ say

$$\therefore\quad \tan\phi=-\cot\theta=-\frac{1+e}{2}$$

$$\therefore\quad \phi=-\tan^{-1}\frac{1+e}{2}$$

the negative sign shows that the angle is obtuse.

Ex. 10. *Three small equal spheres ore projected simultaneously from the corners of an equilateral triangle with equal velocities towards the centre of the triangle and meet near the centre. Prove that they return to the corners with velocities diminished in the ratio e : 1.*

Sol. O is the centre; spheres approach it along the medians DO, EO, FO, each with velocity u and due to symmentry each is reflected back with a velocity say v.

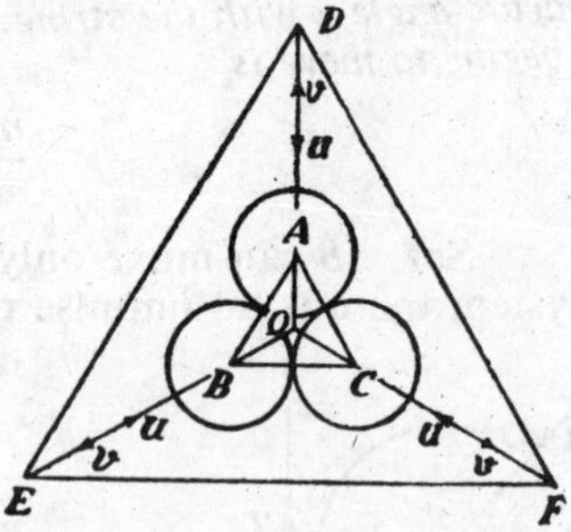

Newton's law for the two spheres A, B, along the line of centre is

$$v \cos 30 + v \cos 30$$
$$= -e(-u \cos 30 - u \cos 30)$$

$$\therefore \quad \frac{v}{u} = \frac{e}{1}.$$

Ex 11. *Two equal balls of radius a are in contact and are struck simultaneously by a ball of radius c moving in the direction of their common tangent. If all the balls be of the same material, the co-efficient of elasticity being e, find the velocities of the balls after impact and prove that the impinging ball will be reduced to rest if*

$$2e = \frac{c^3 (a+c)^2}{a^3 (2a+c)}.$$

Sol. Let the ball A (c) strike the other two balls B and C each of radius a, with velocity u and let its velocity after impact be u' and those of B and C be each v along AB or AC, due to symmetry. Since on the system there is no impulse, a momentum in any direction is conserved and therefore the equation of momentum along the direction of the common tangent is

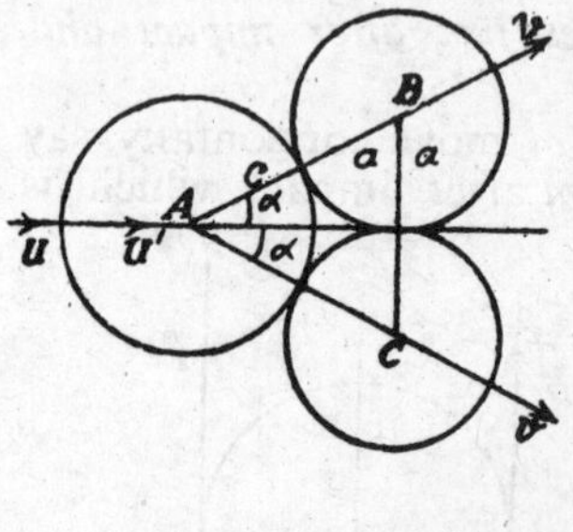

$$m'u = m'u' + 2mv \cos \alpha$$

where m' is the mass of A, m that of B or C.

Newton's law for A, B along AB is

$$u' \cos \alpha - v = -e(u \cos \alpha - 0)$$

If $u' = 0$ then $m'u = 2mv \cos \alpha$

and $$v = eu \cos \alpha$$

$$\therefore \quad m'u = 2meu \cos^2 \alpha$$

or $$2e = \frac{m'}{m \cos^2 \alpha} = \frac{c^3}{a^3(1-\sin^2\alpha)}$$

$$=\frac{c^3}{a^3\left\{1-\frac{a^2}{(a+c)^2}\right\}}=\frac{c^2(a+c)^2}{a^3(2a+c)}$$

Ex. 12. *A smooth spherical ball of mass m′ is tied to a fixed point by a light inextensible string and another spherical ball of mass m impinges directly on it with velocity u in a direction making an acute angle α with the string. Prove that the velocity with which m begins to move is*

$$\frac{m'(1+e)u\sin\alpha}{m+m'\sin^2\alpha}.$$

Sol. B can move only perpendicularly to the string. On the system there is no impulse perp. to the string, so momentum in that direction is conserved.

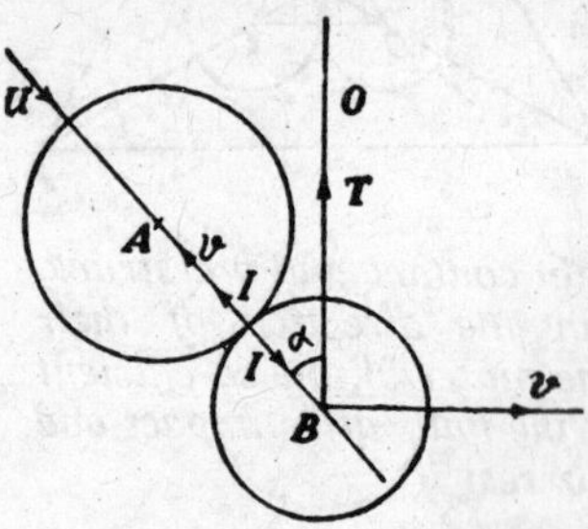

$$\therefore\quad m'u\sin\alpha=m'v\sin\alpha+mv'$$

Newton's law along the line of centres is

$$v'\sin\alpha-v=-e(0-u)=eu$$

$$\therefore\quad mv'=m'u\sin\alpha-m'\sin\alpha\,(v'\sin\alpha-eu)$$

$$\therefore\quad (m+m'\sin^2\alpha)v'=m'\sin\alpha(1+e)u.$$

Ex. 13. *Two equal spheres of masses m′ are suspended by vertical strings so that they are in contact with their centres at the same level. A third equal sphere of mass m falls vertically and strikes the other two simultaneously so that their centres at the instant of impact form an equilateral triangle in a vertical plane. If u is the velocity of m just before impact, find the velocities just after impact and the impulsive tension of the strings.*

Sol. After impact, spheres of masses m' move horizontally, say with velocity v and let u' be the velocity of m after impact which is vertical. Let I be the impulse between the upper and either of the lower sphere, T the impulsive tension of the strings.

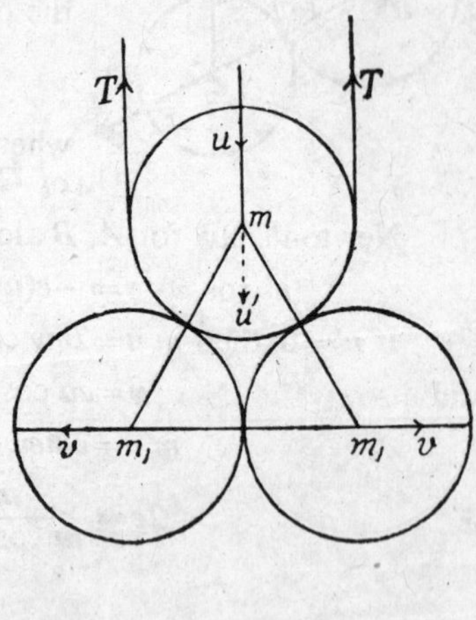

By Newton's law between the upper and one of the lower spheres, along the line of centres, we have

$$u'\cos 30°-v\cos 60°=-eu\cos 30°$$

i.e., $\quad u'\sqrt{3}-v=-eu\sqrt{3}\qquad\ldots(1)$

Also $I\cos 60°=m'v$

and $2I\cos 30°=m(u-u')$

$\therefore\quad m(u-u')=2\sqrt{3}\,m'v.\qquad\ldots(2)$

Hence from (1) and (2) we have by solving

$$u'=\frac{u(m-6em')}{m+6m'} \text{ and } v=\frac{\sqrt{3}\, um\,(1+e)}{m+6m'}.$$

$$\text{Also}\quad T=I\cos 30°=\sqrt{3}\ m'v=\frac{3mm'u(1+e)}{m+6m'}.$$

Ex. 14. *Two equal balls moving with equal velocities collide simultaneously with an equal ball at rest and with each other, their directions of motion being inclined at equal angles α to their common tangents at the moment of impact. Prove that after impact their directions are inclined at an angle* $2\tan^{-1}\left(\frac{3e\tan\alpha}{2-e}\right)$ *to each other and that the fraction* $(1-e^2)(1+2\sin^2\alpha)/3$ *of the original kinetic energy disappears at the impact, e being the co-efficient of restitution between all the balls.*

Sol. Let *A, B, C* be centres of the three balls. They form an equilateral triangle. Let *J* and *K* be the impulses between *A* and *C* and *A* and *B*, *U*, *V* be the components of velocity of *A* or *B*; *U′* be the velocity of *C* after the impact.

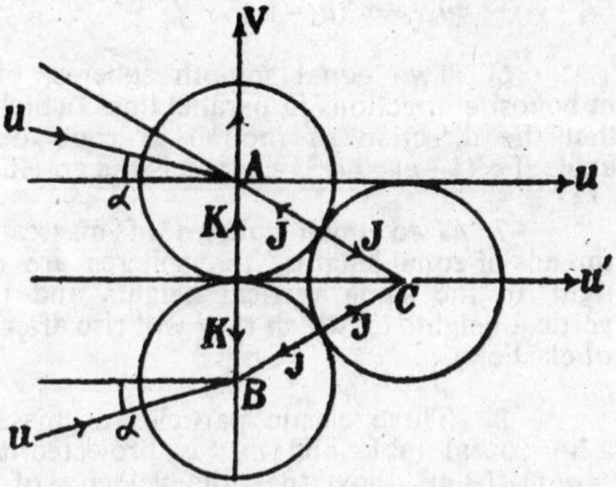

For *A*,

$$m\,(U-u\cos\alpha)=-J\cos 30$$

$$m\,(V+u\sin u)=K+J\sin 30.$$

For *C*, $mU'=2J\cos 30$

Newton's law between *A* and *C* gives

$$U'\cos 30-U\cos 30+V\sin 30=cu\cos(30-\alpha).$$

Newton's law between *A* and *B* gives

$$2V=2eu\sin\alpha.$$

These equations give

$$U=\frac{2-e}{3}u\cos\alpha,\ V=eu\sin\alpha,\ U'=\tfrac{2}{3}(1+e)\,u\cos\alpha$$

$$\therefore\ 2\tan^{-1}\left(\frac{V}{U}\right)=2\tan^{-1}\left(\frac{3e\tan\alpha}{2-e}\right).$$

$$\frac{K.E.\text{ lost}}{\text{Orig. }K.E.}=\frac{mu^2-m\,(U^2+V^2)-\frac{1}{2}mU^2}{mu^2}$$

$$=\frac{(1-e^2)(1+2\sin^2\alpha)}{3}.$$

EXAMPLES V (A)

1. Prove that if two equal and perfectly elastic spheres impinge at right angles, their directions after impact will still be at right angles.

2. A ball impinges on another at rest; prove that if the co-efficient of restitution be equal to the ratio of their masses the balls will move in directions at right angles to each other. What will happen in the case of direct collision ?

3. The masses of three spheres A, B, C are $7m, 7m, m$; their co-efficient of restitution is unity, their centres are in a straight line and C lies between A and B. Initially A and B are at rest and C is given a velocity along the line of centres towards A. Show that it strikes A twice and B once and that the final velocities of A, B, C are proportional to 21, 12, 1.

4. Two equal spheres A, B lie on a smooth horizontal circular groove at opposite ends of a diameter, A is projected along the groove and at the end of time T impinges on B. Show that the second impact will occur after a further time $\frac{2T}{e}$ where e is the co-efficient of elasticity.

5. Two equal balls of elasticity e impinge, having before impact resolved velocities u_1, v_1 in the direction of the common normal and u_2, v_2 perpendicular to it. If their motions after impact are at right angles, prove that $(u_1+v_1)^2+4u_2v_2=e^2(u_1-v_1)^2$.

6. Two equal smooth spheres, of radius r, move with the same speed in opposite directions in parallel lines which are at a distance c apart; prove that the direction of motion of each deviates, on impact, through a right angle if $c^2(1+e)=4er^2$, where e is the co-efficient of restitution.

7. Two small spheres of masses m, m', hang just in contact by two threads of equal lengths, the spheres are drawn apart, each thread remaining tight to the same vertical heights and then let go simultaneously; find the vertical heights to which they will rise after the collision, e being the co-efficient of elasticity.

8. Three elastic particles of masses m_1, m_2, m_3 lie in a straight line on a horizontal table and m_1 is projected towards m_2. If $(m_1+m_2)(m_2+m_3)\,e$ $=m_1m_2(1+e)^2$ show that the velocity of m_1 after striking m_2 will be equal to that of m_2 after striking m_3.

9. A smooth ball strikes another which is at rest, the angle between the line of centres and the direction of motion of the first ball being α. This ball moves off in a direction perpendicular to its original direction. Show that the masses of the balls are in the ratio of $e\cos^2\alpha-\sin^2\alpha : 1$, where e is the co-efficient of restitution.

10. Two equal spheres, centres A and B, lie in contact on a smooth table and A is struck directly by a third sphere, centre C moving with velocity V in a direction making an acute angle θ with AB. Show that after impact A moves in a direction making an angle $\cot^{-1}(\cot\theta+2\tan\theta)$ with CA produced.

11. Two small balls of equal mass can move inside a rough endless horizontal tube of length l. One ball impinges with velocity u on the other at rest. If the friction of the tube produces a retardation f in either ball, and if after impact the balls just meet again, prove that $2lf=u^2e$ where e is the co-efficient of restitution.

12. A smooth sphere of mass m is let fall and when its velocity is u it strikes another smooth sphere of mass M suspended at the end of a fine inextensible string, the line of centres making an angle θ with the vertical at the instant of impact. Show that the horizontal and vertical components of velocity of the impinging ball after impact are

$$\frac{Mu \sin\theta \cos\theta\,(1+e)}{M+m\sin^2\theta},\quad u\left(1-\frac{M(1+e)\cos^2\theta}{M+m\sin^2\theta}\right).$$

5·4. Direct impact of a solid on a fixed smooth surface. Let a solid of mass m impinge perpendicularly on a fixed surface and let its velocities before and after impact be u and v respectively.

The fixed surface may be imagined to be another solid whose velocity before and after impact is zero, then by Newton's law

$$v-0=-e(u-0)$$

or
$$v=-eu,$$

that is, after impact against a fixed smooth surface the solid moves back with a velocity which is e times its velocity before impact. Impulse of the blow on the plane

=impulse of blow on the solid in opposite direction

=change of momentum of the solid

$=m(u-v)=mu\,(1+e)$.

5·5. Oblique impact of a perfectly smooth sphere on a perfectly smooth fixed plane

Let the velocities of the sphere before and after impact be u and v, which make angles α and θ with the line AN, perp to the plane at the point of contact A.

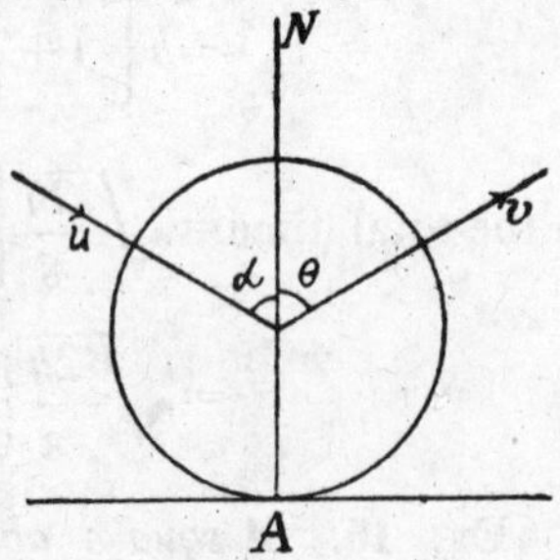

The plane being smooth, there is no force of friction between the sphere and the plane along the plane. There is no other force along the plane, impulse of blow being perp. to the plane, the velocity of the sphere is unaltered along the plane.

$\therefore\ v\sin\theta=u\sin\alpha$ ..(1)

Also by Newton's law

$$v\cos\theta-0=-e(-u\cos\alpha-0)$$

or
$$v\cos\theta=eu\cos\alpha. \quad \text{...(2)}$$

Squaring and adding we get

$$v^2=u^2(\sin^2\alpha+e^2\cos^2\alpha)$$

and dividing
$$\cot\theta=e\cot\alpha.$$

Impulse of the blow = change of momentum of the sphere along the normal

$$= mu \cos \alpha - (-mv \cos \theta) = m(1+e)\ u \cos \alpha.$$

Ex. 15. *A ball dropped from a certain height h on a horizontal plane, bounces up and down and finally comes to rest. Discuss the motion.*

If u be the velocity of fall then $u^2 = 2gh$. If v be the velocity of rebound after the first impact then $v = eu = e\sqrt{2gh}$. The ball goes up with this velocity to a height h', say, so that $h' = \frac{v^2}{2g} = e^2 h$ and it takes a time t', say, when $t' = \frac{v}{g} = e\sqrt{\frac{2h}{g}}$.

Sol. The ball again drops and strikes the plane with velocity v, so rebounds with velocity ev and goes to a height $\frac{e^2 v^2}{2g} = e^4 h$ after a time $\frac{ev}{g} = e^2\sqrt{\frac{2h}{g}}$ and so on.

Hence the total distance described

$$= h + 2e^2 h + 2e^4 h + 2e^6 h + \dots$$

$$= h\left[1 + 2e^2 \frac{1}{1-e^2}\right] = h\,\frac{1+e^2}{1-e^2}$$

and the total time $= \sqrt{\frac{2h}{g}}\left[1 + 2e + 2e^2 + 2e^3 + \dots\right]$

$$= \sqrt{\frac{2h}{g}}\left[1 + \frac{2e}{1-e}\right] = \frac{1+e}{1-e}\sqrt{\frac{2h}{g}}.$$

Ex. 16. *A squa e able ABCD whose side is a has raised edges. A particle of elasticity e is projected from a point P in AB and hits the sides BC, CD, DA in Q, R, S ; prove that PQ, RS are parallel. If α be the angle QPB and BP=x, prove that if the particle returns to P, x (1−e)=a(1−e cot α).*

Sol. Let the angles of reflection at Q, R, S be β, γ, δ.

Then $\cot \beta = e \cot \alpha$, $\cot \gamma = e \tan \beta = \tan \alpha$,

$\cot \delta = e \tan \gamma = e \cot \alpha$

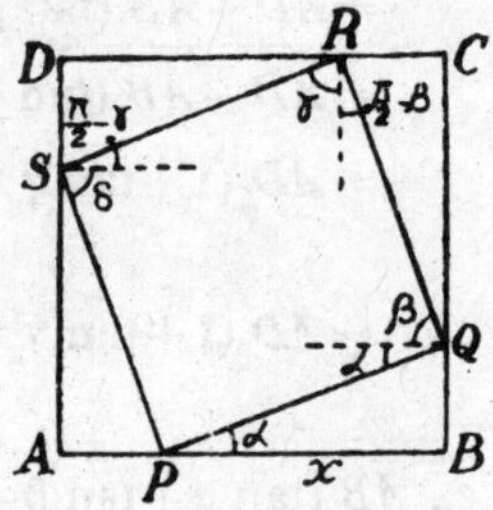

$\therefore \quad \gamma = \pi/2 - \alpha \quad \therefore \quad \angle DRS = \alpha = \angle BPQ$

$\therefore \quad PQ$ is parallel to RS.

Also $\cot \delta = e \cot \alpha = \cot \beta$

$\therefore \quad \delta = \beta \quad \therefore \quad SP$ is parallel to QR.

Hence $PQRS$ is a parallelogram.

$\therefore \quad SP = QR$

Now $SP = AP \sec \delta = (a - x) \sec \beta$

$QR = CQ \operatorname{cosec} \beta = (a - BQ) \operatorname{cosec} \beta$

$= (a - x \tan \alpha) \operatorname{cosec} \beta.$

Hence $\quad (a-x) \sec \beta = (a - x \tan \alpha) \operatorname{cosec} \beta$

or $\quad (a-x) \tan \beta = a - x \tan \alpha$

or $$(a-x) \frac{\tan \alpha}{e} = a - x \tan \alpha$$

or $$x\left(\frac{\tan \alpha}{e} - \tan \alpha\right) = a\left(\frac{\tan \alpha}{e} - 1\right)$$

or $$x(1-e) = a(1 - e \cot \alpha).$$

Ex. 17. *From one corner A of a rectangular billiard table ABCD a ball is projected in a direction making an angle α with the side AB ; it strikes first in the side BC, then AD, then DC, then BC again and then returns to A.*

Prove that

$$\frac{AB}{AD} = \frac{e^2 \cot \alpha}{1 + e^2}.$$

Sol. Let the path of the ball be $APQRSA$, α is the angle of incidence at P, β the angle of reflection at P, γ that at Q, δ at R and ψ at S.

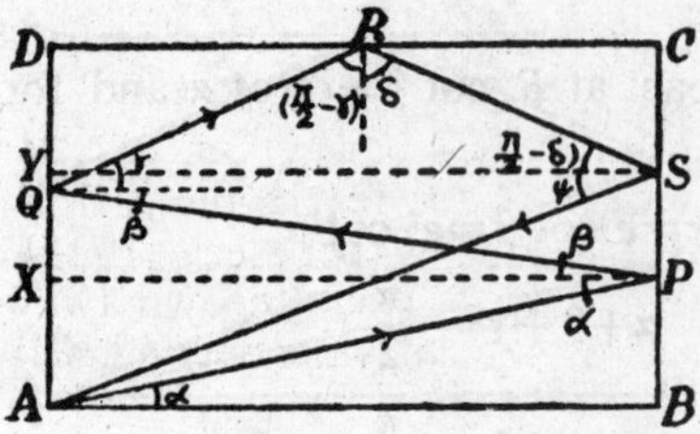

Then $\cot \beta = e \cot \alpha$.

$\cot \gamma = e \cot \beta = e^2 \cot \alpha$,

$\cot \delta = e \tan \gamma = \dfrac{1}{e} \tan \alpha$,

$\cot \psi = e \tan \delta = e^2 \cot \alpha$.

Now, $AB \tan \alpha = BP = BC - CP$

$= AD - XD = AD - XQ - QD$

$= AD - PX \tan \beta - DR \tan \gamma$

$= AD - AB \tan \beta - (DC - RC) \tan \gamma$

$= AD - AB \tan \beta - AB \tan \gamma + CS \tan \delta \tan \gamma$

$$=AD-AB\,(\tan\beta+\tan\gamma)+(BC-BS)\tan\gamma\tan\delta$$

$$=AD-AB\,(\tan\beta+\tan\gamma)+AD\tan\gamma\tan\delta-AY\tan\gamma\tan\delta$$

$$=AD\,(1+\tan\gamma\tan\delta)-AB\,(\tan\beta+\tan\gamma)$$
$$-SY\tan\psi\tan\gamma\tan\delta$$

$$=AD\,(1+\tan\gamma\tan\delta)-AB\,(\tan\beta+\tan\gamma)-AB\tan\gamma$$
$$\tan\delta\tan\psi$$

i.e., $AB\,(\tan\alpha+\tan\beta+\tan\gamma+\tan\gamma\tan\delta\tan\psi)=AD(1+\tan\gamma\tan\delta)$

i.e., $AB\left(\tan\alpha+\dfrac{1}{e}\tan\alpha+\dfrac{1}{e^2}\tan\alpha+\dfrac{1}{e^2}\tan\alpha.e\cot\alpha\,\dfrac{1}{e^2}\tan\alpha\right)$

$$=AD\left(1+\frac{1}{e^2}\tan\alpha.e\cot\alpha\right)$$

i.e., $$AB\tan\alpha(1+e+e^2+e^3)=AD.e^2(1+e)$$

$$\frac{AB}{AD}=\frac{e^2\cot\alpha}{1+e^2}.$$

Ex. 18. *A smooth circular table is surrounded by a smooth rim whose interior surface is vertical. Show that a ball, projected along the table from a point on the rim in a direction making an angle α with the radius through the point, will return to the point of projection after*

(*i*) *two impacts if* $\tan\alpha=\dfrac{e^{\frac{3}{2}}}{\sqrt{1+e+e^2}}$ (*ii*) *three impacts if* $\tan\alpha=e^{\frac{3}{2}}$.

Sol. A is the point of projection and B, C are the points of impact.

(*i*) Let $\angle OAB=\alpha$, $\angle OBC=\beta$ and $\angle OCA=\gamma$

so that $\angle OBA=\alpha$, $\angle OCB=\beta$

and $\angle OAC=\gamma$.

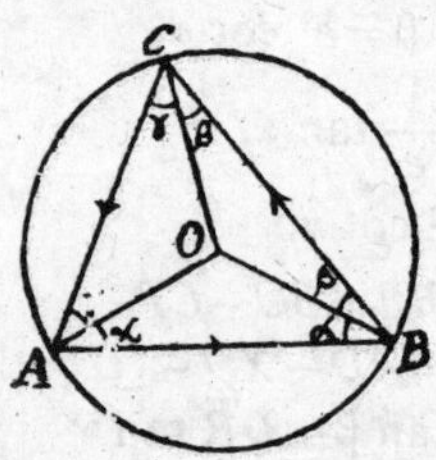

For the impact at B, $\cot\beta=e\cot\alpha$ and for the impact at C,

$$\cot\gamma=e\cot\beta=e^2\cot\alpha$$

But $$\alpha+\beta+\gamma=\frac{\pi}{2}$$

$\therefore$ $$\alpha+\beta=\frac{\pi}{2}-\gamma.$$

Hence $$\frac{\tan\alpha+\tan\beta}{1-\tan\alpha\tan\beta}=\cot\gamma=e^2\cot\alpha$$

or $$\frac{\cot\alpha+\cot\beta}{\cot\alpha\cot\beta-1}=e^2\cot\alpha$$

or $$\frac{\cot\alpha+e\cot\alpha}{e\cot^2\alpha-1}=e^2\cot\alpha$$

or $$e^3\cot^2\alpha-e^2=e+1 \quad \therefore \quad \cot^2\alpha=\frac{1+e+e^2}{e^3}$$

or $$\tan^2\alpha=\frac{e^3}{1+e+e^2}.$$

u, v, w be the velocities of the ball along AB, BC, CA then $v^2=u^2(\sin^2\alpha+e^2\cos^2\alpha)$ and $w^2=v^2(\sin^2\beta+e^2\cos^2\beta)$.

Now $$\tan\alpha=\frac{e^{\frac{3}{2}}}{\sqrt{1+e+e^2}} \quad \therefore \tan\beta=\frac{\tan\alpha}{e}=\sqrt{\frac{e}{1+e+e^2}}$$

$$\therefore \quad \sin^2\beta=\frac{e}{1+2e+e^2}$$

and $$\cos^2\beta=\frac{1+e+e^2}{1+2e+e^2}$$

Hence $$w^2=v^2\left(\frac{e+e^2+e^3+e}{1+2e+e^2}\right)$$

$$\therefore \quad w=v\sqrt{\frac{e(1+e^2)}{1+e}}$$

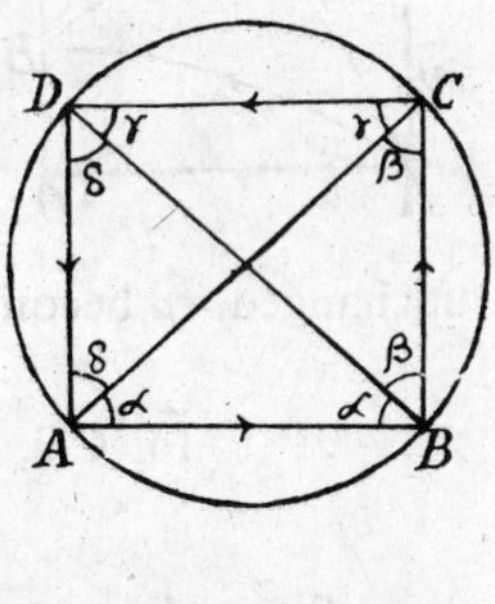

Similarly $$v=eu\sqrt{\frac{1+e}{1+e^2}}$$

$$\therefore \quad \frac{w}{u}=e^{\frac{3}{2}}$$

(*ii*) $\cot\beta=e\cot\alpha$, $\cot\gamma=e\cot\beta=e^2\cot\alpha$, $\cot\delta=e\cot\gamma=e^3\cot\alpha$.

Now $$\alpha+\beta+\gamma+\delta=\pi \quad \therefore \quad \tan(\alpha+\beta)=-\tan(\gamma+\delta)$$

or $$\frac{\tan\alpha+\tan\beta}{1-\tan\alpha\tan\beta}=-\frac{\tan\gamma+\tan\delta}{1-\tan\gamma\tan\delta}\cdot$$

or $$\frac{\tan\alpha\left(1+\frac{1}{e}\right)}{1-\frac{\tan^2\alpha}{e}}=-\frac{\frac{1}{e^2}\tan\alpha\left(1+\frac{1}{e}\right)}{1-\frac{\tan^2\alpha}{e^3}}$$

or $$\tan^2\alpha=e^3 \quad \therefore \quad \tan\alpha=e^{3/2}$$

Also $\quad \tan\alpha \tan\delta = \tan\beta \tan\gamma = \dfrac{\tan^2\alpha}{e^3} = 1.$

Hence $\quad \tan(\alpha+\delta) = \infty$

$\therefore \quad \alpha+\delta = \dfrac{\pi}{2} = \beta+\gamma \quad \therefore \quad \angle A = \dfrac{\pi}{2} = \angle C.$

Ex. 19. *A particle is projected from a point at the foot of one of two smooth vertical walls, so that after three impacts, it may return to the point of projection. If the last impact be direct, show that*

$$e^3+e^2+e=1.$$

Sol. Let the particle be projected from A with components of velocity (u, v) and B, C, D be the points of impact. Let t_1, t_2, t_3, t_4 be the times for AB, BC, CD, DA.

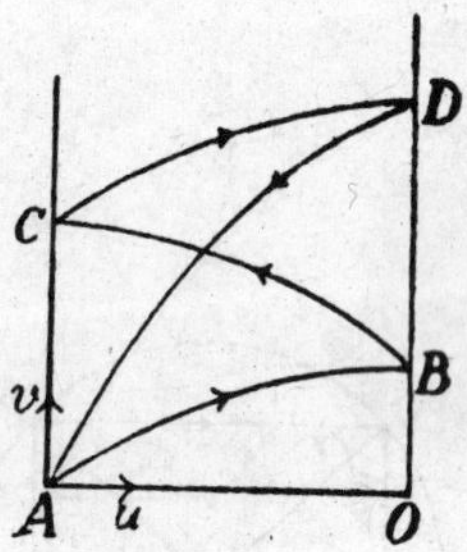

For AB, u remains unchanged so that if $AO=a$, $a=ut_1 \quad \therefore \quad t_1=\dfrac{a}{u}$. At B, vertical velocity remains unchanged, u becomes eu after impact hence for BC, eu remains unchanged.

$$\therefore \quad a=eut_2 \quad \therefore \quad t_2=\frac{a}{eu}.$$

At C, again vertical velocity remains unchanged, eu becomes e^2u after impact.

$\therefore \quad$ for CD, $a=e^2ut_3 \quad \therefore \quad t_3=\dfrac{a}{e^2u}.$

Similarly $\quad t_4=\dfrac{a}{e^3u}.$

Now v becomes zero at D since the impact there is direct *i.e.*, v is destroyed in time $(t_1+t_2+t_3)$ and in moving from D to A, the vertical velocity again becomes v. $\quad \therefore \quad t_1+t_2+t_3=t_4$

i.e., $\quad \dfrac{a}{u}\left(1+\dfrac{1}{e}+\dfrac{1}{e^2}\right)=\dfrac{a}{e^3u}$ i.e, $e^3+e^2+e=1.$

Ex. 20. *A ball is projected from a point in a horizontal plane; after one rebound from the plane it strikes directly against a vertical wall; after two more rebounds it returns to the point of projection.*

Prove that $\quad e=2(1-e^3).$

Sol. A is the point of projection successive points of impact are B, C, D, E, A. Let u, v be the horizontal and vertical components of the velocity of the projection.

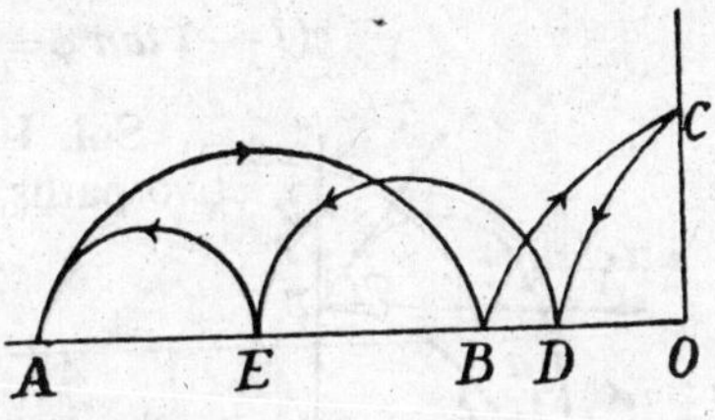

If t_1 be the time from A to B,

$$t_1=\frac{2v}{g} \quad \text{and} \quad AB=ut_1=\frac{2uv}{g}.$$

At B, u remains unchanged, v becomes ev after impact $\therefore$ for BC, if t_2 be time, t_2 = half the time of flight since the impact at C is direct.

$$\therefore \qquad t_2=\frac{ev}{g} \quad \text{and} \quad BO=ut_2=\frac{euv}{g}.$$

At C, u becomes eu after impact and since the impact at C is direct, the time for CD is the same as that for BC.

$$\therefore \qquad OD=eu.t_2=\frac{e^2uv}{g}.$$

At D, vertical velocity and horizontal velocity just before impact are ev and eu and after impact, ev becomes e^2v, eu remains unchanged.

$\therefore$ If t_3 be the time for DE,

$$t_3=\frac{2e^2v}{g} \quad \text{and} \quad DE=eu.t_3=\frac{2e^3uv}{g}.$$

At E, e^2v becomes e^3v and eu remains unchanged

$\therefore$ If t_4 be the time for EA,

$$t_4=\frac{2e^3v}{g},\ EA=eut_4=\frac{2e^4uv}{g}.$$

Now $$AB+BO=OD+DE+EA$$

$$\therefore \qquad \frac{2uv}{g}+\frac{euv}{g}=\frac{e^2uv}{g}+\frac{2e^3uv}{g}+\frac{2e^4uv}{g}$$

or $$2+e=e^2+2e^3+2e^4$$

or $$(1+e)(2e^3+e-2)=0$$

$$2e^3+e-2=0$$

Ex. 21. *A particle is projected at an angle α to the horizontal to strike a smooth vertical wall and after rebounding it passes through* *the point of projection. If ϕ be the angle of inclination to the horizontal at which the particle rebounds from the wall and e the co-efficient of restitution, show that*

$$e(1+e)\tan\phi=(1-e)\tan\alpha.$$

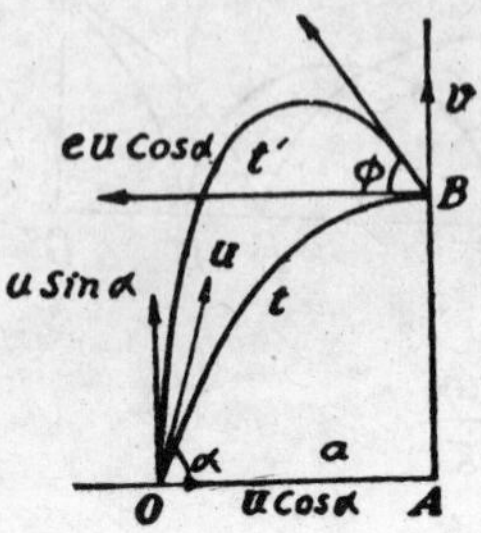

Sol. If $AO=a$, and t, t' be the times of the two paths, then

$$a=u\cos\alpha\, t,$$

$$t+t'=\text{time of flight}=\frac{2u\sin\alpha}{g}.$$

Also at B, horizontal velocity towards left after rebound is $eu\cos\alpha$

$$\therefore\quad a=eu\cos\alpha\, t'$$

Hence $$\frac{a}{u\cos\alpha}+\frac{a}{eu\cos\alpha}=\frac{2u\sin\alpha}{g}$$

or $$a(1+e)=\frac{2eu^2}{g}\sin\alpha\cos\alpha \qquad ...(1)$$

Now at B, the vertical velocity is given by

$$v=u\sin\alpha-gt=u\sin\alpha-\frac{ag}{u\cos\alpha}=\frac{u^2\sin\alpha\cos\alpha-ag}{u\cos\alpha}$$

Hence $$\tan\varphi=\frac{v}{eu\cos\alpha}=\frac{u^2\sin\alpha\cos\alpha-ag}{eu^2\cos^2\alpha}$$

i.e., $$u^2(\sin\alpha\cos\alpha-e\cos^2\alpha\tan\phi)=ag$$

Hence from (1)

$$a(1+e)=2e\sin\alpha\cos\alpha.\frac{a}{\sin\alpha\cos\alpha-e\cos^2\alpha\tan\varphi}$$

$$=\frac{2ae}{1-e\tan\varphi\cot\alpha}$$

$$\therefore\quad (1+e)(1-e\tan\varphi\cot\alpha)=2e$$

or $$(1+e)(\tan\alpha-e\tan\varphi)=2e\tan\alpha$$

or $$e(1+e)\tan\varphi=(1-e)\tan\alpha$$

Ex. 22. *Two parallel vertical walls of height h stand on a horizontal plane. A ball is projected from the foot of one wall directly towards the top of the other and after impact just clears the top of the* *first wall Prove that the point of impact is at a depth* $\frac{h}{(1+e)^2}$ *below the top of the wall.*

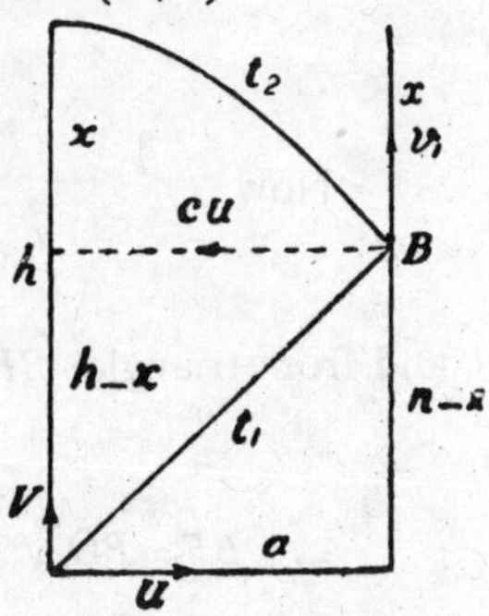

Sol. Let A be the top of the first wall, B the point where the ball strikes the second wall, and it clears the first wall at A horizontlly.

$$t_2+t_2=\text{half the time of flights}=\frac{v}{g}$$

and $$v^2=2gh$$

also $$a=ut_1=eut_2 \qquad \therefore \quad t_1=et_2$$

$$\therefore \quad \frac{v}{g}=t_2\,(1+e)$$

If v' be the vertical velocity at B, $t_2=\frac{v'}{g}$

and $$v'^2=v^2-2g(h-x)$$

$$\therefore \quad \frac{v}{g}=\frac{v'}{g}\,(1+e) \quad \text{or} \quad v=v'\,(1+e)$$

or $$\frac{v^2}{(1+e)^2}=v'^2=v^2-2g\,(h-x)$$

or $$v^2\left\{1-\frac{1}{(1+e)^2}\right\}=2g\,(h-x)$$

or $$2gh\left\{1-\frac{1}{(1+e)^2}\right\}=2g\,(h-x) \quad \therefore \quad x=\frac{h}{(1+e)^2}$$

Ex. 23. *A ball is projected along a horizontal plane from the middle point of one of the sides of an isosceles right angled triangle so as, after reflections at the hypotenuse and the remaining side, to return to the same point, prove that the cotangents of the angles of reflection are e+1 and e+2 respectively.*

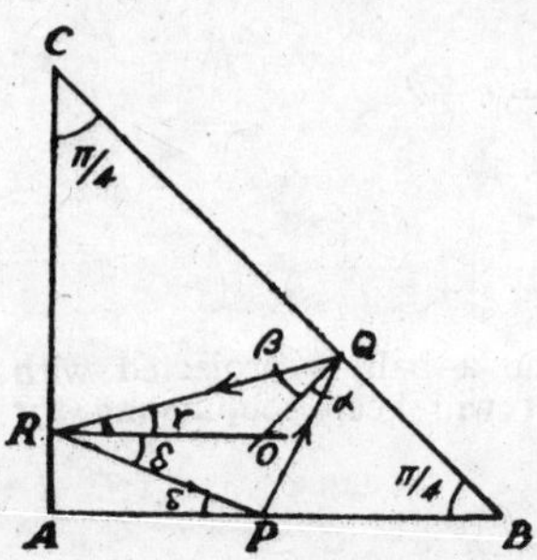

Sol. The ball projected from P, midpoint of AB, strikes BC at Q at angle of incidence α, β the angle of reflection, then it strikes CA at R at angle of incidence γ, δ be the angle of reflection. Evidently $\angle APR=\delta$. Let the normals at Q and R meet at O.

Then at Q, $\cot\beta=e\cot\alpha$

at R, $\cot\delta=e\cot\gamma$

Also $\quad \angle ROQ+\angle C=\pi \quad \therefore \ \pi-(\beta+\gamma)+\dfrac{\pi}{4}=\pi$

$$\therefore \qquad \beta+\gamma=\frac{\pi}{4}$$

Now $\qquad AP=RP\cos\delta$

and from triangle BPQ, $\dfrac{BP}{PQ}=\dfrac{\sin\left(\dfrac{\pi}{4}-\alpha\right)}{\sin\dfrac{\pi}{4}}=\sqrt{2}\cos\alpha$

$\therefore \quad BP=PQ\sqrt{2}\cos\alpha$

But $AP=BP \qquad \therefore \quad RP\cos\delta=PQ.\sqrt{2}\cos\alpha$

or $\qquad \dfrac{RP}{PQ}=\dfrac{\sqrt{2}\cos\alpha}{\cos\delta}$

from triangle PQR, $\dfrac{RP}{PQ}=\dfrac{\sin(\alpha+\beta)}{\sin(\gamma+\delta)}=\dfrac{\sin\alpha\sin\beta\,(\cot\alpha+\cot\beta)}{\sin\gamma\sin\delta\,(\cot\gamma+\cot\delta)}$

$$=\frac{\sin\alpha\sin\beta\,(1+e)\cot\alpha}{\sin\gamma\sin\delta\,(1+e)\cot\gamma}=\frac{\sin\beta\cos\alpha}{\sin\delta\cos\gamma}$$

Hence $\qquad \dfrac{\sqrt{2}\cos\alpha}{\cos\delta}=\dfrac{\sin\beta\cos\alpha}{\sin\delta\cos\gamma} \quad$ or $\quad \dfrac{\sqrt{2}}{\cot\delta}=\dfrac{\sin\beta}{\cos\gamma}$

or $\qquad \dfrac{\sqrt{2}}{e\cot\gamma}=\dfrac{\sin\beta}{\cos\gamma}$ or $e\sin\beta=\sqrt{2}\sin\gamma$

or $\qquad e\sin\beta=\sqrt{2}\sin\left(\dfrac{\pi}{4}-\beta\right)=\cos\beta-\sin\beta$

$$\therefore \qquad \cot\beta=e+1$$

also $\qquad \cot\delta=e\cot\gamma=e\cot\left(\dfrac{\pi}{4}-\beta\right)=\dfrac{e}{\tan\left(\dfrac{\pi}{4}-\beta\right)}$

$$=\frac{e(1+\tan\beta)}{1-\tan\beta}=\frac{e\left[1+\dfrac{1}{e+1}\right]}{1-\dfrac{1}{e+1}}=e+2.$$

EXAMPLES V (B)

1. From a point in a smooth horizontal plane a ball is projected with velocity u at an angle α to the horizon. Show that it will keep rebounding for time $\dfrac{2u\sin\alpha}{g(1-e)}$ and will have a range $\dfrac{u^2\sin 2\alpha}{g(1-e)}$.

2. A particle projected with velocity u and in a direction inclined to the vertical at an angle α, impinges against a vertical wall whose plane is perpendicular to the plane of its path, and distant h from the point of projection.

Prove that, if after impact the particle returns to the point of projection $e=\dfrac{gh}{u^2 \sin 2\alpha - g}$.

3. A particle is projected from a point in a horizontal plane so as to strike a vertical wall at right angles, and after rebounding from the wall, and once from the horizontal plane, returns to the point of projection. Prove that $e=\frac{1}{2}$.

4. A particle is projected with velocity u from a point A on the ground so as to strike a smooth vertical wall distant h and after impact to strike the ground at a point midway between A and the wall; show that the angle of projection is α where $\sin 2\alpha = \dfrac{(2e+1)\,gh}{2eu^2}$.

5. A ball is dropped from the top of a tower of height h and at the same time another ball of equal weight is projected upwards with velocity $\sqrt{2gh}$ from the base of the tower and collides directly with the falling ball. Prove that the falling ball will, in the rebound, rise to a height short of the top of the tower by $\dfrac{h}{4}(1-e^2)$.

6. Particles are projected horizontally from different points in a tower of height h, each with a velocity due to the height of the tower above the point of projection. Prove that they will all cease rebounding from the horizontal plane through the foot of the tower within or on a circle of radius $\dfrac{1+e}{1-e}h$ with its centre at the foot of the tower, where e is the co-efficient of elasticity for all the particles.

7. A ball is projected from a point in one of two smooth parallel vertical walls against the other in a plane perpendicular to both, and after being reflected at each wall impinges again on the second at a point in the same horizontal plane as it started from; show that $be^2=a(1+e+e^2)$, where b is the free range on horizontal plane and a the distance between the planes, e the co-efficient of elasticity.

8. A smooth straight tube of length $(a+b)$ is closed at both ends and contains a particle which is initially at a distance a from one end. The particle is projected with velocity v towards the other end. Prove that it is next moving through the same point relative to the tube and in the same direction after time $\left(\dfrac{a}{e}+b\right)\left(1+\dfrac{1}{e}\right)\Big/ v$ where e is the co-eff. of restitution between the particle and either end of the tube.

9. A smooth ball is projected from the ground with velocity u so as to hit a smooth vertical wall and bounce back again. Show that it cannot bounce further than a distance of $e\left(\dfrac{u^2}{g}-d\right)$ from the wall, where d is the distance of the point of projection from the wall.

10. An imperfectly elastic sphere of mass m, moving with velocity u impinges on another sphere of mass m' at rest. The second sphere afterwards strikes a vertical cushion at right angles to its path ; show that there will be no further impact of the spheres if $(1+e'+ee')<em'$, where e, e' are the co-efficients of restitution between the spheres and the wall respectively.

11. A ball moving with a velocity v impinges directly on a ball at rest ; the latter ball moves with the velocity communicated to it and after a direct rebound from the wall comes into collision with the first ball. If the first ball is reduced to rest, prove that the ratio of the mass of the second ball to that of first is λ where λ is a root of the equation $e^2\lambda^2-e\lambda\,(3+2e+e^2)+1=0$.

12. A particle is projected from a given point in a horizontal plane at an angle α to the horizon and after one rebound at a vertical wall returns to the point of projection ; prove that the point of impact must lie on the straight line $y(1+e)=x \tan \alpha$, x, y being measured horizontally and vertically from the point of projection. When the velocity of projection and not the direction is given, the locus of the point of impact is the ellipse

$$x^2+y^2\,(1+e)^2=4ehy$$

where h is the space due to the velocity of projection.

13. A ball is projected from the middle point of one side of a billiard table, so as to strike first an adjacent side, and the middle point of the side opposite to that from which it started ; show that the ball will hit the other adjacent side at a distance $\dfrac{be}{1+e}$ from the end nearest the opposite side, b being the adjacent side.

5·6. Impact on an inclined plane. A ball is projected from the foot of a plane inclined at an angle β to the horizon, with a velocity u at an angle α to the plane. To discuss the motion.

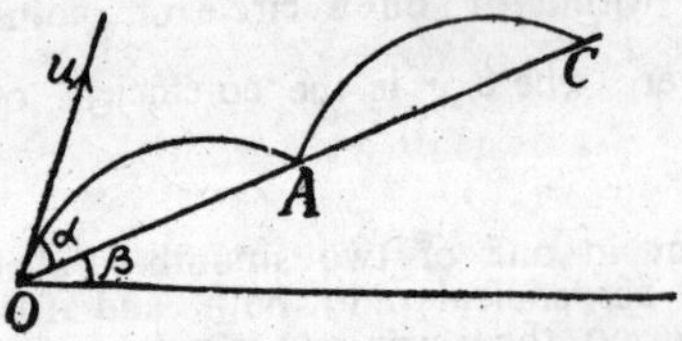

Components of u parallel and perpendicular to the plane are $u \cos \alpha$, $u \sin \alpha$ and components of g in these directions—$g \sin \beta$, $-g \cos \beta$.

Let A, B etc. be the points of impact with the plane,

time for $OA=\dfrac{2u \sin \alpha}{g \cos \beta}$.

At A, velocity parallel to the plane remains unchanged while the velocity perpendicular to the plane, $u \sin \alpha$ is changed to $eu \sin \alpha$ after impact

$$\therefore \quad \text{time for } AB=\frac{2eu \sin \alpha}{g \cos \beta} \text{ and so on.}$$

Hence the total time of rebound $=\dfrac{2u \sin \alpha}{g \cos \beta}(1+e+e^2+\ldots\text{to inf.})$

$$=\frac{2u \sin \alpha}{g \cos \beta\,(1-e)}=T \text{ say.}$$

The distance moved up the plane $= u \cos\alpha . T - \frac{1}{2} g \sin\beta . T^2$

$$= \frac{2u^2 \sin\alpha}{g \cos^2\beta\ (1-e)^2} [\cos(\alpha+\beta) - e \cos\alpha \cos\beta].$$

After this there will be no more rebound but the ball will still move up the plane till $u \cos\alpha$ is destroyed. If t be the total time of moving up the plane

$$0 = u \cos\alpha - g \sin\beta . t \quad \therefore \ t = \frac{u \cos\alpha}{g \sin\beta}.$$

If the ball returns to the point of projection after r impacts, there will be $(r+1)$ parabolas and the time taken is

$$t' = \frac{2u \sin\alpha}{g \cos\beta} (1 + e + e^2 + \ldots + e^r)$$

$$= \frac{2u \sin\alpha}{g \cos\beta} \frac{1 - e^{r+1}}{1-e}.$$

Since the ball returns to the point of projection, distance moved is zero

$$\therefore \quad 0 = u \cos\alpha . t' - \tfrac{1}{2} g \sin\beta . t'^2$$

i.e., $\quad t' = \dfrac{2u \cos\alpha}{g \sin\beta}$ *i.e.*, $\dfrac{2u \sin\alpha}{g \cos\beta} \dfrac{1 - e^{r+1}}{1-e} = \dfrac{2u \cos\alpha}{g \sin\beta}$

or $\quad \tan\alpha \tan\beta\, (1 - e^{r+1}) = 1 - e$

If the ball strikes the plane *normally* at the rth impact, the velocity $u \cos\alpha$ along the plane is destroyed there so that the time upto the rth impact must be equal to the time of returning to the point of projection.

Now time up to rth impact $= \dfrac{2u \sin\alpha}{g \cos\beta} (1 + e + e^2 + \ldots \text{to } r \text{ terms})$

$$= \frac{2u \sin\alpha}{g \cos\beta} \frac{1 - e^r}{1-e}.$$

If the ball returns to the point of projection at the nth impact, the ball will describe $(n-r)$ parabolas after the rth impact,

the time for which is $\dfrac{2u \sin\alpha}{g \cos\beta} \left(e^r + e^{r+1} + \ldots \text{to } (n-r) \text{ terms} \right)$

$$= \frac{2u \sin\alpha}{g \cos\beta} e^r \left[1 + e + e^2 + \ldots\ldots \text{to } (n-r) \text{ terms} \right]$$

$$= \frac{2u \sin\alpha}{g \cos\beta} e^r . \frac{(1 - e^{n-r})}{1-e}.$$

Since the time up the plane is equal to the time down, we have

$$\frac{2u \sin \alpha}{g \cos \beta} \cdot \frac{1-e^r}{1-e} = \frac{2u \sin \alpha}{\cos \beta} e^r . \frac{1-e^{n-r}}{1-e}$$

or $$1-e^r=e^r-e^n$$

or $$e^n-2e^r+1=0.$$

Also since the ball strikes the plane normally at the rth impact the time upto that instant is equal to the time in which $u \cos \alpha$ is destroyed.

i.e., $$\frac{2u \sin \alpha}{g \cos \beta}\frac{1-e^r}{1-e}=\frac{u \cos \alpha}{g \sin \beta}$$

or $$2 \tan \alpha \tan \beta \,(1-e^r)=1-e.$$

Ex. 24. *A particle of elasticity e is dropped from a vertical height a upon the highest point of a plane, which is of length b and is inclined at an angle α to the horizon and descends to the bottom in three jumps.*

Prove that $b=4ae\,(1+e)(1+e+e^2)(1+e^2) \sin \alpha$.

Sol. If u be the velocity at the top O. $u^2=2ga$.

A, B, C are the points of impact, C being the bottom of the plane; components of u along and perp. to the plane $u \sin \alpha$, $u \cos \alpha$ and those of g are $g \sin \alpha$, $g \cos \alpha$.

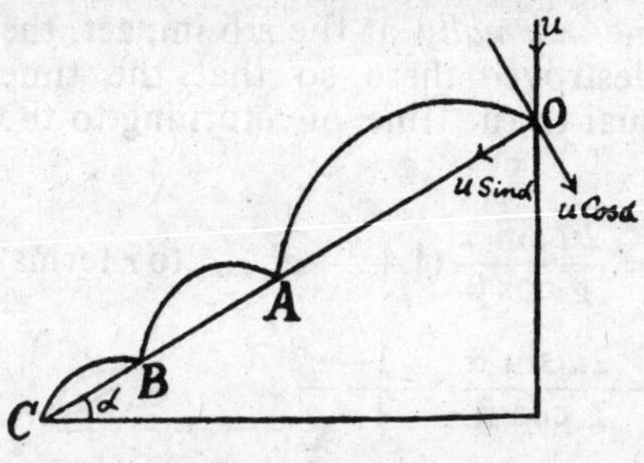

After impact at O, velocity perp. to the plane becomes $eu \cos \alpha$ and velocity parallel to the plane $u \sin \alpha$ remains unchanged.

$\therefore$ if t_1 be the time for OA,

$$t_1=\frac{2eu \cos \alpha}{g \cos \alpha}=\frac{2eu}{g}$$

Similarly if t_2, t_3 be the times for AB, BC,

$$t_2=\frac{2e^2u}{g},\quad t_3=\frac{2e^3u}{g}$$

$\therefore$ $$t=\text{total time}=t_1+t_2+t_3=\frac{2u}{g}\,e(1+e+e^2).$$

Now $$b=u \sin \alpha . t+\tfrac{1}{2}g \sin \alpha . t^2$$

$$=\frac{2u^2 \sin \alpha}{g} e\,(1+e+e^2)+\frac{2u^2 \sin \alpha}{g}\,[e(1+e+e^2)]^2$$

$$=4a \sin \alpha e(1+e+e^2)(1+e+e^2+e^3)$$

$$=4a \sin \alpha e(1+e+e^2)(1+e)(1+e^2).$$

EXAMPLES V (C)

1. A smooth elastic sphere is projected at an elevation α up a plane which makes an angle β with the horizon. If on striking the plane it rebounds vertically, prove that

$$(e+1) \sin \beta \sin (\alpha-\beta)=\cos \alpha.$$

2. An elastic particle is projected from the foot of an inclined plane of angle α and hits the plane at right angles. If β be the angle of projection with the plane, prove that $2 \tan \alpha \tan \beta=1$. If the particle returns to the point of projection in two jumps, prove that

$$e=\frac{\sqrt{5}-1}{2}.$$

3. A particle after falling from rest through a distance h strikes a smooth plane inclined at an angle α to the horizon. Show that the distance between the first two points of impact is $4he\,(1+e) \sin \alpha$ and the whole range on the inclined plane when the particle ceases to rebound is

$$\frac{4he \sin \alpha}{(1-e)^2}.$$

4. A ball is projected with velocity v from a point of a plane inclined at an angle α to the horizontal; the direction of projection is at right angles to the plane. Prove that before ceasing to rebound, it will have described a length $\frac{2v^2 \sin \alpha}{g \cos^2 \alpha\,(1-e)^2}$ along the plane.

5·7. Kinetic Energy lost by impact. *Two spheres of given masses moving with given velocities impinge; to show that there is always a loss of K.E.*

(*i*) *Direct impact.*

The equations are $mv+m'v'=mu+m'u'$

$$v-v'=-e(u-u')$$

$$\therefore\quad (mv+m'v')^2+mm'(v-v')^2=(mu+m'u')^2+mm'e^2\,(u-u')^2$$
$$=(mu+m'u')^2+mm'\,(u-u')^2 - mm'(1-e^2)(u-u')^2$$

i.e.,
$$(m+m')(mv^2+m'v'^2)=(m+m')(mu^2+m'u'^2) - mm'(1-e^2)(u-u')^2$$

i.e., $$mv^2+m'v'^2=mu^2+m'u'^2-\frac{mm'}{m+m'}(1-e^2)(u-u')^2$$

$$\text{Loss in } K.E.=(\tfrac{1}{2}mu^2+\tfrac{1}{2}m'u'^2)-(\tfrac{1}{2}mv^2+\tfrac{1}{2}mv'^2)$$

$$=\frac{1}{2}\frac{mm'}{m+m'}(1-e^2)(u-u')^2.$$

(ii) *oblique impact*

The equations are $u \sin \alpha = v \sin \theta$

$$u' \sin \beta = v' \sin \phi.$$

$$mv \cos \theta + m'v' \cos \phi = mu \cos \alpha + m'u' \cos \beta$$

$$v \cos \theta - v' \cos \phi = e(u \cos \alpha - u' \cos \beta)$$

$$\therefore \quad (mv \cos \theta + m'v' \cos \phi)^2 + mm'(v \cos \theta - v' \cos \phi)^2$$

$$=(mu \cos \alpha + m'u' \cos \beta)^2 + mm'e^2(u \cos \alpha - u' \cos \beta)^2$$

$$=(mu \cos \alpha + m'u' \cos \beta)^2 + mm'(u \cos \alpha - u' \cos \beta)^2 - (1-e^2)mm'(u \cos \alpha - u' \cos \beta)^2$$

i.e., $$(m+m')(mv^2\cos^2\theta + m'v'^2\cos^2\phi) = (m+m')(mu^2 \cos^2\alpha + m'u' \cos^2\beta) - (1-e^2)mm'(u \cos \alpha - u \cos \beta)^2$$

or $$(mv^2 \cos^2 \theta + m'v'^2 \cos^2 \phi) = (mu^2 \cos^2 \alpha + m'u^2 \cos^2 \beta) - (1-e^2)\frac{mm'}{m+m'}(u \cos \alpha - u' \cos \beta)^2$$

$$\text{Loss in } K.E. = (\tfrac{1}{2}mu^2 + \tfrac{1}{2}m'u'^2) - (\tfrac{1}{2}mv^2 + \tfrac{1}{2}m'v^2)$$

$$= \tfrac{1}{2}m(u^2 \cos^2 \alpha + u^2 \sin^2 \alpha) + \tfrac{1}{2}m'(u'^2 \cos^2 \beta + u'^2 \sin^2 \beta) - \tfrac{1}{2}m(v^2 \cos^2 \theta + v^2 \sin^2 \theta) - \tfrac{1}{2}m'(v'^2 \cos^2 \phi + u'^2 \sin^2 \phi)$$

$$= (\tfrac{1}{2}mu^2 \cos^2 \alpha + \tfrac{1}{2}m'u'^2 \cos^2 \beta) - (\tfrac{1}{2}mv^2 \cos^2 \theta + \tfrac{1}{2}m'v'^2 \cos^2 \phi)$$

$$= \frac{1}{2}\frac{mm'}{m+m'}(1-e^2)(u \cos \alpha - u' \cos \beta)^2.$$

Since $e<1$ this is positive and there is always a loss of *K.E.*

This vanishes when $e=1$ *i.e.*, when the spheres are perfectly elastic, there is no loss of *K.E.*

Ex. 25. *Two equal smooth balls are placed in contact and at rest on a horizontal table. A third ball exactly similar to the other two impinges symmetrically upon them. Show that the impact diminishes the whole K.E of the balls in the ratio $(2+3e^2)$ to 5.*

Sol. u is the original velocity of A ; B, C at rest.

'u' the velocity of A, v that of B or C after impact. J is the impulse between A and B or A and C and K the impulse between B and C.

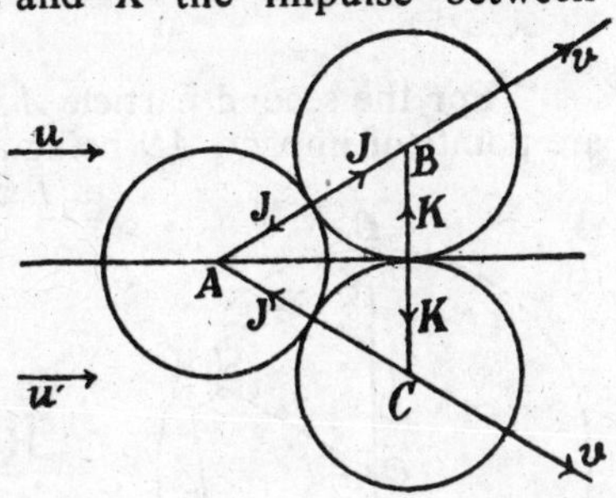

For B or C, $mv = J + K \cos 60°$,

$$K \sin 60 = 0.$$

For A, $m(u' - u) = -2J \cos 30°$

Since on the three spheres taken together, there is no impulse, momentum remains unchanged in any direction.

$\therefore$ momentum equation is

$$mu = mu' + 2mv \cos 30°$$

or $$u = u' + \sqrt{3}v$$

Newton's law between A and B is

$$v - u' \cos 30° = eu \cos 30°$$

These give

$$u' = \frac{2-3e}{5} u, \quad v = \frac{(e+1)\, u\sqrt{3}}{5}$$

$$\therefore \quad \frac{\frac{1}{2}mu'^2 + \frac{1}{2}mv^2}{\frac{1}{2}mu^2} = \frac{3e^2+2}{5}.$$

EXAMPLES V (D)

1. If two inelastic spheres have direct impact, show that the $K.E.$ lost by the impact is that of a body whose mass is half the harmonic mean between those of the spheres whose velocity is equal to their relative velocity before impact.

2. Prove that the energy lost in the collision of two smooth spheres is

$$\frac{1}{2} \frac{e-1}{1+e} \left(\frac{1}{m} + \frac{1}{m'} \right) I^2$$

where I is the momentum transferred from ball to ball.

3. If two particles of masses m, m' moving with velocities v, v' impinge directly, prove that the condition that each loses the same amount of K.E. is $(3+e)(mv+m'v') + (1-e)(mv'+m'v) = 0$.

5·8. Some more worked-out examples.

Ex. 26. *Two particles start from the same point in the circumference of a smooth horizontal circular table with raised edges. After each has been reflected twice at the circumference, the one returns to the point of projection and the other arrives at a point diametrically opposite. Prove that if* α *and* β *be the inclinations of their directions of projection to the radius to the point,* $\tan \alpha \tan \beta = e^2$.

Sol. For the first ball we have by Ex. 18, $\tan^2 \alpha = \dfrac{e^3}{1+e+e^2}$.

For the second particle A is the point of projections, B, C, D are points of impact, AD being a diameter. Let angles of reflection at B and C be γ, δ.

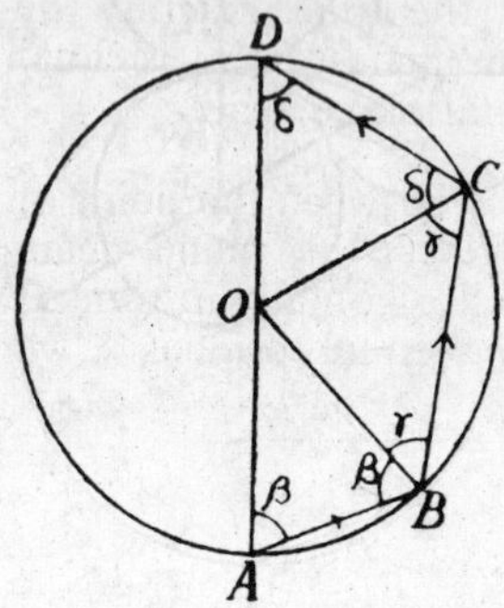

Then $\cot \gamma = e \cot \beta$.

$\cot \delta = e \cot \gamma = e^2 \cot \beta.$

Also $\beta+\gamma+\delta=\pi \qquad \beta+\gamma=\pi-\delta.$

Hence $\dfrac{\cot \beta+\cot \gamma}{\cot \beta \cot \gamma-1} = -\dfrac{1}{\cot \delta}$

or $\dfrac{\cot \beta+e \cot \beta}{e \cot^2 \beta-1} = -\dfrac{1}{e^2 \cot \beta}$

$\therefore \qquad \cot^2 \beta = \dfrac{1}{e+e^2+e^3}$

i.e., $\qquad \tan^2 \beta = e+e^2+e^3.$

Hence $\tan^2 \alpha \tan^2 \beta = \dfrac{e^3}{1+e+e^2} \cdot e(1+e+e^2) = e^4$

$\therefore \qquad \tan \alpha \tan \beta = e^2.$

Ex. 27. *Two equal elastic balls are projected towards each other at the same instant in the same vertical plane, v being the velocity and α the elevation in each case. Show that after impact they will return to the points of projection if ga (1+e)=ev² sin 2α, where 2a is the distance between the points of projection.*

Sol. A, B are the points of projection. The balls will describe equal parabolas and will meet on a vertical plane passing through O, the middle point of AB. The case is therefore, the same as the reflection of a ball from a vertical plane at a distance a from the point of projection. Let them meet at P.

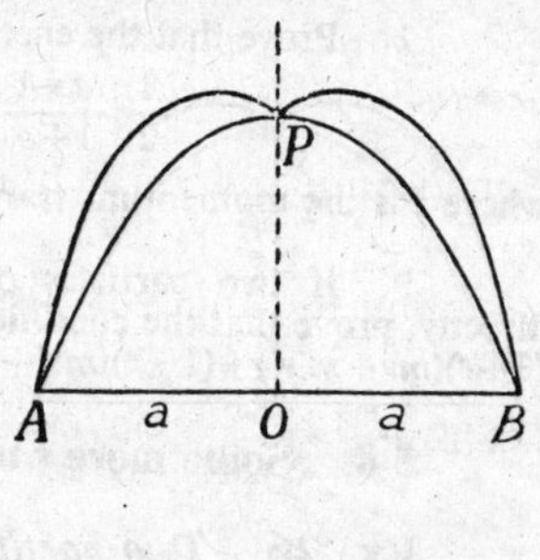

For one ball, the time from A to P and back to A is the total time of flight and therefore, equal to $\dfrac{2v \sin \alpha}{g}$.

Now the time for AP is given by

$a = v \cos \alpha . t_1 \qquad \therefore \quad t_1 = \dfrac{a}{v \cos \alpha}.$

At P, the horizontal velocity $v \cos \alpha$ becomes $ev \cos \alpha$ after impact

$\therefore$ time for PA is $\dfrac{a}{ev\cos\alpha}$

Hence $$\frac{2v\sin\alpha}{g}=\frac{a}{v\cos\alpha}+\frac{a}{ev\cos\alpha}$$

i.e., $$2v^2\sin\alpha\cos\alpha=ag\left(1+\frac{1}{e}\right)$$

or $$ev^2\sin 2\alpha=ag(1+e).$$

Ex. 28. *Two elastic spheres equal in all respects are moving towards each other with equal velocities, their centres being on two parallel lines whose distance apart is d_1. Prove that after impact they will move away from each other with equal velocities so that their centres are on two parallel lines, whose distance apart d_2 is given by $d_2^2[e^2d^2+(1-e^2)d_1^2]=d^2d_1^2$ where d is the diameter of either sphere and d_1 is $<d$.*

Sol. Let the spheres move with velocities u at an angle α to the line of centres and let their velocities after impact be v, v' at angle θ, ϕ to the line of centres.

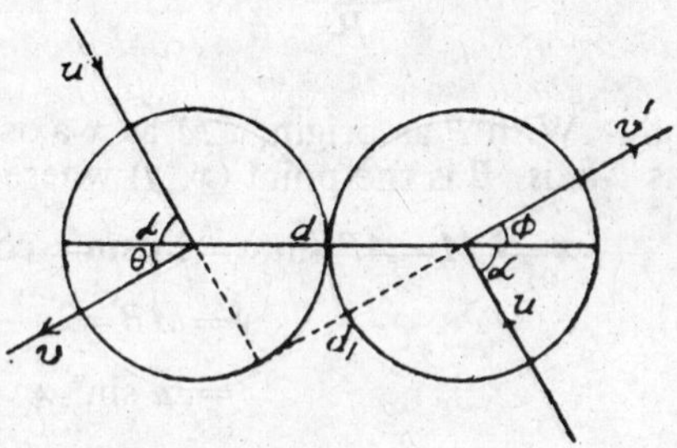

Then $\sin\alpha=\dfrac{d_1}{d}\cdot$

The four equations are

$v\sin\theta=u\sin\alpha,$

$v\sin\phi=u\sin\alpha$

$=v\sin\theta$

$v'\cos\phi-v\cos\theta$

$=u\cos\alpha-u\cos\alpha=0$

$\therefore\quad v'\cos\phi=v\cos\theta$

$\therefore\quad \tan\phi=\tan\theta$

$\therefore\quad \phi=\theta$ hence in parallel lines.

Also $v'^2(\sin^2\phi+\cos^2\phi)=v^2(\sin^2\theta+\cos^2\theta)\qquad\therefore\ v'=v.$

Also $v'\cos\phi+v\cos\theta=-e(-u\cos\alpha-u\cos\alpha)=2eu\cos\alpha$

$$v\cos\theta=eu\cos\alpha$$

also $$v\sin\theta=u\sin\alpha$$

$\therefore$ $$e\tan\theta=\tan\alpha$$

But $$\sin\alpha=\frac{d_1}{d}\text{ and }\sin\theta=\frac{d_2}{d}$$

$\therefore$ $$e\frac{d_2}{\sqrt{d^2-d_2^2}}=\frac{d_1}{\sqrt{d^2-d_1^2}}$$

Hence $$d_2^2[e^2d^2+(1-e^2)d_1^2]=d^2d_1^2.$$

Ex. 29. *A heavy particle, perfectly elastic, falls down a chord from the highest point of a vertical circle, and after reflection at the curve passes through the lowest point. Prove that the inclination of the chord to the vertical is* $\frac{1}{2}\cos^{-1}\left(\frac{\sqrt{3}-1}{4}\right)$.

Sol. A is the highest and B is the lowest point. A particle goes along AP, is reflected at P along PT and moves in a parabola which passes through B. Since the particle is perfectly elastic,

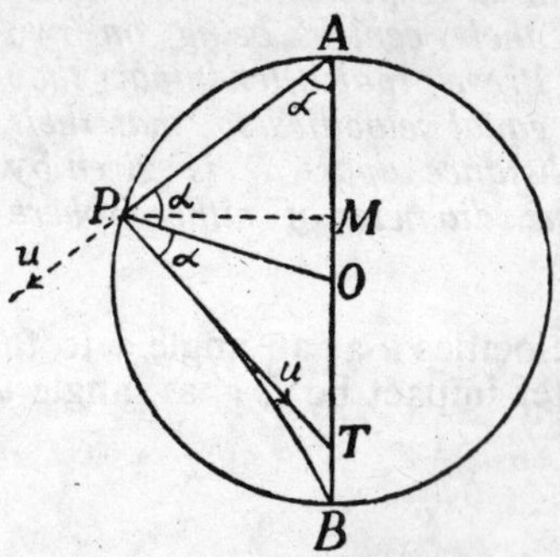

$\angle OPT=\angle OPA=\angle OAP$

$=\alpha$ say, O being the centre.

Also velocity along PT=velocity along $AP=u$ say.

Now $AP=2a\cos\alpha$, a being the radius of the circle.

$\therefore$ If PM be perp. from P on AB, $u^2=2g.\ AM=4ag\cos^2\alpha$

$$\angle TPM=\frac{\pi}{2}-(\pi-3\alpha)=3\alpha-\pi/2,$$

$$=\theta \text{ say.}$$

With P as origin, PM as x-axis, vertical downward through P as y-axis, B is the point (x, y) where

$$x=PM=AP\sin\alpha=2a\sin\alpha\cos\alpha.$$

$$y=MB=2a-2a\cos^2\alpha$$

$$=2a\sin^2\alpha.$$

The equation to the parabola is

$$y=x\tan\theta+\tfrac{1}{2}g\frac{x^2}{u^2\cos^2\theta}=-x\cot 3\alpha+\tfrac{1}{2}g\frac{x^2}{4ag\cos^2\alpha\sin^2 3\alpha}$$

Hence for the point B,

$$2a\sin^2\alpha=-2a\sin\alpha\cos\alpha\cos 3\alpha+\tfrac{1}{2}g\frac{4a^2\sin^2\alpha\cos^2\alpha}{4ag\cos^2\alpha\sin^2 3\alpha}$$

or $\quad 4\sin^2\alpha\sin^2 3\alpha=-4\sin\alpha\cos\alpha\sin 3\alpha\cos 3\alpha+\sin^2\alpha$

or $\quad \sin^2\alpha(1-4\sin^2 3\alpha)=\sin 2\alpha\sin 6\alpha$

or $\quad \sin^2\alpha[1-2(1-\cos 6\alpha)]=\sin 2\alpha\sin 6\alpha$

or $\quad \sin^2 2\alpha(2\cos 6\alpha-1)=\sin 2\alpha\sin 6\alpha$

or $\quad (1-\cos 2\alpha)\cos 6\alpha-\sin^2\alpha=\sin 2\alpha\sin 6\alpha$

or $\quad \cos 6\alpha-\sin^2\alpha=\cos 6\alpha\cos 2\alpha+\sin 6\alpha\sin 2\alpha$

or $$\cos 6\alpha - \sin^2 \alpha = \cos 4\alpha$$

or $$\cos 6\alpha - \cos 4\alpha = \sin^2 \alpha$$

or $$-2 \sin 5\alpha = \sin \alpha$$

or $$-2(16 \sin^5 \alpha - 20 \sin^3 \alpha + 5 \sin \alpha) = \sin \alpha$$

or $$16 \sin^4 \alpha - 20 \sin^2 \alpha + 5 = -\tfrac{1}{2}$$

or $$4(1-\cos 2\alpha)^2 - 10(1-\cos 2\alpha) + 5 = -\tfrac{1}{2}$$

or $$8 \cos^2 2\alpha + 4 \cos 2\alpha - 1 = 0$$

$$\therefore \quad \cos 2\alpha = \frac{-1 \pm \sqrt{3}}{4} = \frac{\sqrt{3}-1}{4}.$$

Ex. 30. *A wire ABC in the form of an equilateral triangle is fixed on a horizontal table. A particle is projected from a point in BC in a direction parallel to BA. If the point of projection divided BC in the ratio 2e : 3e−1, prove that the particle will return to it after impinging on AC and AB.*

Sol. P is the point of projection, Q, R the points of impact.

If a be the side of the triangle, let $BP = x$ so that $PC = a - x$.

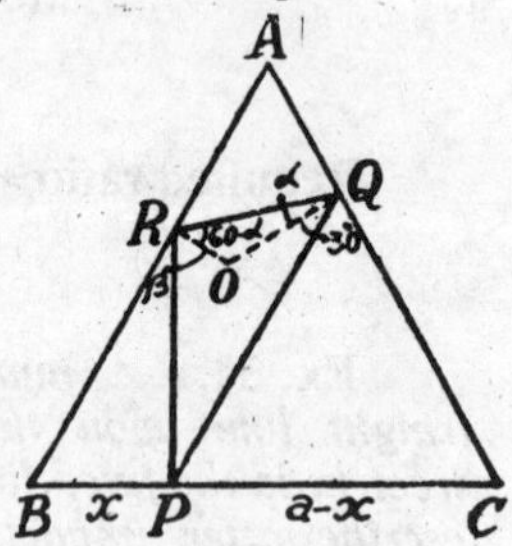

QO and RO be the perpendiculars at Q and R.

α, β be the angles of reflection at Q and R.

Then $\angle PQO = 30°$ since PQ is parallel to AB and $\angle QRO = 60° - \alpha$.

Hence $\cot \alpha = e \cot 30 = \sqrt{3}\, e$

and $$\cot \beta = e \cot (60-\alpha) = e \frac{\cot 60 \cot \alpha + 1}{\cot \alpha - \cot 60} = e \cdot \frac{1/\sqrt{3} . \sqrt{3}.e + 1}{\sqrt{3}.e - 1/\sqrt{3}}$$

$$= e \frac{e+1}{(3e-1)/\sqrt{3}} = \frac{\sqrt{3}e(1+e)}{3e-1}.$$

Now $PQ = a - x$ and $\dfrac{PR}{\sin 60} = \dfrac{x}{\cos \beta}$ from triangle BPR

$$\therefore \quad PR = \frac{\sqrt{3}x}{2 \cos \beta}$$

From triangle PQR, $$\frac{PR}{\sin (\alpha+30)} = \frac{PQ}{\sin (\beta + 60 - \alpha)}$$

i.e., $$\frac{\sqrt{3}x}{2 \cos \beta \sin (\alpha+30)} = \frac{a-x}{\cos (\alpha - \beta + 30)}$$

or $$\sqrt{3}x \cos (\beta - \alpha + 30) = 2(a-x) \cos \beta \sin (\alpha + 30).$$

or $x[\sqrt{3} \cos \beta \cos (\alpha+30) + \sqrt{3} \sin \beta \sin (\alpha+30) + 2 \cos \beta \sin (\alpha+30)]$

$$=2a\cos\beta\sin(\alpha+30)$$

or $x[3\cos\alpha\cos\beta-\sqrt{3}\sin\alpha\cos\beta+3\sin\alpha\sin\beta+\sqrt{3}\sin\beta\cos\alpha$
$+2\sqrt{3}\sin\alpha\cos\beta+2\cos\alpha\cos\beta]$

$$=2a\cos\beta(\sqrt{3}\sin\alpha+\cos\alpha)$$

or $x[5\cot\alpha\cot\beta+\sqrt{3}\cot\beta+\sqrt{3}\cot\alpha+3]$

$$=2a(\sqrt{3}\cot\beta+\cot\alpha\cot\beta)$$

or $$x\left[15.\frac{e^2(1+e)}{3e-1}+\sqrt{3}.\frac{\sqrt{3}.e(1+e)}{3e-1}+3e+3\right]$$

$$=2a\left[\frac{3e(1+e)}{3e-1}+\frac{3e^2(1+e)}{3e-1}\right]$$

or, $x[15e^2(1+e)+3e(1+e)+3(1+e)(3e-1)]$

$$=2a[3e(1+e)+3e^2(1+e)]$$

or $$x(5e^2+4e-1)=2ae(1+e)$$

or $$x(5e-1)=2ae$$

$$\text{Required ratio}=\frac{a}{a-x}=\frac{\frac{2ae}{5e-1}}{a-\frac{2ae}{5e-1}}=\frac{2ae}{3ae-a}=\frac{2e}{3e-1}.$$

Ex. 31. *A number of particles are let fall from a horizontal straight line upon the convex arc of a fixed vertical parabola whose directrix is the fixed line ; show that the parabolas which the particles describe after impact on the curve have a common directrix at a distance a $(1-e^2)$ below the given line where 4a is the latus rectum of the given parabola and e the co-efficient of restitution.*

Sol. Let parabola be $y^3=4ax$. Let P be the point of impact and let the tangent PT at P make an angle θ with the axis of the parabola, so that its slope $m=\tan\theta$.

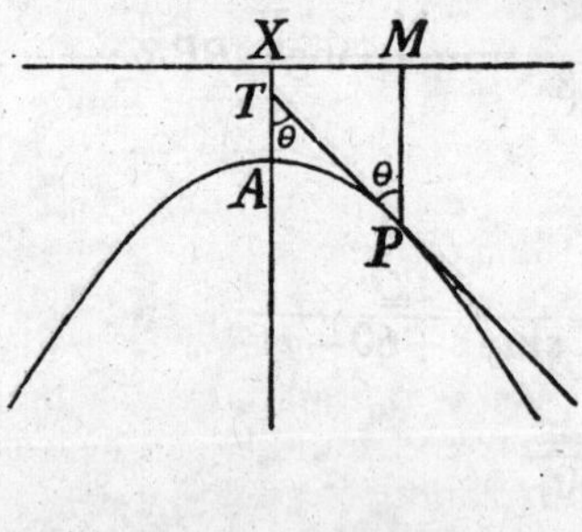

Equation to the tangent is

$$y=mx+\frac{a}{m}$$

or $$y=x\tan\theta+a\cot\theta.$$

P is the point $(a\cot^2\theta, 2a\cot\theta)$

$\therefore\ PM=a+a\cot^2\theta=a\operatorname{cosec}^2\theta$

$\therefore$ Velocity due to fall through PM

$$=\sqrt{2ga\operatorname{cosec}^2\theta}=\sqrt{2ga}\operatorname{cosec}\theta.$$

Its components along TP and normally are $\sqrt{2ga}\operatorname{cosec}\theta\cos\theta$ and

$$\sqrt{2ga}\ \text{cosec}\ \theta \sin \theta \quad i.e., \sqrt{2ga} \cot \theta \text{ and } \sqrt{2ga}.$$

By impact the velocity along the tangent is unaltered and the velocity along the normal becomes e-times the original velocity. Hence after impact components of velocity along the tangent and normal are $\sqrt{2ga} \cot \theta$ and $e\sqrt{2ga}$, $\therefore$ the velocity of rebound $=\sqrt{2ga}\sqrt{(e^2+\cot^2 \theta)}$.

$\therefore$ height of directrix of the subsequent path above P

$$=\frac{(\text{vel. at } P)^2}{2g}=a(e^2+\cot^2 \theta)$$

$\therefore$ its depth below $XM=a \text{ cosec}^2 \theta-a(e^2+\cot^2 \theta)=a(1-e^2)$.

Ex. 32. *A perfectly elastic ball is at the focus of an elliptic billiard table ; show that the ball, in whatever manner struck, will ultimately be moving along the major axis.*

Sol. Let the ball be projected along $S'P'$ from S'. As the elasticity is perfect, the velocity of rebound is equal to the velocity of approach and the angle of reflection with the normal is equal to the angle of incidence. Also the normal at any point is equally inclined to the focal radii.

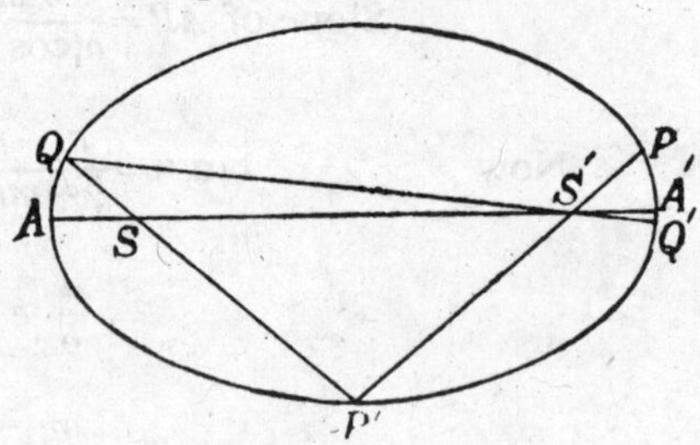

Thus the ball along $S'P'$ will be reflected along $P'S$ striking at Q and then rebounding passes through S' again and strikes at Q'.

But $\angle A'S'Q'$ is less than $\angle A'S'P'$.

After rebounding from Q', the ball will go to S and after another impact will pass through S' again striking at R' say.

Again $\angle A'S'R'$ is less than $\angle A'S'Q'$ and so on.

Thus the angle which the direction of ball makes with $A'S'$ goes on decreasing each time it passes through S'. Ultimately it will be moving along $S'A'$.

Ex. 33. *A ball at the focus of an ellipse whose eccentricity is e, receives a blow and after one impact on the elliptic periphery, passes through the other end of the major axis. Find the point of impact on the ellipse and show that the co-efficient of restitution cannot be greater than* $\dfrac{2e}{1+e^2}$.

Sol. O is the centre of the ellipse, S the focus point of projection, P the point of impact and PG the normal at P.

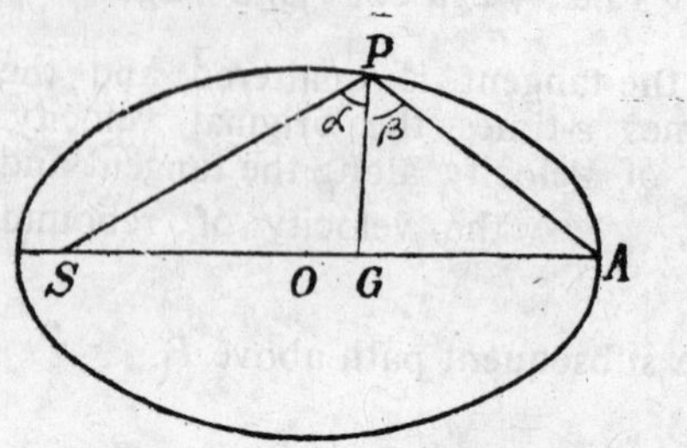

α, β be the angles of incidence and reflection at P.

$\therefore \cot \beta = e' \cot \alpha$ where e' = co-efficient of restitution.

If ϕ be the eccentric angle of P, $P \to a \cos \phi, b \sin \phi$

Equation to PG is

$$ax \sec \phi - by \operatorname{cosec} \phi = a^2 - b^2$$

$$\text{its slope} = \frac{a}{b} \tan \phi = m$$

$$\text{Slope of } SP = \frac{b \sin \phi}{a(\cos \phi + e)} = m_1$$

$$\text{Slope of } AP = \frac{b \sin \phi}{a(\cos \phi - 1)} = m_2$$

Now
$$\tan \alpha = \frac{m - m_1}{1 + mm_1} = \frac{a^2 \sin \phi\, e(e \cos \phi + 1)}{ab(1 + e \cos \phi)}$$

$$= \frac{a}{b} e \sin \phi$$

$$\tan \beta = \frac{m_2 - m}{1 + m_2 m} = \frac{a^2 \sin \phi (1 - e^2 \cos \phi)}{ab(1 - \cos \phi)}$$

Hence
$$\frac{ab(1 - \cos \phi)}{a^2 \sin \phi\, (1 - e^2 \cos \phi)} = e' \frac{b}{ae \sin \phi}$$

or
$$e(1 - \cos \phi) = e'(1 - e^2 \cos \phi)$$

$$e' = \frac{e(1 - \cos \phi)}{1 - e^2 \cos \phi}$$

Its maximum value is obtained when $\phi = \pi$.

Hence the max. value of $e' = \dfrac{2e}{1 + e^2}$.

6

Circular and Harmonic Motions

6·1. Uniform circular motion. If a particle of mass m describes a circle of radius r with uniform speed v or angular velocity ω the only acceleration is towards the centre and is equal to $\frac{v^2}{r}$ or $\omega^2 r$. The effective force on the particle is, therefore, $m\frac{v^2}{r}$ towards the centre. To produce this effective force some external force will be required. If F be the external force acting on the particle towards the centre, the equation of motion is

$$F=m\frac{v^2}{r}$$

or
$$F=m\omega^2 r.$$

If the particle makes n revolutions per second, then

$$\omega=2\pi n$$

$$F=4m\pi^2 n^2 r$$

6.2. Conical Pendulum. *In a conical pendulum a particle tied to one end of a light string, the other end of which is kept fixed, is made to move in a horizontal circle whose axis passes through the fixed end, the string generating a cone.*

Let OP $(=l)$ be the string, O being fixed and the particle being attached at P. The particle describes a circle of radius r, h being the depth of the circle below O. If the string be inclined at an angle θ to the vertical then $r=l\sin\theta$, $h=l\cos\theta$. If ω be the angular velocity then the central acceleration is $\omega^2 r$ or $\dfrac{v^2}{r}$. If T be the tension of the string then resolving vertically and horizontally we have

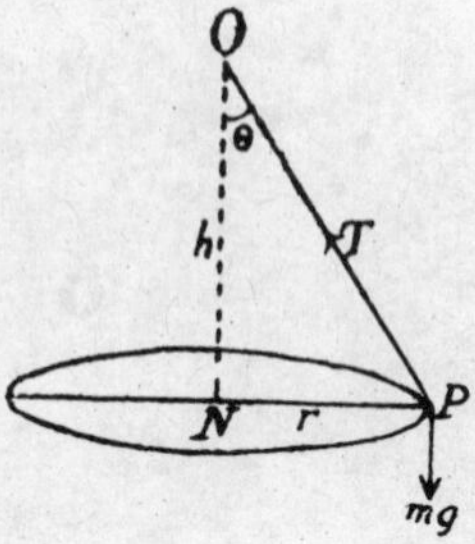

$$T\cos\theta - mg = 0 \quad \text{(statical equation)}$$

$$T\sin\theta = m\omega^2 r \quad \text{(dynamical equation)}$$

$$= m\omega^2\, l\sin\theta$$

$\therefore$ eliminating T, we get

$$m\omega^2 l\cos\theta = mg \text{ or } \omega^2 h = g \quad \therefore \quad \omega = \sqrt{\frac{g}{h}}$$

$\therefore$ time of one complete revolution

$$= \frac{2\pi}{\omega} = 2\pi\sqrt{\frac{h}{g}}$$

Further $\cos\theta = \dfrac{h}{l} = \dfrac{g}{\omega^2 l}$; hence $g < \omega^2 l$

or
$$\omega > \sqrt{\frac{g}{l}}.$$

Ex. 1. *One end of a string 2l ft. long is fastened to a point A on a fixed smooth vertical rod, the other to a small ring P of mass m lbs. which slides on the rod, another mass Q of m′ lbs. is fastened to the middle point of the string and revolves with velocity v in a horizontal circle so that the angle AQP is a right angle ; prove that*

$$v^2 = \frac{lg}{\sqrt{2}}, \quad \frac{2m+m'}{m'}.$$

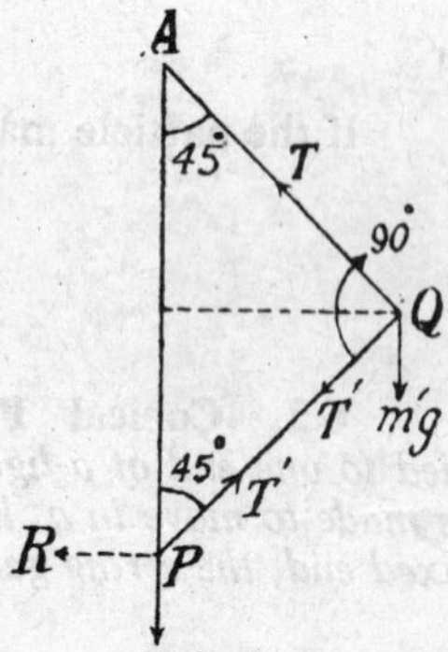

Sol. Since the angle APQ is maintained at right angle, P is at rest.

For P, the equations are

$$mg = T'\cos 45° = \frac{T'}{\sqrt{2}}$$

$$R = \frac{T'}{\sqrt{2}}$$

For Q, the equations are

$$\frac{T}{\sqrt{2}} = m'g + \frac{T'}{\sqrt{2}} = (m+m')g$$

and $$\frac{T}{\sqrt{2}} + \frac{T'}{\sqrt{2}} = m'\,\frac{v^2}{r} = m'\,\frac{v^2\sqrt{2}}{l}$$

since $AQ=l=PQ$

$$\therefore \quad (m+m')g+mg=m'\frac{v^2\sqrt{2}}{l}$$

$$\therefore \quad v^2=\frac{(2m+m')gl}{m'\sqrt{2}}.$$

Ex. 2. *A smooth cup in the form of a paraboloid is set whirling with angular velocity ω about its axis which is vertical. Show that a particle will remain in equilibrium on any part of its surface if* $\omega^2=\frac{g}{2a}$ *where 4a is the latus rectum of parabolic section of the cup.*

Sol. $P(x, y)$ be the position of the particle so that the particle describes a horizontal circle of radius x. If R be the reaction on the particle, the equations are

$$R\cos\psi=mg$$
$$R\sin\psi=m\omega^2 x$$

$$\therefore \quad \tan\psi=\frac{\omega^2 x}{g}.$$

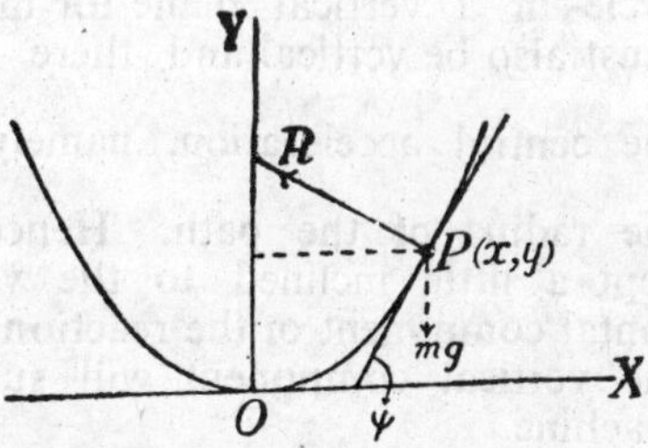

Equation to the parabola is $x^2=4ay$

$$\therefore \quad \tan\psi=\frac{dy}{dx}=\frac{x}{2a}$$

Hence $\frac{\omega^2 x}{g}=\frac{x}{2a}$; $\therefore \quad \omega^2=\frac{g}{2a}$.

This is independent of the position of P ∴ The particle will remain in any position.

EXAMPLES VI (A)

1. Prove that a particle can rest in relative equilibrium on the lower half of a hollow smooth sphere, revolving with uniform angular velocity ω about the vertical diameter unless $\omega<\sqrt{\frac{g}{a}}$, in which case it will rest at the lowest point, where a is the radius of the sphere.

2. An elastic string of natural length l is extended by an amount x when it supports a mass m at rest and is extended by an amount y when it is rotating as a conical pendulum carrying a particle of the same mass. Show that $gy=\omega^2 x$.

3. A particle of mass m is attached to a point O by an inextensible string of length l. Prove that it can describe a horizontal circle about a point vertically below O with uniform angular velocity ω, provided $l\omega^2>g$.

Prove that if the string is elastic of natural length l_0 and modulus λ the above condition is replaced by

$$l_0\lambda\frac{\omega^2}{g} > \lambda\, m l_0\omega^2 > 0.$$

4. A string of length l has its ends fastened to two points A and B in the same vertical line, A above B. A bead C on the string rotates about AB with uniform angular velocity ω. Prove that

(*i*) if the string is not vertical $\cos A - \cos B = \dfrac{gl}{ab\omega^2}$;

(*ii*) BC is horizontal, if $\lambda^2 = \dfrac{2gl}{c(l^2 - c^2)}$.

5. A body is moving in a circle making n revolutions per sec. on the surface of a smooth hollow cone of vertical angle 2α. Show that the distance of the body from the axis of the cone is

$$\frac{g \cot \alpha}{4\pi^2 n^2}.$$

6·3. Motion in a nearly circular path. A cyclist running in a nearly circular path cannot do so if he keeps himself and the cycle in a vertical plane for in that case the reaction of the ground must also be vertical and there is no horizontal force to produce the central acceleration, namely, $\dfrac{v^2}{r}$ where v is the velocity and r the radius of the path. Hence the man and the machine must be kept a little inclined to the vertical so that in that case the horizontal component of the reaction will produce the acceleration and the vertical component will support the weight of the man and the machine.

In the case of a car or a railway carriage moving in a circular path, the friction on the wheels across the track produces the necessary acceleration. Let P_1 and P_2 be the vertical reactions at points A and B where the wheels are in contact with the rails and F_1, F_2 be the horizontal frictions at those points. Resolving vertically and horizontally we have

$$P_1 + P_2 - mg = 0 \text{ (statical equation)}$$

and $$\frac{mv^2}{r} = F_1 + F_2 \text{ (dynamical equation)}$$

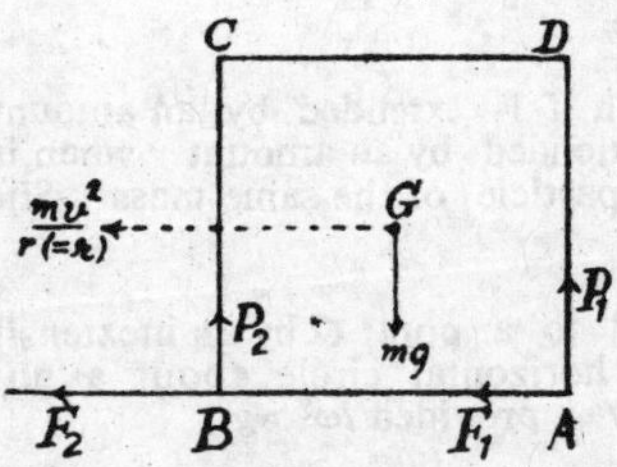

Moments about A and B give

$$m\frac{v^2}{r}h = mga - P_2\, 2a$$

and $$m\,\frac{v^2}{r}\,h = P_1.2a - mga$$

where $AB = 2a$ and h is the height of the C.G. above AB.

These give $$P_1=\tfrac{1}{2}m\left(g+\frac{v^2h}{ar}\right)$$

and $$P_2=\tfrac{1}{2}m\left(g-\frac{v^2h}{ar}\right).$$

Thus if $v^2>\dfrac{gar}{h}$, P_2 is negative which is impossible hence the carriage will overturn outwards.

Here we see that the presence of the frictions at A and B is necessary in order to produce the central acceleration $\dfrac{v^2}{r}$ but these frictions are also responsible for the wear and tear of the rails hence are undesirable. In order to avoid these undesirable effects, the outer rails are raised so that the horizontal components of the reactions R_1 and R_2 at A and B are sufficient to produce the necessary $\dfrac{v^2}{r}$ and the vertical components support the weight. If AB is inclined at angle θ with the horizontal then supposing there is no friction the equations are

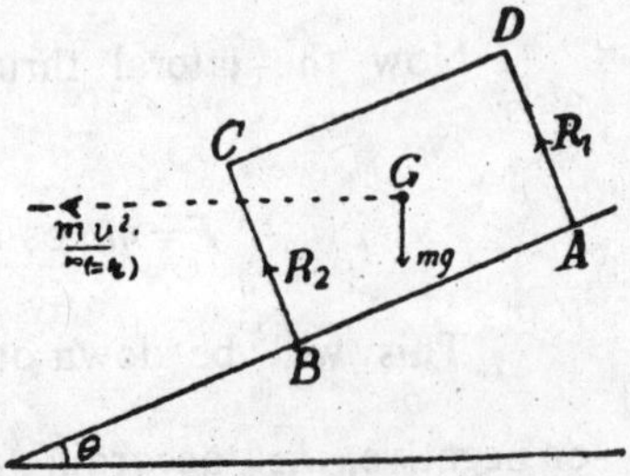

$$(R_1+R_2)\sin\theta=m\frac{v^2}{r}$$

and $$(R_2+R_1)\cos\theta=mg$$

so that $$\tan\theta=\frac{v^2}{gr} \qquad \text{...(1)}$$

This is the case when the track is called *banked up* for velocity v. If, however, the actual velocity is V, different from v, then there will be some friction. In this case the equations will be

$$(R_1+R_2)\sin\theta+(F_1+F_2)\cos\theta=m\frac{V^2}{r}$$

and $$(R_1+R_2)\cos\theta-(F_1+F_2)\sin\theta=mg$$

giving $$F_1+F_2=m\frac{V^2}{r}\cos\theta-mg\sin\theta$$

$$=m\cos\theta\left(\frac{V^2}{r}-g\tan\theta\right)$$

$$=m\cos\theta\frac{V^2-v^2}{r}.$$

Ex. 3. *If the maximum and minimum speeds of trains on a certain curve of radius r are v and u respectively, show that if the track is banked up so that the sleepers are inclined at an angle* β *to the horizon such that* $\tan\beta=\dfrac{u^2+v^2}{2gr}$, *then the outer lateral thrust of the fastest train is equal to the inner lateral thrust of the slowest train, assuming their weights to be equal.*

Sol. The proper elevations for the speeds v and u, when the lateral thrusts are to be avoided, are given by

$$\tan\alpha=\frac{v^2}{gr},\quad \tan\alpha'=\frac{u^2}{gr}.$$

Now the lateral thrust when the speed is V and inclination is β is

$$F=m\cos\theta\left(\frac{V^2}{r}-g\tan\beta\right).$$

This will be down or up the sleepers according as F is positive or negative, *i.e.*, according as $\dfrac{V^2}{gr}>$ or $<\tan\beta$.

But $$\tan\beta=\frac{u^2+v^2}{2gr}\text{ where } v>u$$

$$\therefore\quad \frac{v^2}{gr}>\tan\beta>\frac{u^2}{gr}.$$

Thus the train running with speed v will get a lateral thrust down the sleepers and the trains with speed u a lateral thrust up the sleepers.

When the speed is v, the downward thrust $F=m\cos\beta\left(\dfrac{v^2}{r}-g\tan\beta\right)$ and when the speed is u, the upward thrust $F=m\cos\beta\left(g\tan\beta-\dfrac{u^2}{r}\right)$ but since $\tan\beta=\dfrac{u^2+v^2}{2gr}$, we get $F=F'$.

Ex. 4. *A car takes banked corner of a racing track at a speed V, the lateral gradient* θ *being designed to reduce the tendency to side-slip to zero for a lower speed U. Show that the co-efficient of friction to prevent side-slip for the greater speed V must be at least*

$$\frac{(V^2-U^2)\sin\theta\cos\theta}{V^2\sin^2\theta+U^2\cos^2\theta}$$

Sol. If R_1 and R_2 be the normal reactions and F the lateral pressure for speed V then

$$(R_1+R_2)\sin\theta+F\cos\theta=m\frac{V^2}{r}$$

$$(R_1+R_2)\cos\theta-F\sin\theta=mg.$$

$$\therefore\ F=m\cos\theta\left(\frac{V^2}{r}-g\tan\theta\right)=m\cos\theta\,\frac{V^2-U^2}{r}$$ if U is the speed for which F is zero.

Also $$R_1+R_2=m\left(g\cos\theta+\frac{V^2}{r}\sin\theta\right)$$

$\therefore$ the co-efficient of friction necessary to prevent side-slipping must be at least

$$\frac{F}{R_1+R_2}=\frac{m\cos\theta\,(V^2-U^2)}{m(V^2\sin\theta+gr\cos\theta)}=\frac{(V^2-U^2)\sin\theta\cos\theta}{V^2\sin^2\theta+gr\sin\theta\cos\theta}$$

$$=\frac{(V^2-U^2)\sin\theta\cos\theta}{V^2\sin^2\theta+U^2\cos^2\theta}\text{ for }\tan\theta=\frac{U^2}{r}.$$

EXAMPLES VI (B)

1, A train of mass M is standing on a railway curve of radius r which is banked up to suit a speed v. Show that there is a lateral thrust on the rails of magnitude $\dfrac{v^2Mg}{\sqrt{v^4+r^2g^2}}$.

6.4. Constrained Motion. If a particle is forced to move along a given curve or surface (whether smooth or rough), then the motion of the particle is said to be the constrained motion.

6·5. Motion on a smooth curve in a vertical plane. A heavy particle is made to move on a smooth curve in a vertical plane; to discuss the motion.

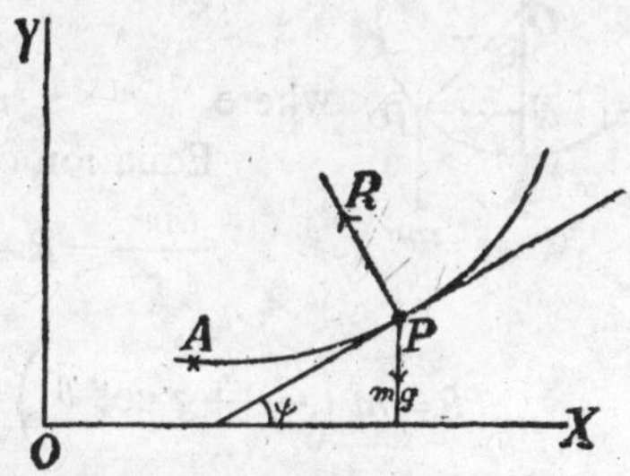

The external forces are the weight mg of the particle downwards and the normal reaction R of the curve. If P be the particle at time t, the equations of motion (tangential and normal) are

$$mv\frac{dv}{ds}=-mg\sin\psi$$

$$m\frac{v^2}{\rho}=R-mg\cos\psi$$

if s be measured from a fixed point A down the curve below P, ρ the radius of curvature of the curve at P.

Since $\sin\psi=\dfrac{dy}{ds}$ we have from the first equation

$$mv\frac{dv}{ds}=-mg\frac{dy}{ds}$$

Integrating $\frac{1}{2}mv^2=-mgy+C$ where C is a constant. If the particle be projected from a point where $y=y_1$ with a velocity u then

$$\tfrac{1}{2}mu^2=-mg\,y_1+C$$

Hence $\frac{1}{2}mv^2-\frac{1}{2}mu^2=-mg(y-y_1)=-mgh$.

where h is the height of P above the point of projection. This is in fact the equation of energy and follows at once from the principle of energy namely the change in kinetic energy is equal to the work done.

Thus for upward projection

$$v^2=u^2-2gh.$$

and for downward projection $v^2=u^2+2gh$

where h is the vertical distance between the two points. These equations hold good whatever be the paths followed. The second equation gives R, the reaction on the particle. The particle will leave the curve when $R=0$.

6·6. Motion on the inside of a smooth vertical circle.

A particle is projected from the lowest point with velocity u and moves along the inside of a smooth vertical circle ; to discuss the motion.

Let A and B be the lowest and highest points of the circle, P the position of the particle at time t where $\angle AOP=\theta$, O being the centre of the circle. Let v be the velocity at P and PN perpendicular to AB.

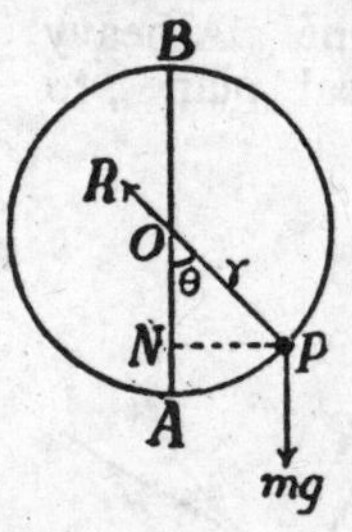

By the equation of energy

$$v^2=u^2-2g\,AN$$
$$=u^2-2gr(1-\cos\theta)$$

where $r=$ radius of the circle.

Equation of motion along the normal is

$$\frac{mv^2}{r}=R-mg\cos\theta.$$

$$\therefore\ R=m\left(\frac{v^2}{r}+g\cos\theta\right)=m\left[\frac{u^2}{r}-2g(1-\cos\theta)+g\cos\theta\right]$$

$$=m\left[\frac{u^2}{r}-2g+3g\cos\theta\right].$$

v vanishes when $u^2=2gr(1-\cos\theta)$ or $\cos\theta=1-\dfrac{u^2}{2gr}$ ...(1)

R vanishes when $\dfrac{u^2}{r}=2g-3g\cos\theta$ or $\cos\theta=\dfrac{1}{3}\left(2-\dfrac{u^2}{gr}\right)$ ...(2)

If v vanishes before R vanishes, *i.e.*, if θ from (1) is less than θ from (2) *i.e.*, if $\cos\theta$ from (1) is greater than $\cos\theta$ from (2) then

$$1-\frac{u^2}{2gr}>\frac{1}{3}\left(2-\frac{u^2}{gr}\right)$$

i.e., $$3(2gr-u^2)>2(2gr-u^2)$$

i.e., $$u^2<2gr.$$

If $u^2=2gr$, v and R will vanish simultaneously and the particle will rise up to the horizontal diameter of the circle.

Therefore when $u^2<2gr$, v will vanish in some point in the first quadrant, R will not vanish, therefore, the particle will stop there and will come down, pass through the lowest point and will go up in the fourth quadrant and rise up to the same height and will then come down and so on. Thus the particle will oscillate about the lowest point. When $u^2=2gr$, the particle will go up to the end of the horizontal diameter and will oscillate through a quadrant on each side of the vertical.

If R vanishes before v vanishes, we will have

$$1-\frac{u^2}{2gr}<\frac{1}{3}\left(2-\frac{u^2}{gr}\right) \qquad \therefore\ u^2>2gr.$$

The particle will rise above the horizontal diameter.

The particle will go right round the circle if both v and R do not vanish at the highest point when $\theta=\pi$, *i.e.*, if $u^2>4gr$ and if $u^2>5gr$ *i.e.*, if $u^2>5gr$, the particle will describe the complete circle.

Thus if $u^2>2gr$ but $<5gr$, the particle will go above the horizontal diameter but will not go up to the highest point *i.e.*, the particle will leave the circle at some point in the second quadrant and will afterwards move in a parabola.

Thus we obtain the following results ;—

(*i*) *when* $u^2<2gr$, *the particle oscillates on each side of the lowest point,*

(*ii*) *when* $u^2>2gr$ *but* $<5gr$, *the particle leaves the circle and describes a parabola.*

(iii) *when $u^2 > 5gr$, the particle makes complete revolutions.*

The reaction at A is

$$R_0 = m\left[\frac{u^2}{r} - 2g + 3g\right] = m\left(\frac{u^2}{r} + g\right)$$

Hence when $u^2 > 5gr$, $R_0 > 6mg$, *i.e., the pressure at the lowest point must be greater than 6 mg, if the particle is to go right round the circle.*

In case (*ii*) we will have to find the point when the particle after leaving the circle cuts it again.

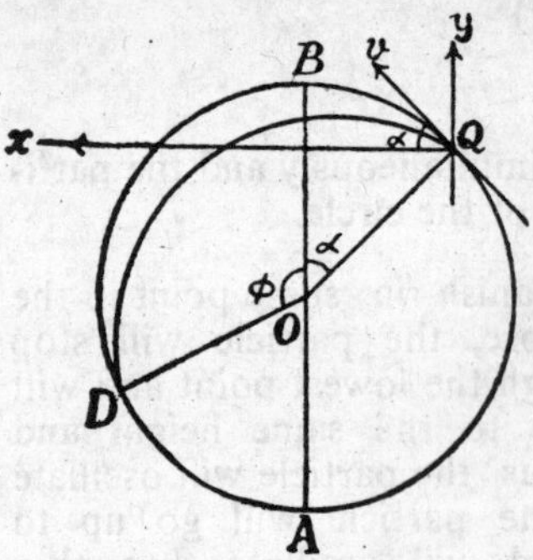

Let the particle leave the circle at some point Q where $R=0$

i.e., $$\cos\theta = \frac{1}{3}\left(2 - \frac{u^2}{gr}\right)$$

If $\angle QOB = \alpha$, $\cos\alpha = \frac{1}{3}\left(\frac{u^2}{gr} - 2\right)$

If v be the velocity at Q

$$v^2 = u^2 - 2gr\left\{1 - \frac{1}{3}\left(2 - \frac{u^2}{gr}\right)\right\}$$

$$= (u^2 - 2gr) - \tfrac{2}{3}(u^2 - 2gr)$$

$$= \frac{u^2 - 2gr}{3} = gr\cos\alpha.$$

Latus rectum of the parabola it describes after leaving is given by

$$\frac{2v^2\cos^2\alpha}{g} = 2r\cos^3\alpha = \frac{2(u^2 - 2gr)^3}{27\,g^3\,r^2}$$

Take Q as origin, the vertical upwards as y-axis and the horizontal to the left as x-axis. Then the equation to the parabolic path of the particle after leaving at Q is

$$y = x\tan\alpha - \tfrac{1}{2}g\,\frac{x^2}{v^2\cos^2\alpha}$$

$$= x\tan\alpha - \frac{x^2}{2r\cos^3\alpha}.$$

The equation to the circle with Q as origin is

$$(x - r\sin\alpha)^2 + (y + r\cos\alpha)^2 = r^2$$

i.e., $$y = x\tan\alpha - \frac{x^2 + y^2}{2r\cos\alpha}.$$

The point D where the two curves cut again be (x', y') so that

$$\frac{x'^2}{2r\cos^3\alpha}=\frac{x'^2+y'^2}{2r\cos\alpha}$$

or $$x'^2\sin^2\alpha=y'^2\cos^2\alpha$$

$\therefore$ $$y'=\pm x'\tan\alpha$$

when $y'=x'\tan\alpha$, we get $x'=0=y'$, *i.e.*, the point Q.

when $y'=-x'\tan\alpha$, we get $x'=4r\cos^2\alpha\sin\alpha$.

$$y'=-4r\cos\alpha\sin^2\alpha.$$

Let angle $BOD=\phi$, then $\tan\phi=\dfrac{x'-r\sin\alpha}{y'-(-r\cos\alpha)}$

$$=\frac{4r\cos^2\alpha\sin\alpha-r\sin\alpha}{-4r\cos\alpha\sin^2\alpha+r\cos\alpha}$$

$$=\tan\alpha.\frac{4\cos^2\alpha-1}{1-4\sin^2\alpha}$$

$$=\tan\alpha.\frac{4-(1+\tan^2\alpha)}{1+\tan^2\alpha-4\tan^2\alpha}$$

$$=\tan\alpha.\frac{3-\tan^2\alpha}{1-3\tan^2 a}=\tan 3a$$

$\therefore$ $$\phi=3\alpha$$

Hence arc $BD=3$ arc BQ.

Height reached by the particle above the horizontal line through Q

$$=\frac{v^2\sin^2\alpha}{2g}=\frac{gr\cos\alpha\sin^2\alpha}{2g}=\tfrac{1}{2}r\cos\alpha\sin^2\alpha$$

$\therefore$ Height above $A=\frac{1}{2}r\cos\alpha\sin^2\alpha+r\cos\alpha+r.$

To find at what point the particle crosses the vertical diameter, we see that the equation to OB is

$$x=r\sin\alpha$$

Putting $x=r\sin\alpha$ in the equation to the parabola

$$y=x\tan\alpha-\frac{x^2}{2r\cos^3\alpha}$$

we get $$y=r\left[\frac{\sin^2\alpha}{\cos\alpha}-\frac{\sin^2\alpha}{2\cos^3\alpha}\right]$$

hence the height of the point above A

$$r+r\cos\alpha+y=r+r\cos\alpha+r\left[\frac{\sin^2\alpha}{\cos\alpha}-\frac{\sin^2\alpha}{2\cos^3\alpha}\right]$$

$$=r\,\frac{2\cos^3\alpha+2\cos^4\alpha+2\sin^2\alpha\cos^2\alpha-\sin^2\alpha}{2\cos^3\alpha}$$

$$=\frac{r}{2\cos^3\alpha}\left\{2\cos^3\alpha\,(1+\cos\alpha)+\sin^2\alpha\,(2\cos^2\alpha-1)\right\}$$

$$=\frac{r(1+\cos\alpha)}{2\cos^3\alpha}\left\{2\cos^3\alpha+(1-\cos\alpha)\,(2\cos^2\alpha-1)\right\}$$

$$=\frac{r(1+\cos\alpha)^2}{2\cos^3\alpha}\,(2\cos\alpha-1).$$

Cor. In the case of a bead moving in a circular wire or a particle moving inside a circular tube, the condition for a complete revolution is that $v>0$ when $\theta=\pi$, *i.e.*, $u^2>4gr$.

Ex. 5. *A particle is free to move on a smooth vertical circular wire of radius r. It is projected from the lowest point with velocity just sufficient to carry it to the highest point. Show that the reaction between the particle and the wire is zero after a time*

$$\sqrt{\frac{r}{g}}\,\log\,(\sqrt{5}+\sqrt{6}).$$

Sol. Here $u^2=4gr$

$\therefore$ $v^2=u^2-2gr\,(1-\cos\theta)=2gr\,(1+\cos\theta)$

$=4gr\cos^2\theta/2$

$$R=\frac{mv^2}{r}+mg\cos\theta$$

$\therefore$ where $R=0,\ v^3=-gr\cos\theta$

If at $\theta=\theta_1$, $R=0$ then $-gr\cos\theta_1=2gr(1+\cos\theta_1)$

$\therefore$ $\cos\theta_1=-\frac{2}{3}$

Also $v=2\sqrt{gr}\cos\theta/2$

Hence $r\dot{\theta}=2\sqrt{gr}\cos\theta/2$

or $$\sqrt{\frac{g}{r}}\,t=\int_0^{\theta_1}\frac{d\theta}{2\cos\theta/2}$$

$$=\log(\tan\theta/2+\sec\theta/2\Big]_0^{\theta_1}$$

$$=\log\left(\tan\frac{\theta_1}{2}+\sec\frac{\theta_1}{2}\right)$$

Now $\cos\theta_1=-\dfrac{2}{3}\quad\therefore\quad 2\cos^2\theta_1/2-1=-\dfrac{2}{3}$

$\therefore\quad \cos^2\dfrac{\theta_1}{2}=\dfrac{1}{6}\quad\therefore\ \sec\dfrac{\theta_1}{2}=\sqrt{6},\ \tan\dfrac{\theta_1}{2}=\sqrt{5}$

Hence $$t=\sqrt{\frac{r}{g}}\log(\sqrt{5}+\sqrt{6}).$$

Ex. 6. *Find the velocity with which a particle must be projected along the interior of a smooth vertical hoop of radius r, from the lowest point in order that it may leave the hoop at an angular distance of 30° from the vertical. Show that it will strike the hoop again at an extremity of the horizontal diameter.*

Show also that if velocity of projection be $\sqrt{\tfrac{7}{2}gr}$, the particle will leave the hoop and return to the lowest point.

Sol. The particle leaves the circle at an angle α from the upward drawn vertical where $\cos\alpha=\dfrac{1}{3}\left(\dfrac{u^2}{gr}-2\right)$

Hence $\alpha=30$, $\therefore\ \dfrac{\sqrt{3}}{2}=\dfrac{1}{3}\left(\dfrac{u^2}{gr}-2\right)$

$$\therefore\qquad u^2=\tfrac{1}{2}gr(3\sqrt{3}+4).$$

Since $\angle BOQ=30°\quad\therefore\quad\angle BOD=3\cdot30°=90°$ *i.e.*, D is at the extremity of the horizontal diameter.

Where $u^2=\frac{7}{2}gr$, $\cos\alpha=\frac{1}{3}(\frac{7}{2}-2)=\frac{1}{2}\quad\therefore\quad\alpha=60$

i.e., $\angle BOQ=60°$ Hence $\angle BOD=180°$ *i.e.*, D coincides with A.

Ex. 7. *A particle is projected along the inside of a smooth vertical circle of radius r, from the lowest point. Show that the velocity of projection so that after leaving the circle the particle may pass through the centre is $\sqrt{\tfrac{1}{2}gr[\sqrt{3}+1]}$.*

Sol. The particle will leave the circle at some point Q at an angle α with the verticle with velocity v

where $$v^2=\frac{u^2-2gr}{3}=gr\cos\alpha$$

and $$\cos\alpha=\frac{1}{3}\left(\frac{u^2}{gr}-2\right)$$

With Q origin, the centre is the point $(r\sin\alpha, -r\cos\alpha)$ and by the question it lies on the parabola

$$y=x\tan\alpha-\frac{1}{2}\,\frac{gr^2}{v^2\cos^2\alpha}$$

i.e., $$y=x\tan\alpha-\frac{1}{2}\,\frac{gx^2}{gr\cos^3\alpha}.$$

Hence $$-r\cos\alpha=r\sin\alpha\tan\alpha-\frac{1}{2}\,\frac{r^2\sin^2\alpha}{r\cos^3\alpha}$$

i.e., $$\frac{\sin^2\alpha}{\cos\alpha}+\cos\alpha=\tfrac{1}{2}\frac{\sin^2\alpha}{\cos^3\alpha}\quad\text{or}\quad\tan^2\alpha=2$$

$\therefore$ $$\sec^2\alpha=3.\quad\text{Hence }\cos\alpha=\frac{1}{\sqrt{3}}$$

Therefore $$\frac{1}{\sqrt{3}}=\frac{1}{3}\left(\frac{u^2}{gr}-2\right)$$

$\therefore$ $$u^2=gr(2+\sqrt{3})$$

$\therefore$ $$u=\sqrt{\frac{gr}{2}}(\sqrt{3}+1).$$

Ex. 8. *A heavy particle, hanging vertically from a fixed point by a light inextensible cord of length l, is struck by a horizontal blow which imparts to it a velocity $2\sqrt{gl}$; prove that the cord becomes slack when the particle has risen to a height $\frac{2}{3}l$ above the fixed point and find the highest point of the parabola subsequently described.*

Sol. The cord becomes slack at an angle α from the vertical here

$$\cos\alpha=\frac{1}{3}\left(\frac{u^2}{gl}-2\right)=\frac{1}{3}\left(\frac{4gl}{gl}-2\right)=\frac{2}{3}$$

i.e., at a height above the fixed point (centre of the circle)

$$=l\cos\alpha=\tfrac{2}{3}l.$$

The velocity at that point is $v^2=gl\cos\alpha=\frac{2}{3}gl.$

The greatest height above this point

$$=\frac{v^2\sin^2\alpha}{2g}=\frac{\frac{2}{3}gl.(1-\frac{4}{9})}{2g}=\frac{l}{3}\cdot\frac{5}{9}=\frac{5l}{27}$$

$\therefore$ height above the fixed point $\frac{2}{3}l+\frac{5l}{27}$.

Ex. 9 *A particle is slightly displaced from the position of rest from the highest point of a vertical circle and falls under gravity; determine where the particle will leave the circle.*

Sol. AOC is the vertical diameter of the circle of radius r. P is the position of the particle at time t and PN is perpendicular to AB. $\angle AOP=\theta$.

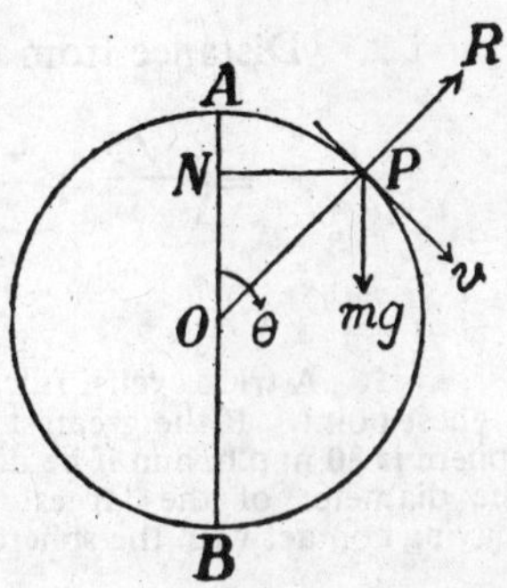

From the equation of energy, if v be the velocity at P

$$v^2=2g.AN=2gr(1-\cos\theta)$$

The normal equation of motion is

$$\frac{mv^2}{r}=mg\cos\theta-R$$

$$\therefore\quad R=m\left(g\cos\theta-\frac{v^2}{r}\right)$$

$$=m[g\cos\theta-2g(1-\cos\theta)]$$

$$=mg(3\cos\theta-2).$$

The particle will leave the circle when $R=0$, *i.e.*, when $\cos\theta=\frac{2}{3}$

$$\therefore\quad AN=r(1-\cos\theta)=\tfrac{1}{3}r.$$

Thus the particle leaves the circle at an angle θ down the horizontal where $\cos\theta=\frac{2}{3}$ with a velocity v such that

$$v^2=2gr(1-\tfrac{2}{3})=\tfrac{2}{3}gr.$$

The latus rectum of the parabola subsequently described

$$=\frac{2v^2\cos^2\theta}{g}=\frac{2.\frac{2}{3}gr.\frac{4}{9}}{g}=\frac{16}{27}r.$$

If P be the point where the particle leaves the circle, take P as origin, vertical downwards as y-axis and horizontal to the right as x-axis, then the equation to the parabola described is

$$y=x\tan\theta+\tfrac{1}{2}\frac{gx^2}{v^2\cos^2\theta}=\frac{\sqrt{5}}{2}x+\frac{27}{16r}x^2$$

The equation to the horizontal line through B is

$$y=r+r\cos\theta=\tfrac{5}{3}r.$$

Therefore, the particle will strike this plane at a point where

$$\frac{\sqrt{5}}{2}x+\frac{27}{16r}x^2=\frac{5}{3}r$$

or
$$x^2+\frac{8\sqrt{5}}{27}rx-\frac{80r^2}{81}=0$$

$$\therefore \quad x=\frac{-4\sqrt{5}+20\sqrt{2}}{27}r$$

$\therefore$ Distance from the vertical diameter $=r\sin\theta+x$

$$=\left(\frac{\sqrt{5}}{3}+\frac{-4\sqrt{5}+20\sqrt{2}}{27}\right)r=\frac{5\sqrt{5}+20\sqrt{2}}{27}r.$$

EXAMPLES VI (C)

1. A trick cyclist rides inside a sphere, riding head downwards past the highest point. If the greatest velocity with which he can reach half-way up the sphere is 30 m.p.h. and if he does not pedal in the upper hemisphere, prove that the diameter of the largest sphere in which he can perform his feat without leaving contact with the sphere is 40·3 ft.

Also show that the least speed the cyclist must have when upside down at the highest point of the sphere of radius 21 ft. is 24 f.p.s. if the mass of the machine and rider is supposed concentrated at a point 3 ft. from the track.

2. An aeroplane describes a vertical circle of 150 yards radius. Prove that the least speed it must have when upside down at the highest point in order that no force may be necessary to keep the pilot in his seat is 81·82 m.p.h.

Show also that the force required to keep the pilot in his seat when the aeroplane is moving upside down at the highest point of the circle with a velocity which is just sufficient to take it to the highest point is 3 times the weight of the pilot.

3. Show that the greatest angle through which a person can oscillate on a swing, the ropes of which can support twice the person's weight when at rest is 120°.

If the ropes are strong enough and he can swing through 180° and if v is his speed at any point, prove that the tension in the rope at that point is $\frac{3mv^2}{2l}$ where m is the mass of the person and l the length of the rope.

4. A heavy particle is suspended by a string from a fixed point and rotates in a vertical circle. Show that the sum of tension of the string when the particle is at the opposite ends of a diameter is the same for all diameters.

5. A particle slides outside a fixed smooth circle in a vertical plane starting at rest at the top and another is projected from the lowest point inside the circle with just sufficient velocity to carry it to the top; prove that both leave the circle at the same point and proceed to describe parts of the same parabola.

6. A heavy particle hangs from a point O by a string of length a. It is projected horizontally with a velocity v such that $v^2=(2+\sqrt{3})ag$. Show that the string becomes slack when it has described an angle

$$\cos^{-1}\left(-\frac{1}{\sqrt{3}}\right).$$

7. A particle is hanging from a fixed point by a light cord 3 feet long and is started moving with an initial horizontal velocity such that the cord slackens when the particle is 5 feet above the lowest point. Show that it will rise further through $\frac{5}{9}$ feet.

8. A heavy particle hangs by an inextensible string of length a from a fixed point and is then projected horizontally with a velocity $\sqrt{2gh}$. If $\frac{5}{2}a>h>a$, prove that the circular motion ceases when the particle has reached the height $\frac{1}{3}(a+2h)$. Prove also that the greatest height ever reached by the particle above the point of projection is $\dfrac{(4a-h)\,(a+2h)^2}{27a^2}$.

9. A particle moves on the inner side of smooth circular hoop of radius a whose plane is vertical. It is projected from the bottom with a velocity insufficient to carry it round the hoop so that it comes off at an angular distance α from the highest point. Show that it crosses the vertical diameter of the hoop at a distance $\frac{1}{2}$ a $\sec^3\alpha(1-\cos\alpha)^2\,(1+2\cos\alpha)$ below the top of the hoop.

10. A particle attached to a fixed peg O by a string of length l, is lifted up with the string horizontal and let go. Prove that when the string makes an angle θ with the horizontal, the resultant acceleration is $g\sqrt{1+3\sin^2\theta}$.

If the particle is let fall from a point in the horizontal through O at a distance $l\cos\theta$ from O, show that the velocity when it is vertically below O is

$$\sqrt{2gl(1-\sin^3\theta)}$$

11. A particle is projected from the lowest point inside a smooth sphere of radius a with a velocity due to a height h above the centre. Find the point at which the particle will leave the sphere and show that it will afterwards pass

(*i*) through the centre if $h=\dfrac{\sqrt{3}}{2}a$.

(*ii*) through the lowest point if $h=\frac{3}{4}a$.

12. A body is projected along the surface of a smooth sphere of radius r from the highest point with velocity $\frac{1}{2}\sqrt{gr}$: find where it will leave the surface and prove that it will strike a horizontal plane through the centre of the sphere at a distance from the centre $\dfrac{9\sqrt{39}+7\sqrt{7}}{64}r$.

13. A particle is hanging from a fixed point O by a light string of length a. There is a small smooth nail at O' in the same horizontal line with O at a distance b ($<a$) from O. Find the minimum velocity with which the particle should be projected from its lowest position in order that it may make a complete revolution round the nail without the string becoming slack.

14. A heavy bead slides on a smooth fixed circular wire of radius a. If it be projected from the lowest point, with a velocity just sufficient to carry it to the highest point, prove that the radius through the bead, in time t, will turn through an angle $2 \tan^{-1}\left(\sinh t \sqrt{\frac{g}{a}}\right)$.

15. A heavy particle slides under gravity down the inside of a smooth circular tube held in a vertical plane. It starts from the highest point with a velocity $\sqrt{2ga}$ where a is the radius of the circle. Prove that where the vertical component of its acceleration is a maximum, the pressure on the curve is equal to twice the weight of the particle.

6.7. Simple Pendulum. *A heavy particle is fixed to one end of a light inextensible string the other end of which is fixed and is allowed to oscillate in vertical plane through a small angle. Such a system is called a Simple Pendulum.*

O is the fixed point, P the particle at any time t when the string OP is inclined at an angle θ to the vertical OA, A being the lowest position of the particle. If arc $OP=s$, $s=l\theta$, l being the length of the string. The only forces are the weight mg downwards and the tension T of the string.

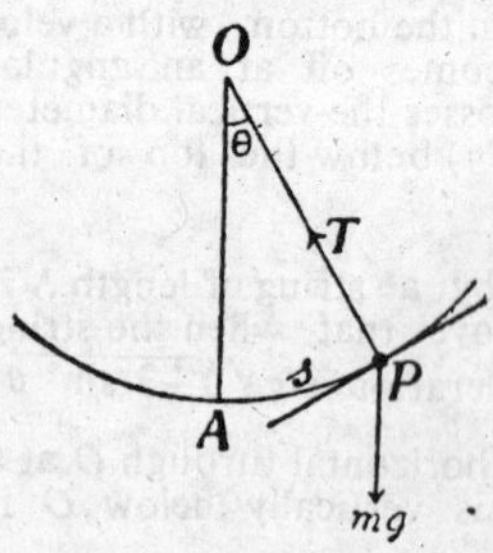

The tangential and normal equations of motion are

$$ml\ddot{\theta} = -mg \sin \theta$$

and $$ml\dot{\theta}^2 = T - mg \cos \theta.$$

From the first equation we get when θ is small, $\ddot{\theta} = -\frac{g}{l}\theta$

Thus the motion is *simple harmonic* and period $= 2\pi\sqrt{\frac{l}{g}}$.

If θ is not small, we have by integrating the equation

$$l\ddot{\theta} = -g \sin \theta$$

$$l\dot{\theta}^2 = 2g \cos \theta + C$$

when C is constant.

If the amplitude is $\theta = \alpha$ *i.e.*, when $\theta = \alpha$, $\dot{\theta} = 0$

then $$C = -2g \cos \alpha$$

Hence $$l\dot{\theta}^2 = 2g (\cos \theta - \cos \alpha)$$

or $\sqrt{\frac{2g}{l}} \cdot dt = -\frac{d\theta}{\sqrt{\cos \theta - \cos \alpha}}$, minus since θ is decreasing.

If t be the time from one extremity to the lowest position,

$$\sqrt{\frac{2g}{l}}\; t = -\int_\alpha^0 \frac{d\theta}{\sqrt{\cos\theta-\cos\alpha}}=\int_0^\alpha \frac{d\theta}{\sqrt{\cos\theta-\cos\alpha}}$$

$$=\int_0^\alpha \frac{d\theta}{\sqrt{2\sin^2\frac{\alpha}{2}-2\sin^2\frac{\theta}{2}}}.$$

$$\therefore\ \textit{i.e.,}\ 2\sqrt{\frac{g}{l}}\,.\,t=\int_0^\alpha \frac{d\theta}{\sqrt{\sin^2\frac{\alpha}{2}-\sin^2\frac{\theta}{2}}}$$

Putting $\sin\frac{\theta}{2}=\sin\frac{\alpha}{2}\sin\phi$, so that

$$\tfrac{1}{2}\cos\frac{\theta}{2}\; d\theta=\sin\frac{\alpha}{2}\cos\phi d\phi$$

and when $\theta=0,\ \phi=0$

$\theta=\alpha,\ \phi=\pi/2$

we get $$2\sqrt{\frac{g}{l}}\; t=\int_0^{\pi/2}\frac{2\sin\frac{\alpha}{2}\cos\phi d\phi}{\cos\frac{\theta}{2}\sin\frac{\alpha}{2}\cos\phi}$$

$$=\int_0^{\pi/2}\frac{2d\phi}{\cos\frac{\theta}{2}}$$

$$=\int_0^{\pi/2}\frac{2d\phi}{\sqrt{1-\sin^2\frac{\alpha}{2}\sin^2\phi}}$$

$$=2\int_0^{\pi/2}\left[1+\tfrac{1}{2}\sin^2\frac{\alpha}{2}\sin^2\phi+\tfrac{3}{8}\sin^4\frac{\alpha}{2}\sin^4\phi+\ldots.\right]d\phi$$

by expanding the Binomial

$$=2\left[\frac{\pi}{2}+\tfrac{1}{2}\sin^2\frac{\alpha}{2}\,.\,\frac{\pi}{4}+\ldots.\right]\ \text{since}\int_0^{\pi/2}\sin^n\phi d\phi=\frac{\sqrt{\pi}}{2}\cdot\frac{\Gamma\left(\frac{n+1}{2}\right)}{\Gamma\left(\frac{n+2}{2}\right)}$$

$$\therefore\quad t=\frac{\pi}{2}\sqrt{\frac{l}{g}}\left[1+\tfrac{1}{4}\sin^2\frac{\alpha}{2}+\ldots.\right]$$

If T be the time for complete oscillation *i.e.*, period

then $$T=2\pi\sqrt{\frac{l}{g}}\left[1+\tfrac{1}{4}\sin^2\frac{\alpha}{2}+\dots\right]$$

$$=2\pi\sqrt{\frac{l}{g}}\left[1+\frac{\alpha^2}{16}\right]$$ if higher powers of α be neglected.

To find whether the clock will go fast or slow if g or l or both change slightly.

Now time for one swing of the pendulum is $\pi\sqrt{\frac{l}{g}}$ hence if there be n swings in a given interval T, we have

$$T=n\pi\sqrt{\frac{l}{g}} \text{ or } n=\frac{T}{\pi}\sqrt{\frac{g}{l}}.$$

From the theory of errors, if δg, δl be slight changes in g and l and if δn be the corresponding change in n, we have

$$\delta n=\frac{\partial n}{\partial g}\delta g+\frac{\partial n}{\partial l}\delta l=\frac{T}{2\pi}\frac{\delta g}{\sqrt{gl}}-\frac{T}{2\pi}\sqrt{\frac{g}{l}}\cdot\frac{\delta l}{l}$$

$$=\frac{T}{\pi}\sqrt{\frac{g}{l}}\left[\frac{1}{2}\frac{\delta g}{g}-\frac{1}{2}\frac{\delta l}{l}\right]$$

$$\therefore \quad \frac{\delta n}{n}=\frac{1}{2}\frac{\delta g}{g}-\frac{1}{2}\frac{\delta l}{l}.$$

If by some means the length of the pendulum changes, g remaining unchanged, then we get

$$\frac{\delta n}{n}=-\frac{1}{2}\frac{\delta l}{l}.$$

Now when l increases, δl is positive and when l decreases δl is negative, thus if the length is increased, δn becomes negative which means that in a given interval there is less number of swings hence the clock will go slow. Similarly we can show that when a pendulum of a given length is taken from one place to another so that g increases, the clock will go fast.

If a clock be taken to the top of a mountain of height h, then since in this case force $F=\frac{\mu}{r^2}$ where r is the distance from the centre of the earth and μ a constant.

We have $\delta F=-\frac{2\mu}{r^3}\delta r$. But when $r=a$, a being the radius of the earth, $g=\frac{\mu}{a^2}$, $\delta r=h$

then $$\delta g=(\delta F)_{r=a}=-\frac{2ga^2}{a^3}h=-\frac{2hg}{a}$$

$$\therefore \quad \frac{\delta n}{n}=\frac{1}{2}\frac{\delta g}{g}=-\frac{h}{a} \text{ which is negative.}$$

Hence the clock becomes slow.

If the clock be taken inside a mine at depth h, then in this case

$$F=\mu r$$

so that $$\delta F=\mu\delta r. \quad \text{Also } g=\mu a,\ \delta r=-h$$

$$\therefore \quad \delta g=(\delta F)_{r=a}=\frac{g}{a}\delta r=-\frac{gh}{a}$$

Hence $$\frac{\delta n}{n}=\frac{1}{2}\frac{\delta g}{g}=-\frac{1}{2}\frac{h}{a} \text{ also negative.}$$

The clock again goes slow.

Ex. 10. *A pendulum is carried to the top of a mountain half a mile high, how many seconds will it lose per day? By how much must its present length be shortened so that it may beat seconds at the top of the mountain.*

Sol. If h be the height of mountain

$$\frac{\delta n}{n}=-\frac{h}{a}.$$

Hence $$h=\tfrac{1}{2} \text{ mile},\ a=4000 \text{ miles},\ n=86400$$

$$\therefore \quad \delta n=-\frac{86400\times\frac{1}{2}}{4000}=-10{\cdot}8.$$

$\therefore$ number of seconds lost per day $=10{\cdot}8$.

Let l be the present length and suppose an increment δl in l will keep the pendulum correct at the top of the mountain,

then from $$\frac{\delta n}{n}=\frac{1}{2}\frac{\delta g}{g}-\frac{1}{2}\frac{\delta l}{l}$$

we have here $$\delta n=0, \quad \therefore \quad \delta l=\frac{l\delta g}{g}=l\times\left(-\frac{2h}{a}\right)$$

$$=-l.\frac{2\times\frac{1}{2}}{4000}=-\frac{l}{4000}$$

$\therefore$ the present length must be decreased by $\frac{1}{4000}$ of itself.

Ex. 11. *A second's pendulum was too long on a given day by a quantity α ; it was then over-corrected so as to be too short by α during the next day ; prove that l being the correct length, the number of minutes gained in the two days was 1080 $\frac{\alpha^2}{l^2}$ nearly.*

Sol. One second $=\pi\sqrt{\frac{l}{g}}$.

If t be the time of one swing when l becomes $l+\alpha$,

then $$t=\pi\sqrt{\frac{l+\alpha}{g}}.$$

If n be the number of vibrations in a day

$$nt=86400$$

$$\therefore \quad n=\frac{86400}{\pi}\sqrt{\frac{g}{l+a}}=\frac{86400}{\pi}\sqrt{\frac{g}{l}}\sqrt{\frac{l}{l+\alpha}}$$

$$=86400\sqrt{\frac{l}{l+a}} \text{ since } \pi\sqrt{\frac{l}{g}}=1$$

$$=86400\left(1+\frac{\alpha}{l}\right)^{-\frac{1}{2}}$$

$$=86400\left(1-\frac{\alpha}{2l}+\frac{3\alpha^2}{8l^2}+\dots\right)$$

$\therefore$ loss in a day $=86400\left(\frac{\alpha}{2l}-\frac{3\alpha^2}{8l^2}\right)$ vibrations.

Similarly when l becomes $l-\alpha$, the loss in a day is

$$86400\left(-\frac{\alpha}{2l}-\frac{3\alpha^2}{8l^2}\right)$$

i.e., gain in a day $=86400\left(\frac{\alpha}{2l}+\frac{3\alpha^2}{8l^2}\right)$ vibrations.

Therefore, during the two days the gain in the number of vibrations is $\frac{3\alpha^2}{4l^2}$. 86400.

But the gain in one vibration means one second

$$\text{the time gained} = 86400 \times \frac{3\alpha^2}{4l^2} \text{ seconds}$$

$$= 1080 \frac{\alpha^2}{l^2} \text{ minutes.}$$

EXAMPLES VI (D)

1. Show that an incorrect second pendulum of a clock which loses x seconds in a day must be shortened by $\frac{x}{432}$ per cent of its length in order to keep correct time.

2. A pendulum beats seconds accurately at a place where g is 32 ft./sec^2. If taken to a place where the value of g is 32·2 ft./sec.^{2}, prove that it will gain 270 seconds per day.

3. Show that the height of a mountain at the top of which a pendulum which beats seconds at sea level loses 20 sec. a day, is approximately $\frac{25}{27}$ mile.

4. A clock loses 20 seconds a day at a place where $g=32$ ft./sec.^{2} Show that either the pendulum must be shortened by about ·018 in. or it must be taken to a place where $g=32{\cdot}015$ ft. sec.^{2} in order it may give correct time.

5. A second pendulum hangs against a wall inclined at an angle θ to the horizontal. Show that the time of a small oscillation is $\pi \sqrt{\frac{l}{g \sin \theta}}$.

6. A pendulum of length l has one end of the string fastened to a peg on a smooth inclined plane of inclination α. With the string and the weight on the plane, its time of oscillation is t seconds. If pendulum of length l' oscillates in one second when suspended vertically, prove that

$$\sin \alpha = \frac{l}{l' t^2}.$$

7. In an oscillating pendulum the tension of the string when the bob is in its lowest position is n times the weight of the bob. Prove that the angle of swing on each side of the vertical is

$$\cos^{-1}\left(\frac{3-n}{2}\right).$$

8. For a simple pendulum making complete revolutions in a vertical plane, if ω_1, ω_2 be the greatest and the least angular velocities and T_1, T_2 the greatest and the least tension, prove that when the pendulum makes an angle θ with the vertical, the angular velocity is

$$\sqrt{\omega_1^2 \cos^2 \frac{\theta}{2} + \omega_2^2 \sin^2 \frac{\theta}{2}}$$

and that the tension is $T_1 \cos^2 \frac{\theta}{2} + T_2 \sin^2 \frac{\theta}{2}$.

Further if the tension when the bob is in its lowest position is K times the tension when the bob is in its highest position and u_1, u_2 be the velocities in these positions, prove that

$$\frac{u_1^2}{u_2^2} = \frac{5K+1}{K+5}.$$

8·9 Cycloidal pendulum. We have seen that the motion of a simple pendulum is simple harmonic as long as the amplitude is small. For larger amplitudes the motion is not simple harmonic and consequently the period is different and cannot be said to be independent of the amplitude and variations in amplitude will cause irregularities in the time-keeping of a clock.

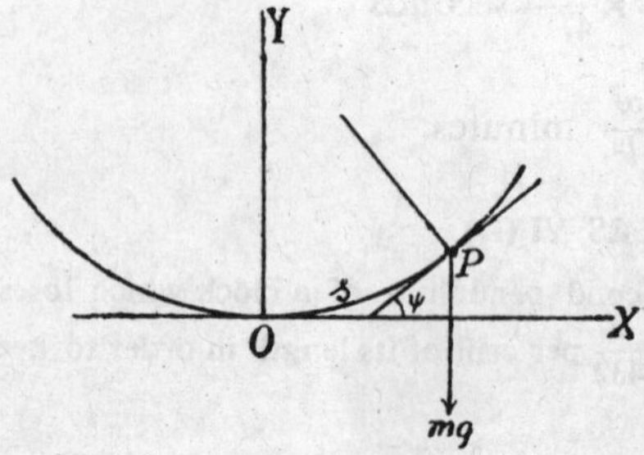

If a particle is constrained to move along the arc of smooth cycloid in a vertical plane, placed with vertex downwards and axis vertical, the tangential and normal equations of motion are

$$m\ddot{s}=-mg\sin\psi$$

and
$$\frac{m\dot{s}^2}{\rho}=R-mg\cos\psi$$

where s is measured from the vertex O, ψ the angle the tangent at any point P makes with the horizontal, R the reaction of the cycloid and ρ its radius of curvature at P. But the equation to the cycloid is $s=4a\sin\psi$.

Hence the first equation can be written as

$$\ddot{s}=-\frac{g}{4a}s,$$

which shows that the *motion is simple harmonic whatever be the amplitude and the time of a complete oscillation is* $2\pi\sqrt{\frac{4a}{g}}$. This property finds its application in the formation of clocks.

Let CAC' be a cycloidal arc and let CA' and $C'A'$ be the evolutes of CAC' which are also cycloids such that arc $C'A'$=arc CA.

If $AA'=4a$, let the upper end of a string of length $4a$ be attached at A' and let P be the bob. When the bob is let fail the string is always normal to CAC' and tangent to either CA or $C'A'$. Hence P describes the involute of CA' and $C'A'$, *i.e.*, CAC'. Thus the string wraps or unwraps itself about CA' or $C'A'$ and the bob moves on the

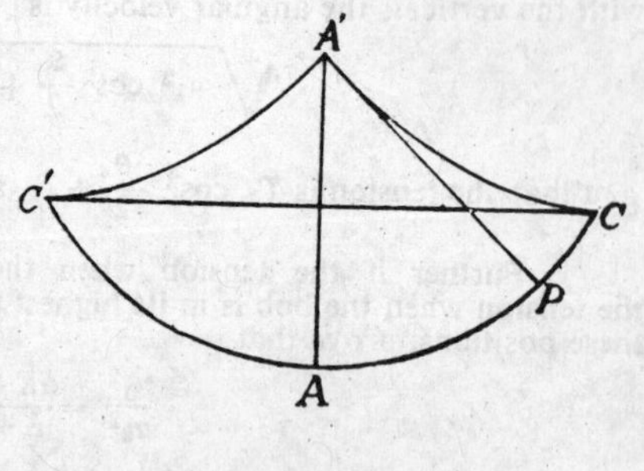

cycloidal arc CAC' and moves in true simple harmonic motion independent of the amplitude. The system is called a *Cycloidal Pendulum.*

Ex. 12, *A heavy particle slides down a smooth cycloid starting from rest at the cusp, the axis being vertical and vertex downwards. Prove that the magnitude of the acceleration of the particle is equal to g at every point of its path.*

Sol. Cycloid $s=4a\sin\psi,\ \rho=4a\cos\psi$

$$\ddot{s}=-g\sin\psi=-\frac{g}{4a}s$$

Integrating

$$\frac{\dot{s}^2}{2}=-\frac{g}{8a}s^2+C$$

when $\qquad s=4a,\ \dot{s}=0 \quad \therefore \quad C=2ag$

$$\therefore \quad \dot{s}^2=-\frac{g}{4a}s^2+4ag=4ag\cos^2\psi$$

If f be the acceleration

$$f^2=(\ddot{s})^2+\left(\frac{\dot{s}^2}{\rho}\right)^2=g^2\sin^2\psi+\left(\frac{4ag\cos^2\psi}{4a\cos\psi}\right)^2$$

$$=g^2\sin^2\psi+g^2\cos^2\psi=g^2$$

$$\therefore \quad f=g.$$

Ex. 13. *A particle is projected with velocity V from the cusp of a smooth inverted cycloid down the arc ; show that the time of reaching the vertex is*

$$2\sqrt{\frac{a}{g}}\tan^{-1}\left(\frac{\sqrt{4ag}}{V}\right).$$

Sol. The equation of motion is $\ddot{s}=-\dfrac{g}{4a}s$

Its solution is $s=A\cos\sqrt{\dfrac{g}{4a}}\,t+B\sin\sqrt{\dfrac{g}{4a}}\,t.$

Initially $t=0,\ s=4a,\ \dot{s}=-V$

Hence $4a=A$

also $\qquad \dot{s}=-A\sqrt{\dfrac{g}{4a}}\sin\sqrt{\dfrac{g}{4a}}\,t+B\sqrt{\dfrac{g}{4a}}\cos\sqrt{\dfrac{g}{4a}}\,t$

$$\therefore \quad -V=B\sqrt{\frac{g}{4a}}. \qquad \therefore \quad B=-V\sqrt{\frac{4a}{g}}.$$

Hence $$s=4a\cos\sqrt{\frac{g}{4a}}\,t-V\sqrt{\frac{4a}{g}}\,\sin\sqrt{\frac{g}{4a}}\,t$$

when it comes to the vertex, $s=0$

$$\therefore\quad \tan\sqrt{\frac{g}{4a}}\,t=\frac{4a}{V\sqrt{\frac{4a}{g}}}=\frac{\sqrt{4ag}}{V}$$

$$\therefore\qquad t=\sqrt{\frac{4a}{g}}\tan^{-1}\left(\frac{\sqrt{4ag}}{V}\right).$$

Ex. 14. *A particle slides down the arc of a smooth cycloid whose axis is vertical and vertex lowest; prove that the time occupied in falling down the first half of the vertical height is equal to the time of falling down the second half.*

Sol. This equation of motion is $\ddot{s}=-\frac{g}{4a}s.$

$$\therefore\quad \text{period}=2\pi\sqrt{\frac{4a}{g}}=4\pi\sqrt{\frac{a}{g}}.$$

$$\therefore\quad \text{time from cusp to vertex}=\pi\sqrt{\frac{a}{g}}.$$

Integrating the above equation, we get

$$\dot{s}^2=-\frac{g}{4a}s^2+C \text{ where } C \text{ is a constant.}$$

Initially $s=4a$, $\dot{s}=0$ at the cusp $\quad\therefore C=4ag$

Hence $$\dot{s}^2=4ag-\frac{g}{4a}s^2=\frac{g}{4a}(16a^2-s^2)$$

$$\therefore\qquad \sqrt{\frac{g}{4a}}\,dt=-\frac{ds}{\sqrt{16a^2-s^2}}$$

when $y=a$, we have since $s^2=8ay$ for a cycloid

$$s^2=8a^2\qquad\therefore\ s=2a\sqrt{2}.$$

Hence time from cusp in falling through the first half of the vertical height is given by

$$\sqrt{\frac{g}{4a}}\,t=-\int_{4a}^{2a\sqrt{2}}\frac{ds}{\sqrt{16a^2-s^2}}=\int_{2a\sqrt{2}}^{4a}\frac{ds}{\sqrt{16a^2-s^2}}$$

$$=\left[\sin^{-1}\frac{s}{4a}\right]_{2a\sqrt{2}}^{4a}=\sin^{-1}1-\sin^{-1}\frac{1}{\sqrt{2}}$$

$$=\frac{\pi}{2}-\frac{\pi}{4}=\frac{\pi}{4}$$

$$\therefore \quad t=\frac{\pi}{2}\sqrt{\frac{a}{g}}=\text{half the time from cusp to vertex.}$$

Therefore, the remaining half of the vertical distance is also described in time $\frac{\pi}{2}\sqrt{\frac{a}{g}}$.

Ex. 15. *A particle started from the vertex of a cycloid whose axis is vertical and vertex upwards. Prove that it will leave the curve when it has fallen half the vertical height of the cycloid and the latus rectum of the parabola subsequently described is equal to the height of the cycloid. Also show that after leaving the curve it falls upon the base at a distance* $\left(\frac{\pi}{2}+\sqrt{3}\right)a$ *from the centre of base, where a is the radius of the generating circle.*

Sol. The tangential and normal equations of the motion are

$$m\ddot{s}=mg\sin\psi$$

$$m\frac{\dot{s}^2}{\rho}=mg\cos\psi-R$$

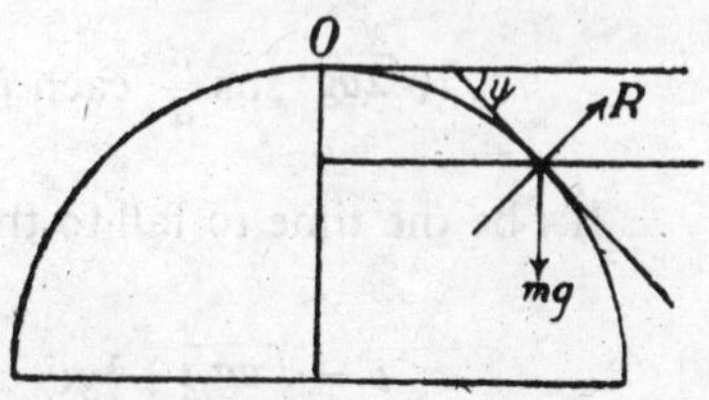

where R is the reaction, ρ the radius of curvature.

But $s=4a\sin\psi$

so that $\rho=4a\cos\psi$

Hence $\quad \ddot{s}=\frac{g}{4a}s$

Integrating $\quad \dot{s}^2=\frac{g}{4a}s^2$, the constant being zero

since $\quad s=0=\dot{s}$

$$\therefore \quad R=mg\cos\psi-\frac{m\dot{s}^2}{\rho}=mg\cos\psi-m\frac{\frac{g}{4a}16a^2\sin^2\psi}{4a\cos\psi}$$

$$=mg\left(\cos\psi-\frac{\sin^2\psi}{\cos\psi}\right).$$

The particle will leave the curve when $R=0$

i.e., when $\cos^2\psi-\sin^2\psi=0 \quad \therefore \cos 2\psi=0$

$$\therefore \quad \psi=\frac{\pi}{4}.$$

At that point $s=4a\sin\frac{\pi}{4}=2\sqrt{2}a$

$$\therefore \quad y=\frac{s^2}{8a}=\frac{8a^2}{8a}=a.$$

Also when leaving $v^2=\dot{s}^2=\frac{g}{4a}\cdot 8a^2=2ag$ at an angle $\frac{\pi}{4}$ to the horizontal

$\therefore$ latus rectum of the parabolic path

$$=\frac{2v^2\cos^2\frac{\pi}{4}}{g}=\frac{4ag.\frac{1}{2}}{g}=2a.$$

Also the depth of the base below the point where the particle leaves is a and the vertical and horizontal components of velocity are

$$\sqrt{2ag}\cos\frac{\pi}{4} \text{ each } i.e., \sqrt{ag} \text{ each.}$$

If t be the time to fall to the base

$$a=\sqrt{ag}.t+\tfrac{1}{2}gt^2 \quad \therefore \quad t=(\sqrt{3}-1)\sqrt{\frac{a}{g}}$$

and the horizontal distance moved since leaving the curve

$$=\sqrt{ag}.t=a(\sqrt{3}-1).$$

$\therefore$ Total horizontal distance from the axis

$$=x+a(\sqrt{3}-1) \text{ where } x=a(\theta+\sin\theta)_{\theta=2\psi=\frac{\pi}{2}}=a\left(\frac{\pi}{2}+1\right)$$

$$=a\left(\frac{\pi}{2}+1\right)+(\sqrt{3}-1)a$$

$$=\left(\frac{\pi}{2}+\sqrt{3}\right)a.$$

Ex. 16. *Two particles are let drop from the cusp of a cycloid down the curve of an interval of time t ; prove that they will meet in a time*

$$2\pi\sqrt{\frac{a}{g}}+\frac{t}{2}.$$

Sol, Equation of motion $\ddot{s}=-\frac{g}{4a}s$

$\therefore$ Solution is $s=4a\cos\sqrt{\frac{g}{4a}}\,t$ since when $t=0$,

$s=4a$ and $\dot{s}=0$.

For the second particle $s=4a\cos\sqrt{\frac{g}{4a}}\,(t-t_1)$ if it starts after interval t_1.

Hence when they meet $\cos\sqrt{\frac{g}{4a}}\,t=\cos\sqrt{\frac{g}{4a}}\,(t-t_1)$.

$$\therefore \quad \sqrt{\frac{g}{4a}}\,(t-t_1)=2\pi-\sqrt{\frac{g}{4a}}\,t$$

or $$t-t_1=4\pi\sqrt{\frac{a}{g}}-t$$

or $$2t=4\pi\sqrt{\frac{a}{g}}+t_1 \quad \therefore \quad t=2\pi\sqrt{\frac{a}{g}}+\frac{t_1}{2}.$$

Ex. 17. If a particle is projected from a point B downwards along the arc of a smooth cycloid with axis vertical and vertex downwards with velocity equal to that with which it reaches the vertex A when started from rest at B, prove that it will now reach A in half the time taken in the first case.

Solution. If particle starts from rest at B, $v=0$ when $s=b$ (say). Then tangential equation of motion $\ddot{s}=\frac{g}{4a}s$, yields to

$$v^2=(g/4a)\,(b^2-s^2)$$

Therefore at A where $s=0$, $v^2=gb^2/4a$ ...(1)

If t_1 be the time from B to A, in this case $t_1=\frac{1}{4}$ time period

$$=\frac{1}{4}\cdot 4\pi\sqrt{a/g}=\pi\sqrt{a/g}.$$

Now if the particle starts from B with velocity given by (1), the initial conditions are at $t=0$ when $s=b$, $v=-b\sqrt{g/4a}$, the negative sign being taken due to velocity in the sense of s decreasing.

Now general solution of $\ddot{s}=-(g/4a)\,s$ is

$$s = A\cos(\sqrt{g/4a}\,t) + B\sin(\sqrt{g/4a}\,t) \quad ...(2)$$

so that $\dot{s} = -A\sqrt{g/4a}\sin(\sqrt{g/4a}\,t) + B(\sqrt{g/4a})\cos(\sqrt{g/4a}\,t)$ ...(3)

Using initial conditions, (2) and (3) give, $A = b$, $B = -b$.

Hence $s = b(\cos\sqrt{g/4a}\,t - \sin\sqrt{g/4a}\,t)$ ...(4)

At A, $s = 0$, $t = t_2$ (say); then from (4),

$$0 = b(\cos\sqrt{g/4a}\,t_2 - \sin\sqrt{g/4a}\,t_2)$$

or $\tan\sqrt{g/4a}\,t_2 = 1$; so that $t_2 = \frac{\pi}{4}\sqrt{4a/g} = \frac{\pi}{2}\sqrt{a/g}$

Hence $$t_2 = \frac{1}{2}t_1.$$

EXAMPLES VI (E)

1. A particle falls down a cycloid starting from the cusp. Show that when it arrives at the vertex the pressure on the curve is equal to twice the weight of the particle.

2. When a particle falls from rest at the cusp down the arc of an erect cycloid, prove that it describes half the path to the lowest point in two-thirds the time to the lowest point.

3. A particle is moving on a smooth curve under gravity and its velocity varies as the distance (measured along the arc) from the highest point. Prove that the curve must be a cycloid.

4. If a particle slides down a smooth cycloid starting from a point whose arcual distance from the vertex is b, prove that its speed at any time t is $\frac{2\pi b}{T}\sin\frac{2\pi t}{T}$ where T is the time of complete oscillation of the particle.

5. A particle oscillates from cusp to cusp in a smooth cycloid whose axis is vertical and vertex lowest. Show that the velocity v at any point P is equal to the resolved part of the velocity V at the vertex along the tangent at P.

6. If a particle starts from rest at a given point of a cycloid with its axis vertical and vertex downwards, prove that it falls through $\frac{1}{n}$ of the vertical distance to the lowest point in the time $2\sqrt{\frac{a}{g}}\sin^{-1}\frac{1}{\sqrt{n}}$ where a is the radius of the generating circle.

7. Two particles connected by a fine string are constrained to move in a fine cycloid tube in a vertical plane, the axis of the cycloid being vertical and the vertex upwards; prove that the tension of the string is constant throughout the motion.

8. A cycloid is placed with its axis vertical and vertex upwards and a heavy particle is projected from the cusp up the concave side of the curve with velocity $\sqrt{2gh}$; prove that the latus rectum of the parabola described after leaving the arc is $\frac{h^2}{2a}$ where a is the radius of the generating circle.

9. Prove that for a particle, sliding down the arc and starting from the cusp of a smooth cycloid, whose vertex is lowest, the vertical velocity is maximum, when it has described half the vertical height.

7

Hodograph, Units and Dimension, Lines of Quickest Descent

7·1. Hodograph. If from a fixed point lines be drawn representing the velocities of a particle moving in a plane curve then the extremities of these lines will be on a certain curve which is called the *hodograph* of the path of the particle. If P and Q be two consecutive positions of the particle at time t and at time $t+\delta t$ and lines are

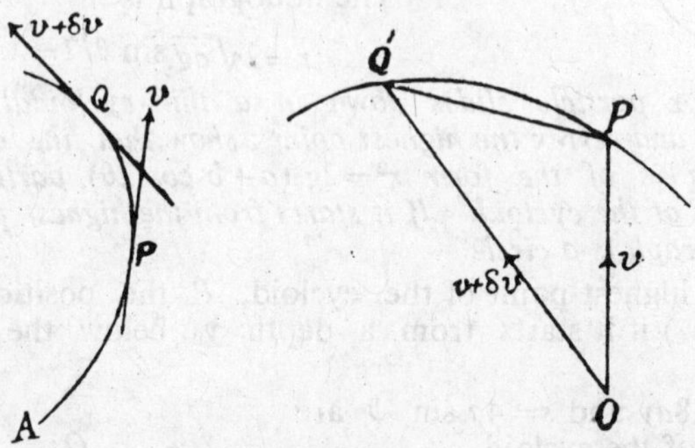

drawn from O, namely OP' and OQ' to represent the velocities v and $v+\delta v$ at P and Q, then the curve $P'Q'$ is the hodograph of the curve PQ. Now $P'Q'$ represents the change in velocity as the particle moves from P to Q and hence in the limit when Q approaches P, the line $P'Q'$ will be the tangent at P' to the hodograph and will become the direction of acceleration of P and the magnitude of the acceleration will be

$$\underset{\delta t \to 0}{\text{Lt}} \frac{P'Q'}{\delta t} = \text{velocity of } P' \text{ in the hodograph.}$$

Hence *the acceleration of P is equal to the velocity of the corresponding point P′ in the hodograph.*

If (x, y) and (ξ, η) be the co-ordinates of the points P and P' referred to axes through O, then

$$\xi = \lambda x, \quad \eta = \lambda y \text{ where } \lambda \text{ is a constant.}$$

Thus to find the hodograph, we first find v in terms of the angle ψ, the angle which the tangent at P makes with a fixed direction. Hence if we get $v = cf(\psi)$, where c is a constant then the hodograph is obtained by writing r for v, θ for ψ, that is, the hodograph is

$$r = cf(\theta).$$

Ex. 1. *A bead moves on the arc of a smooth vertical circle starting from rest at the highest point. Show that the equation to the hodograph is* $r = \lambda \sin \frac{\theta}{2}$.

Sol. *A* is the highest point, *P* the position at time *t*. If *v* be the velocity at *P*,

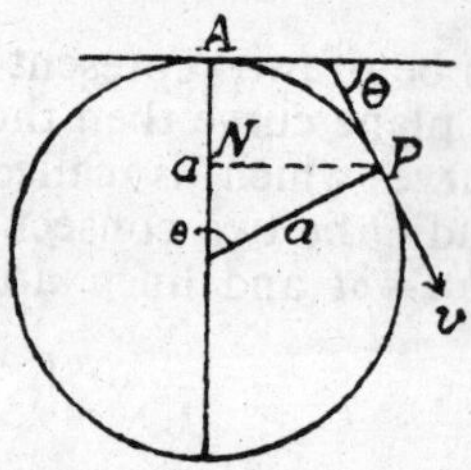

$$v^2 = 2g.AN = 2ga(1 - \cos\theta)$$

$$= 4ag \sin^2 \theta/2$$

or $\quad v = 2\sqrt{ag} \sin \theta/2$

$\therefore$ The hodograph is

$$r = 2\sqrt{ag}.\sin \theta/2 = \lambda \sin \theta/2.$$

Ex. 2. *A particle slides down in a thin cycloidal tube whose axis is vertical and vertex the highest point; show that the equation to the hodograph is of the form* $r^2 = 2g\,(a + b \cos 2\theta)$, *particle starting from any point of the cycloid. If it starts from the highest point, show that the hodograph is a circle.*

Sol. *O* is the highest point of the cycloid, *P* the position at time t, $v^2 = 2g\,(y - y_0)$ if it starts from a depth y_0 below the horizontal through *O*.

But $s^2 = 8ay$ and $s = 4a \sin \psi$ are the equations of the cycloid.

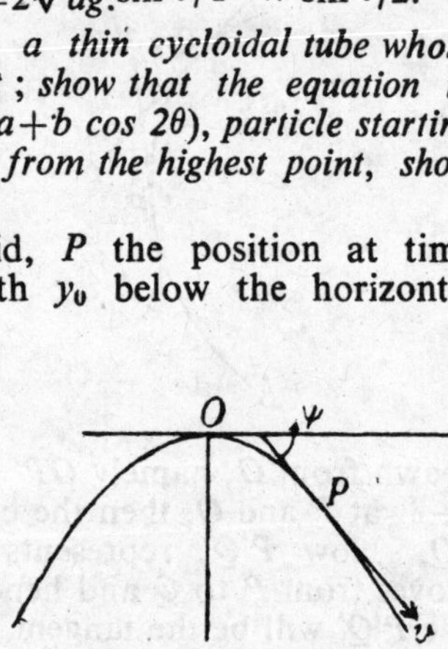

Hence $v^2 = 2g\left(\dfrac{s^2}{8a} - \dfrac{s_0^2}{8a}\right) = \dfrac{g}{4a}(s^2 - s_0^2)$

$$= \frac{g}{4a}(16a^2 \sin^2 \psi - s_0^2)$$

$$= \frac{g}{4a}[8a^2 (1 - \cos 2\psi) - s_0^2]$$

$\therefore$ The hodograph is $r^2 = \frac{g}{4a}[8a^2(1-\cos 2\theta)-s_0^2]$

which is of the form $\quad r^2 = 2g(a+b\cos 2\theta)$.

If the particle starts from O, $s_0=0$.

$$\therefore \quad r^2 = \frac{g}{4a}.\ 16a^2\sin^2\theta = 4ag\sin^2\theta$$

$$\therefore \quad r = \sqrt{4ag}\sin\theta \text{ which is a circle.}$$

Ex. 3. *The hodograph of an orbit is a parabola whose ordinate increases uniformly. Show that the orbit is a semi-cubical parabola.*

Sol. Let (ξ, η) be the point on the hodograph corresponding to the point (x, y) on the path, hence $\xi = k\dot{x}$, $\eta = k\dot{y}$.

Now the hodograph is $\eta^2 = 4a\xi$ where $\dot{\eta}$=constant$=c$ say

$\eta = ct$, the constant of integration supposed to be zero,

and $$\xi = \frac{c^2t^2}{4a}.$$

Hence $$k\dot{x} = \frac{c^2t^2}{4a},\ k\dot{y} = ct$$

$$\therefore \quad kx = \frac{c^2t^3}{12a} + c_1,$$

$$ky = \frac{ct^2}{2} + c_2.$$

Suppose c_1, c_2 are zeros.

Hence $kcy^3 = 18a^2x^2$ which is a semi-cubical parabola.

EXAMPLES VII (A)

1. Prove that the hodograph of the path of the projectile is a vertical line described with uniform velocity.

2. If the hodograph is a circle described under constant angular velocity about a point on its circumference, show that the path is a cycloid.

7·2. Units and Dimensions. Any physical quantity is measured in terms of a unit, thus if u be a quantity, its unit is U, then the measure of the quantity is $\frac{u}{U}$.

The fundamental units are those of mass, length and time, and all quantities can be expressed in terms of these units. Thus a quantity can be expressed in the form

$$L^pM^qT^r$$

where L, M, T are the units of length, mass and time. The numbers p, q, r are known as measure of dimension. Thus we say that the given quantity has dimension p in length, q in mass and r in time. In terms of units we write as follows :

$$[\text{velocity}] = \frac{[\text{length}]}{[\text{time}]} = \frac{L}{T},$$

$$[\text{acceleration}] = \frac{[\text{velocity}]}{[\text{time}]} = \frac{L}{T^2},$$

$$[\text{momentum}] = [\text{mass}]\,[\text{velocity}] = \frac{ML}{T},$$

$$[\text{force}] = [\text{mass}]\,[\text{acceleration}] = \frac{ML}{T^2},$$

$$[\text{energy}] \text{ or } [\text{work}] = [\text{force}]\,[\text{length}] = \frac{ML^2}{T^2},$$

$$[\text{power}] = \frac{[\text{work}]}{[\text{time}]} = \frac{ML^2}{T^3}.$$

As mentioned above, every physical quantity F can be expressed as $kL^pM^qT^r$ where k is a constant, thus

$$[F] = kL^pM^qT^r$$

If the units are L', M', T' then we must have

$$[F] = k'L'^pM'^qT'^r, \text{ where } k' \text{ is also constant.}$$

We note that the dimensions of the same quantity are always the same. The numbers k and k' are the measures of the quantity in the two systems of units used. Thus we have

$$kL^pM^qT^r = k'(L')^p(M')^q(T')^r.$$

When certain physical quantities are connected by a relation in the form of an equation, each term has some dimensions and in order that the equation may be correct. the dimensions of each term must be the same. This is the fundamental idea in the theory of dimensions.

Ex. 4. *If the acceleration caused by gravity be the unit of acceleration and velocity of a mile in 5 minutes that of velocity, find the units of time and space employed.*

$$k.M^pL^qT^r = k'\,(M')^p(L')^q(T')^r$$

when $\dfrac{L}{T^2} = g, \dfrac{(1 \text{ ft.})}{(1 \text{ sec.})^2}, \quad \dfrac{L}{T} = \dfrac{1760 \times 3}{5 \times 60} \cdot \dfrac{(1 \text{ ft.})}{(1 \text{ sec.})}$

since $\quad v = 1 \text{ mile/5 sec.} = \dfrac{1760 \times 3}{5 \times 60} \text{ ft./sec.}$

These give $T=\frac{11}{20}$ sec , $L=9{\cdot}68$ ft.

Ex. 5. *Given that the unit of power is a million ergs per min., that the unit of force is a thousand dynes and the unit of time one-tenth of a second, what are the units of mass and length ?*

Sol. $$\frac{ML^2}{T^3}=\frac{10^6}{60}\cdot\frac{(1\text{ gm.})(1\text{ cm.})^2}{(1\text{ sec.})^3}$$

$$\frac{ML}{T^2}=10^3.\frac{(1\text{ gm.})(1\text{ cm.})}{(1\text{ sec.})^2}$$

$$T=\frac{(1\text{ sec.})}{10}.$$

These give $M=6$ gms., $L=1\frac{2}{3}$ cms.

Ex. 6. *Show that if a body be acted upon by constant force in the direction of its motion, the space passed over in time t from rest must vary as the square of the time.*

Sol. Here s will depend upon f and t.

$$\therefore \quad s\propto f^{\lambda}\ t^{\mu}$$

$$\therefore \quad L\propto\left(\frac{L}{T^2}\right)^{\lambda}T^{\mu}=L^{\lambda}\ T^{\mu-2\lambda}$$

Equating dimensions from both sides, we get

$$\lambda=1, \qquad \mu-2\lambda=0 \qquad \therefore \quad \mu=2.$$

Hence $$s\propto ft^2.$$

Ex. 7. *Find the time of oscillation of a simple pendulum.*

Sol. Let m be the mass of the parrticle, l the lennth of the string. Time of oscillation will depend upon m, l and g.

Let $$t\propto m^p l^q g^r$$

$$\therefore \quad T\propto M^p L^q\left(\frac{L}{T^2}\right)^r=M^p L^{q+r}T^{-2r}.$$

$\therefore$ Equating dimensions, we get $p=0$, $q+r-0$, $-2r=1$

$$\therefore \quad r=-\tfrac{1}{2},\ q=\tfrac{1}{2},\ p=0.$$

Hence $$t\propto l^{\frac{1}{2}}\quad g^{-\frac{1}{2}}=\sqrt{\frac{l}{g}}.$$

EXAMPLES VII (B)

1. Given that 1 lb.=453 gms., 1 ft.=30·5 cms., show that the number of dynes in a poundal is 13816·5 and the number of ergs in a foot pound is $1{\cdot}348\times10^7$.

2. If the unit of velocity be 20 ft. per sec., the unit of acceleration 40 ft./sec.^{2} and the unit of force be 30 poundals, prove that units of mass, length and time are $\frac{3}{4}$ lb., 10 ft. and $\frac{1}{2}$ sec. respectively.

3. Show that 1 watt which is 10^7 C.G.S. units of power is equal to $\frac{1}{746}$ horse power, taking 1 lb.$=453{\cdot}6$ grams.

$$g=32{\cdot}2 \text{ ft. sec.}^2=981 \text{ cm. sec.}^2$$

4. If the unit of kinetic energy be that of a train of mass m tons moving with a velocity v miles per hour, the unit of power that of an engine of horse power h and the unit of force the weight of n tons, prove that unit of mass is $m\left[\dfrac{448\ vn}{75h}\right]^2$ tons.

5. A body moves in a straight line with S.H.M. its acceleration being μx at a distance x from the centre of motion. Show that the time of arriving at the centre varies as $1/\sqrt{\mu}$.

7·3. Lines of quickest and slowest descent.

The time of descent down all chords of a vertical circle from the highest point is the same.

A is the highest point, AC any chord at an angle θ to the vertical diameter AOB, O the centre of the circle.

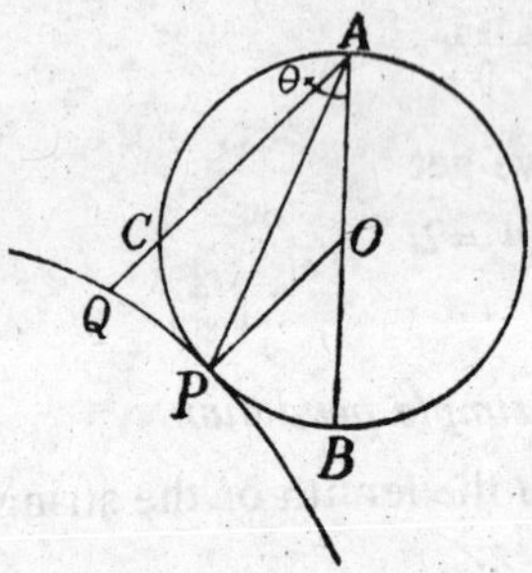

$$f=\text{acc. down } AC=g\cos\theta$$

$$\therefore \quad AC=\tfrac{1}{2}ft^2=\tfrac{1}{2}g\cos\theta.t^2.$$

But $AC=2a\cos\theta$ if a is the radius of the circle.

Hence $2a\cos\theta=\frac{1}{2}g\cos\theta.t^2$

$$\therefore \quad t^2=\frac{4a}{g}.$$

This is independent of θ; hence the times down all chords from A are the same.

Let us now find the line of quickest descent from a given point A to a given curve. Draw a circle with A as highest point and touching the curve at some point P. Then AP will be the line of quickest descent from A to the curve.

Let AQ be any other line from A to the curve cutting the circle at C.

Then $\qquad AQ>AC$

But $\qquad$ time down AC=time down AP

$\therefore \qquad$ time down $AQ>$time down AC

i.e., $\qquad >$time down AP.

Further $\qquad AO=OP$ and $\angle APO=\angle PAO$.

∴ the chord of quickness descent, the normal to the curve at P and the vertical through A form an isosceles triangle. Hence *the chord of quickest descent is equally inclined to the vertical and the normal to the curve at the point where it meets the curve.* Similarly if we want to find out the chord of quickest descent from a curve to a given point, we shall find that the circle described should have the given point as the lowest point and touch the curve at some point and then the line joining the point of contact with the given point is the required line.

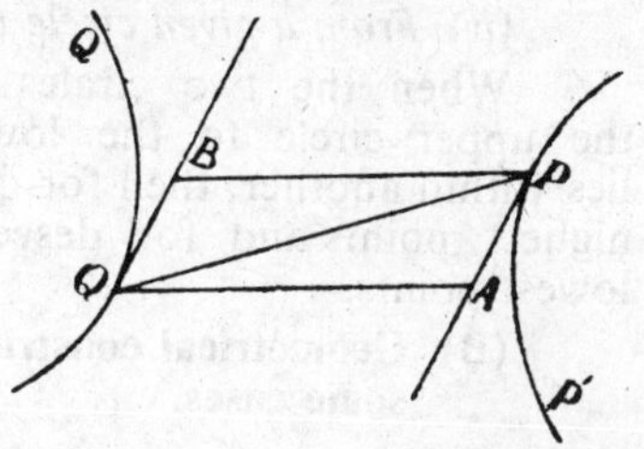

Suppose the chord of quickest descent from the curve PP' to the curve QQ' is the line PQ.

Since PQ is the chord of quickest descent from P to curve QQ',

$$\angle PQA = \angle QPA.$$

Similarly because PQ is the chord of quickest descent from curve QQ to P, $\angle PQB = \angle BPQ$.

Also $\angle APQ = \angle PQB \quad \therefore \quad \angle PQA = \angle BPQ$

∴ the figure $PBQA$ is a rhombus.

Therefore, *the chord of quickest descent from one curve to another is the diagonal of the rhombus formed by the normals and the verticals at the points where it meets the curves.*

(A) Geometrical constructions for lines of quickest descent in some cases.

(*i*) *From a given point P to a given straight line, or vice versa.*

Through P, draw a horizontal straight line cutting the given line in A and take $AQ = AP$ down in first case and up in second case. Then PQ is the required line.

(*ii*) *From a given point P to a given circle, or vice versa.*

In first case when P is outside the circle, join P to the lowest point and when P is inside, join P to the highest point of the circle ; in either case if the joining line cuts the circle in Q then PQ is the required line.

In second case just do the opposite on other case.

(*iii*) *From a given straight line to a given circle, or vice versa.*

In first case, draw a horizontal line through the lowest point B of the circle, and in second case, draw a horizontal through the highest point B, cutting the given line in K and take $KP = BK$ and then join PB to cut the circle in Q. Thus PQ is the required line.

(*iv*) *From a given circle to another circle.*

When the two circles are external, join the highest point of the upper circle to the lowest point of the other ; when one circle lies within another, then for descent from bigger to smaller, join the highest points and for descent from smaller to bigger, join the lowest points.

(B) Geometrical constructions for lines of slowest descent in some cases.

(*i*) *From a given point P to a given circle, P being without the circle and higher than the highest point of the circle.*

Join P to the highest point of the circle and produce it to meet the circle in Q, then PQ is the required line of the slowest descent.

(*ii*) *From a given circle to a point without it.*

Join P to the lowest point of the circle and produce it to meet the circle in Q, then QP is the required line.

(*iii*) *From a given circle to another given circle lying wholly without it, the lowest point of the first circle being higher than the highest point of the second.*

Join the lowest point of the first circle with the highest point of the second and produce it both ways to the circles in P and Q, then PQ is the required line.

Ex. 8. *Show that the line of quickest descent from a given straight line to a given circle is* $\sqrt{\frac{2l}{g}}$ *sec* $\theta/2$ *where* θ *is the inclination of the straight line to the horizontal and l is the shortest distance between the straight line and the circle.*

Sol. B is the lowest point of the circle, AX the given line. Through B draw a horizontal line cutting AX in A. Cut off $AX=AB$ and join XB cutting the cricle at Y Then XY is the required line.

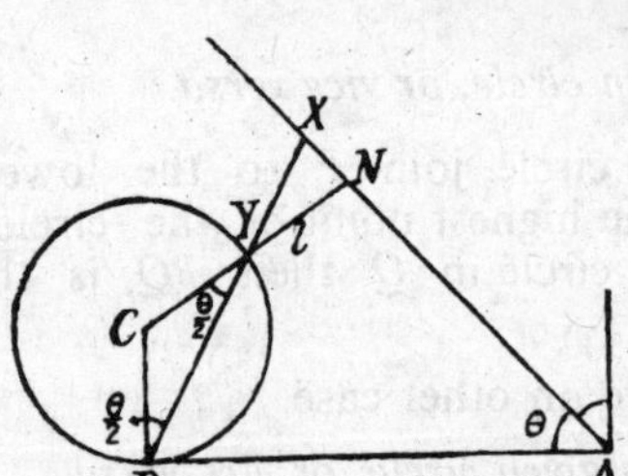

If C be the centre of the circle, join $CB.CY$ and produce CY to cut AX in N, then $YN=l$, and $\angle CYB = \theta/2$.

$$\therefore \quad XY = l \sec \theta/2.$$

Component of g along XY is $g \cos \theta/2$.

Hence $l \sec \theta/2 = XY = \frac{1}{2} g \cos \theta/2 . t^2$

$$\therefore \quad t = \sqrt{\frac{2l}{g}} \sec \theta/2.$$

Ex. 9. *A parabola is placed with its axis vertical and vertex upwards ; find the chord of quickest descent from the focus to the curve and also the focal chord of quickest descent.*

Sol. Let QSP be a focal chord making angle θ with the vertical.

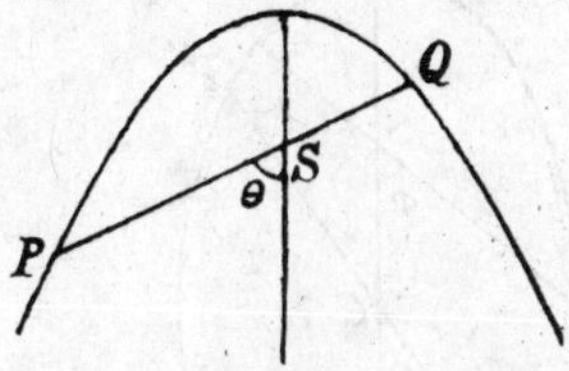

Then $SP=\dfrac{l}{1-\cos\theta}$,

$SQ=\dfrac{l}{1+\cos\theta}$, where l= semi-latus rectum.

Component of g down SP is $g\cos\theta$.

$\therefore \quad \dfrac{l}{1-\cos\theta}=SP=\frac{1}{2}g\cos\theta . t^2$ if t be the time down SP.

or $$t^2=\frac{2l}{g\cos\theta\,(1-\cos\theta)}$$

t is minimum when $\cos\theta\,(1-\cos\theta)$ is maximum

i.e., when $-\sin\theta\,(1-\cos\theta)+\cos\theta\sin\theta=0$

i.e., $\theta=0$ or $\cos\theta=\frac{1}{2}$ *i.e.,* $\theta=60°$.

Discarding $\theta=0$, we get that the required chord is inclined at an angle 60° to the axis.

For the focal chord QSP, we have

$$QSP=\frac{l}{1-\cos\theta}+\frac{l}{1+\cos\theta}=\frac{2l}{\sin^2\theta}.$$

If t be the time down QSP, we have

$$\frac{2l}{\sin^2\theta}=\tfrac{1}{2}g\cos\theta\; t^2$$

$$\therefore \quad t^2=\frac{4l}{g\cos\theta\sin^2\theta}$$

This is minimum when $-\sin^3\theta+2\sin\theta\cos^2\theta=0$

i.e., either $\theta=0$ or $-\sin^2\theta+2\cos^2\theta=0$

i.e., $\sin^2\theta=\frac{2}{3}$.

Discarding $\theta=0$, we get the chord QSP inclined at an angle θ where $\sin\theta=\sqrt{\frac{2}{3}}$ and the length of $QSP=2l/\frac{2}{3}=3l=6a$ where $4a$ is the latus rectum.

Ex. 10. *An ellipse is placed with the major axis vertical ; find diameter of the quickest descent and the limiting value of the eccentri-*

city that this may not be the major axis itself. Also find the chord of the quickest descent from the focus to the curve.

Sol. C is the centre and S the focus of the ellipse. Let PCQ be a diameter of length $2r$ inclined at an angle θ to the vertical.

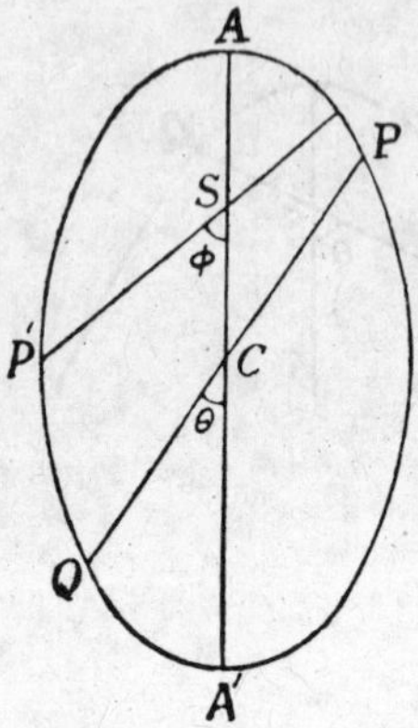

Then $r^2\left(\dfrac{\cos^2\theta}{a^2}+\dfrac{\sin^2\theta}{b^2}\right)=1$

$$\therefore \quad r^2=\frac{a^2b^2}{a^2\sin^2\theta+b^2\cos^2\theta}.$$

If t be the time down PCQ,

$$r=\tfrac{1}{2}g\cos\theta . t^2$$

$$\therefore \quad t^2=\frac{2r}{g\cos\theta}$$

$$\therefore \quad t^4=\frac{4r^2}{g^2\cos^2\theta}$$

$$=\frac{4a^2b^2}{g^2(a^2\sin^2\theta+b^2\cos^2\theta)\cos^2\theta}$$

$$=\frac{4b^2}{g^2[\sin^2\theta+(1-e^2)\cos^2\theta]\cos^2\theta}$$

$$=\frac{4b^2}{g^2(\cos^2\theta-e^2\cos^4\theta)}.$$

This is minimum when $\cos^2\theta-e^2\cos^4\theta$ is maximum.

i.e., when $(4e^2\cos^3\theta-2\cos\theta)\sin\theta=0.$

Giving $\theta=0,\ \pi/2$ or $\cos^2\theta=\dfrac{1}{2e^2}.$

Discarding $\theta=0,\ \pi/2$ we get $\cos\theta=\dfrac{1}{e\sqrt{2}}$

In order that this is possible $e\sqrt{2}>1 \quad \therefore \quad e>\dfrac{1}{\sqrt{2}}.$

In this case $r^2=\dfrac{b^2}{1-e^2\cos^2\theta}=\dfrac{b^2}{1-e^2\dfrac{1}{2e^2}}=2b^2$

$\therefore$ length of the chord $=b\sqrt{2}.$

For the focal chord SP' inclined at an angle ϕ to the vertical we have $SP'=\dfrac{l}{1-e\cos\phi}$ where $l=$ semi-latus rectum.

$\therefore$ If t be the time down SP',

we have $\dfrac{l}{1-e\cos\phi}=\tfrac{1}{2}g\cos\phi\ t^2 \quad \therefore \quad t^2=\dfrac{2l}{g\cos\phi(1-e\cos\phi)}.$

This is maximum when $-\sin\phi(1-e\cos\phi)+e\cos\phi\sin\phi=0$

i.e., either $\phi=0$ or $\cos\phi=\dfrac{1}{2e}$.

Discarding $\phi=0$, we get $\cos\phi=\dfrac{1}{2e}$; hence $2e>1$ or $e>\frac{1}{2}$ and

$$SP'=\frac{l}{1-e.\dfrac{1}{2e}}=2l=\text{latus rectum}.$$

Ex. 11. *A hyperbola is placed in a vertical plane with its transverse axis horizontal; show that the time of descent down a diameter is least when the conjugate diameter is equal to the distance between the foci.*

Sol Let PQ be the diameter inclined at $\angle\theta$ with its axis, its length be $2r$.

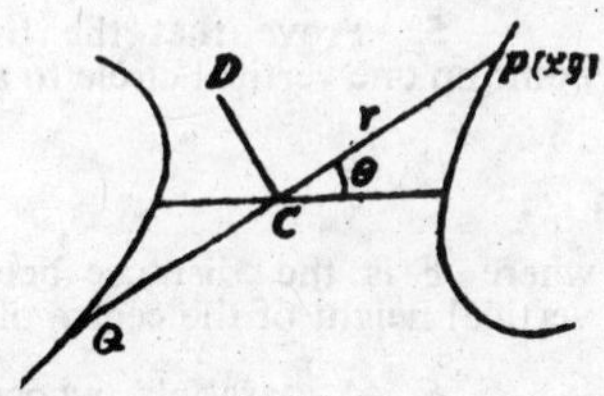

$$\therefore \quad 2r=PQ=\tfrac{1}{2}g\sin\theta t^2$$

$$\therefore \quad t^2=\frac{4r}{g\sin\theta}$$

Now hyperbola is $\dfrac{x^2}{a^2}-\dfrac{y^2}{b^2}=1$

$$\therefore \quad \frac{\cos^2\theta}{a^2}-\frac{\sin^2\theta}{b^2}=\frac{1}{r^2}$$

or
$$\frac{1-\sin^2\theta}{a^2}-\frac{\sin^2\theta}{b^2}=\frac{1}{r^2}$$

$$\therefore \quad \sin^2\theta=\frac{(r^2-a^2)\,b^2}{(a^2+b^2)\,r^2}$$

Hence
$$t^2=\frac{4r^2\sqrt{a^2+b^2}}{gb\sqrt{r^2-a^2}}$$

t is least when $\dfrac{dt}{dr}=0$

$$\therefore \quad 2r\sqrt{r^2-a^2}-\frac{r^2\,.\,r}{\sqrt{r^2-a^2}}=0$$

i.e.,
$$r^2=2a^2$$

If CD be the conjugate semi-diameter to CP,

$$CP^2-CD^2=a^2-b^2$$

$$\therefore \quad CD^2=2a^2-a^2+b^2=a^2+b^2=a^2e^2$$

$$\therefore \quad CD=ae$$

$$\therefore \quad 2CD=2ae \text{ hence.}$$

EXAMPLES VII (C)

1. AC is a given horizontal line and AB a line elevated about it at an angle α, prove that the least time in which it is possible for a body to descend from AB to C along a straight line is $2\left(\frac{AC}{g}\tan\frac{\alpha}{2}\right)^{\frac{1}{2}}$.

2. Show that the shortest time of descent down a straight smooth tube from a circle to a point in the same vertical plane is

$$2\sqrt{\frac{r\tan\alpha\cot(\beta-\alpha)}{g}}$$

where r is the radius of the circle and α, β are the angles which the lines drawn from the given point to the highest and lowest points of the circle make with the vertical.

3. Prove that the chord of quickest descent from a parabola to the focus when the axis of the parabola is horizontal is inclined at an angle of 60° to the axis and falls towards the same side of the focus as the vertex.

4. Prove that the locus of all points from which the time of descent in straight lines to two given points is the same is a rectangular hyperbola.

5. Prove that the time of quickest descent along a straight line from a point on one vertical circle to another in the same plane is

$$\sqrt{\frac{2(c^2-a^2)}{g(a+h)}}$$

where c is the distance between the centres, a is the sum of the radii and h the vertical height of the centre of the former circle above that of the latter.

6. A parabola whose latus rectum p is placed in a vertical plane with its axis horizontal. Prove that the normal terminated by the axis down which a particle would descend in the shortest time is at 45° and the time of descent is

$$\sqrt{\frac{2p}{g}}.$$

8

Central Forces

8·1. *A particle moves in a plane with an acceleration which is always directed to a fixed point O in the plane ; to discuss the motion.*

Take O as origin and a fixed line OX as initial line and let the polar co-ordinates of P, the position of the particle at time t, be (r, θ).

Let P be the acceleration of the particle towards O, *i.e.*, force per unit mass acting on the particle.

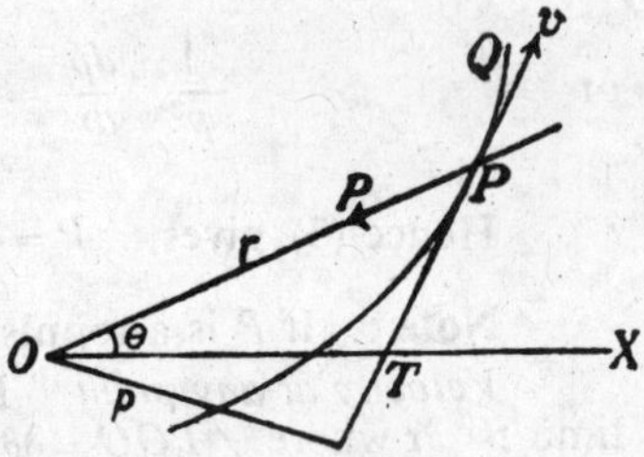

The radial equation of motion is

$$\ddot{r}-r\dot{\theta}^2=-P. \qquad \ldots(1)$$

Since there is no acceleration perp. to OP, the cross-radial equation is

$$\frac{1}{r}\frac{d}{dt}(r^2\dot{\theta})=0$$

i.e.,
$$r^2\dot{\theta}=\text{const.}=h, \text{ say} \qquad \ldots(2)$$

$$\dot{\theta}=\frac{h}{r^2}=hu^2 \text{ if } u=\frac{1}{r}$$

$$\dot{r}=\frac{d}{dt}\left(\frac{1}{u}\right)=\dot{\theta}\frac{d}{d\theta}\left(\frac{1}{u}\right)=-\dot{\theta}\cdot\frac{1}{u^2}\frac{du}{d\theta}$$

$$=-hu^2\frac{1}{u^2}\frac{du}{d\theta}=-h\frac{du}{d\theta}$$

and $\ddot{r}=\frac{d}{dt}\left(-h\frac{du}{d\theta}\right)=\dot{\theta}\frac{d}{d\theta}\left(-h\frac{du}{d\theta}\right)$

$$=-h\dot{\theta}\frac{d^2u}{d\theta^2}=-h^2u^2\frac{d^2u}{d\theta^2}.$$

Hence (1) becomes

$$-h^2u^2\frac{d^2u}{d\theta^2}-\frac{1}{u}h^2u^4=-P$$

i.e., $$P=h^2u^2\left(\frac{d^2u}{d\theta^2}+u\right). \quad ...(3)$$

Equation in pedal form. If p be the perpendicular from C upon the tangent at P,

then $$\frac{1}{p^2}=u^2+\left(\frac{du}{d\theta}\right)^2.$$

Differentiating both sides with respect to θ, we get

$$-\frac{2}{p^3}\frac{dp}{d\theta}=2u\frac{du}{d\theta}+2\frac{du}{d\theta}\frac{d^2u}{d\theta^2}$$

i.e., $$-\frac{1}{p^3}\frac{dp}{dr}\frac{dr}{d\theta}=\left(u+\frac{d^2u}{d\theta^2}\right)\frac{du}{d\theta}$$

i.e., $$-\frac{1}{p^3}\frac{dp}{dr}\times\left(-\frac{1}{u^2}\frac{du}{d\theta}\right)=\left(u+\frac{d^2u}{d\theta^2}\right)\frac{du}{d\theta}$$

or $$\frac{1}{p^3}\frac{dp}{dr}=u^2\left(u+\frac{d^2u}{d\theta^2}\right).$$

Hence (3) gives $$P=\frac{h^2}{p^3}\frac{dp}{dr}. \quad ...(4)$$

Note : If P is a repulsive force, replace P by $-P$ in (3) and (4).

Velocity at any point. If Q be the position of the particle at time $t+\delta t$ where $\angle POQ=\delta\theta$

then $$\text{area } POQ=\tfrac{1}{2}.OP.OQ\sin POQ$$
$$=\tfrac{1}{2}r.(r+\delta r)\sin\delta\theta$$

$\therefore$ $$\text{Areal velocity}=\lim_{\delta t\to 0}\frac{\text{area } POQ}{\delta t}$$

$$=\lim_{\delta t\to 0}\frac{\tfrac{1}{2}r(r+\delta r)\sin\delta\theta}{\delta t}$$
$$=\tfrac{1}{2}r^2\dot{\theta}\text{ in the limit}$$
$$=\tfrac{1}{2}h.$$

Thus areal velocity is also independent of time i.e. in every central orbit, radius vector to the centre of force traces out an area which increases uniformly per unit time.

If v be the velocity at P, it has components $\dot{x}$, $\dot{y}$ parallel to OX and perpendicularly and componets $\dot{r}$, $r\dot{\theta}$ along OP and perpendicularly.

Taking moment of the velocity about O, we get

$$vp = r^2\dot{\theta} = x\dot{y} - \dot{x}y$$

This relation is an identity and holds good for any curve.

When the curve is a central orbit, this relation becomes

$$vp = r^2\dot{\theta} = x\dot{y} - \dot{x}y = \text{const.} = h.$$

Thus
$$v^2 = \frac{h^2}{p^2} = h^2\left[u^2 + \left(\frac{du}{d\theta}\right)^2\right].$$

Again here, areal velocity $\frac{1}{2}\ \dot{\theta} r^2 = \text{const.}$

$\therefore \quad r^2\dot{\theta} = k$ (say)

Hence transverse acceleration $= \frac{1}{r}\ \frac{d}{dt}(r^2\ \dot{\theta}) = \frac{1}{r}\ \frac{d}{dt}(k) = 0$

Thus if radius vector to a point which describes an orbit sweeping out equal areas in equal intervals of time, the acceleration is directed towards the centre.

This determines velocity, if path is given.

Also $v = \frac{h}{p}$ shows that $v \propto \frac{1}{p}$

Hence velocity of a particle at any point tracing a central orbit varies inversely as the length of perpendicular from the centre to the tangent at that point.

Application. (*a*) *Given the orbit in polar co-ordinates, to find the force.*

(1) *If the central orbit is an ellipse, the focus being the centre of force, find the law of force.*

The equation to the ellipse with focus as pole is

$$\frac{l}{r} = 1 + e\cos\theta \quad \therefore \quad u = \frac{1}{l} + \frac{e}{l}\cos\theta$$

$$\therefore \qquad \frac{d^2u}{d\theta^2} = -\frac{e}{l}\cos\theta.$$

Hence
$$P = h^2u^2\left[\frac{d^2u}{d\theta^2} + u\right] = \frac{h^2}{l}u^2 = \frac{\mu}{r^2} \text{ say}$$

where
$$\mu = \frac{h^2}{l} \text{ i.e., } h = \sqrt{\mu l}.$$

Thus the central force varies inversely as the square of the distance from the focus

Also
$$v^2 = h^2\left[u^2 + \left(\frac{du}{d\theta}\right)^2\right]$$

$$=h^2\left[\left(\frac{1}{l}+\frac{e}{l}\cos\theta\right)^2+\left(\frac{e}{l}\sin\theta\right)^2\right]$$

$$=\frac{\mu}{l}\,[1+2e\cos\theta+e^2]$$

$$=\mu\left[2\,\frac{1+e\cos\theta}{l}-\frac{1-e^2}{l}\right]$$

$$=\mu\left[\frac{2}{r}-\frac{1}{a}\right]$$

for l=semi-latus rectum$=\dfrac{b^2}{a}=a(1-e^2)$

$2a$ being the major axis of the ellipse.

If T be the periodic time *i.e.*, the time the particle takes to describe the whole arc of the ellipse, we have

$\frac{1}{2}h.T=$ area of the ellipse$=\pi ab$, since $\frac{1}{2}h$ is the areal velocity *i.e.*, $\dfrac{h}{2}$ is the area described in unit time.

Also $$h^2=\mu l=\mu\,\frac{b^2}{a}$$

$$\therefore \qquad T=\frac{2\pi ab}{h}\quad\frac{2\pi ab}{\sqrt{\mu\dfrac{b^2}{a}}}=\frac{2\pi}{\sqrt{\mu}}\quad a^{\frac{3}{2}}$$

(2) *If the central orbit is $r^n=a^n\cos n\theta$ under a force towards the pole, to find the law of force.*

Here $$u^na^n\cos n\theta=1$$

Taking logarithmic differential, we get

$$\frac{du}{d\theta}=u\tan n\theta$$

$$\therefore \qquad \frac{d^2u}{d\theta^2}=\frac{du}{d\theta}\tan n\theta+nu\sec^2 n\theta=u\,(\tan^2 n\theta+n\sec^2 n\theta)$$

$$\therefore \qquad \frac{d^2u}{d\theta^2}+u=u(n+1)\sec^2 n\theta=(n+1)a^{2n}u^{2n+1}$$

Hence $$P=h^2u^2\left(u+\frac{d^2u}{d\theta^2}\right)=(n+1)h^2a^{2n}u^{2n+3}.$$

Also taking logarithmic differential of $r^n=a^n\cos n\theta$, we get

$$\frac{n}{r}\cdot\frac{dr}{d\theta}=-n\tan n\theta$$

or $\cot\phi=-\tan n\theta$ where ϕ is the angle between the tangent and radius vector

$$\therefore \qquad \phi = \frac{\pi}{2} + n\theta$$

Now $p = r \sin \phi = r \cos n\theta = \dfrac{r^{n+1}}{a^n}$.

Hence $$\frac{dp}{dr} = (n+1)\frac{r^n}{a^n}$$

$$\therefore\ P = \frac{h^2}{p^3}\frac{dp}{dr} = \frac{h^2 a^{3n}}{r^{3n+3}}(n+1)\frac{r^n}{a^n} = \frac{h^2 a^{2n}(n+1)}{r^{2n+3}}.$$

(*b*) *Given the central force as a function of r, to find the orbit.*

(1) *If the central force varies inversely as the square of the distance from a fixed point, to find the orbit.*

Here $$P = \frac{\mu}{r^2}$$

$$\therefore \qquad \frac{d^2u}{d\theta^2} + u = \frac{P}{h^2u^2} = \frac{\mu u^2}{h^2u^2} = \frac{\mu}{h^2}$$

or $$\frac{d^2u}{d\theta^2} = -\left(u - \frac{\mu}{h^2}\right)$$

solution of which is $u - \dfrac{\mu}{h^2} = A \cos \theta + B \sin \theta$

or $$u = \frac{\mu}{h^2} + A_1 \cos(\theta - \alpha)$$

or $$\frac{h^2/\mu}{r} = 1 + C\cos(\theta - \alpha), \text{ putting } \frac{A_1h^2}{\mu} = C,$$

i.e., $$\frac{l}{r} = 1 + C\cos(\theta - \alpha)$$

where $$l = \frac{h^2}{\mu}$$

or $$h^2 = \mu l$$

Thus the orbit is a conic, focus being the pole.

(2) *If the central force varies as the distance from a fixed point, to find the orbit.*

Here $$P = \mu r$$

$$\therefore \qquad \frac{d^2u}{d\theta^2} + u = \frac{P}{h^2u^2} = \frac{\mu}{h^2u^3}$$

Multiplying by $2\dfrac{du}{d\theta}$, we get,

$$2\frac{d^2u}{d\theta^2}\frac{du}{d\theta} + 2u\frac{du}{d\theta} = \frac{2\mu}{h^2u^3}\frac{du}{d\theta}$$

Integrating, we get

$$\left(\frac{du}{d\theta}\right)^2+u^2=-\frac{\mu}{h^2u^2}+C \text{ where } C \text{ is a const.}$$

or

$$\int\frac{udu}{\sqrt{\left(Cu^2-\frac{\mu}{h^2}-u^4\right)}}=\theta+\text{const.}$$

Putting $u^2=z$, we get

$$\int\frac{dz}{\sqrt{\left(Cz-\frac{\mu}{h^2}-z^2\right)}}=2\theta+\alpha \text{ where } \alpha \text{ is a constant}$$

or

$$2\theta+\alpha=\int\frac{dz}{\sqrt{\{(a-z)(z-b)\}}}$$

where

$$a+b=C,\ ab=\frac{\mu}{h^2}$$

Putting $z=a\sin^2\lambda+b\cos^2\lambda$, we get

$$2\theta+\alpha=\int\frac{2(a-b)\sin\lambda\cos\lambda\, d\lambda}{\sqrt{\{(a-b)\cos^2\lambda.(a-b)\sin^2\lambda}}=2\lambda$$

Choosing $\alpha=0$, we get $\theta=\lambda$

Hence $u^2=z=a\sin^2\theta+b\cos^2\theta$

or

$$\frac{1}{r^2}=a\sin^2\theta+b\cos^2\theta$$

or

$$ay^2+bx^2=1$$

Thus the orbit is a conic, centre being the pole.

(3) *If the central force varies inversely as the cube of the distance from a fixed point, to find the orbit.*

Here $P=\frac{\mu}{r^3}$

$$\therefore \quad \frac{d^2u}{d\theta^2}+u=\frac{\mu}{h^2}u$$

or

$$\frac{d^2u}{d\theta^2}=-\left(1-\frac{\mu}{h^2}\right)u$$

If $\frac{\mu}{h^2}>1$, then $\frac{d^2u}{d\theta^2}=\left(\frac{\mu}{h^2}-1\right)u=n^2u$ say

where

$$n^2=\frac{\mu}{h^2}-1$$

Its solution is $u=A\cosh n\theta+B\sinh n\theta$

If $\dfrac{\mu}{h^2}=1$, then $\dfrac{d^2u}{d\theta^2}=0$ so that $u=A\theta+B$.

If $\dfrac{\mu}{h^2}<1$, then $\dfrac{d^2u}{d\theta^2}=-n^2u$ so that $u=A\cos n\theta+B\sin n\theta$.

8·2. **(p, r) equations of some curves.**

(1) *Circle...pole at any point.*

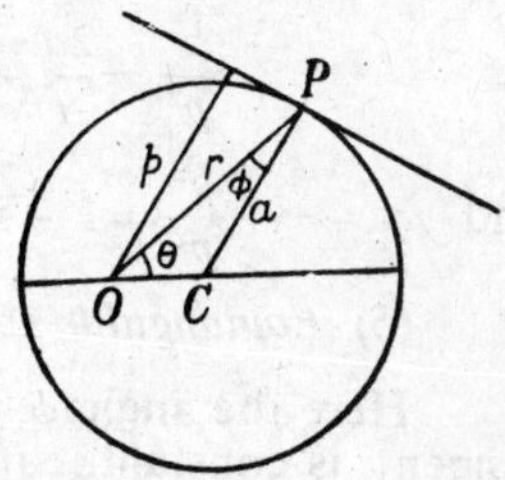

C is centre of the circle, O the pole.

$OC=c$.

$$p=r\sin(90-\phi)=r\cos\phi$$

$$=r.\frac{a^2+r^2-c^2}{2ar}$$

i.e., $$r^2+a^2-c^2=2ap.$$

If pole be on the circumference, $c=a$, then $r^2=2ap$.

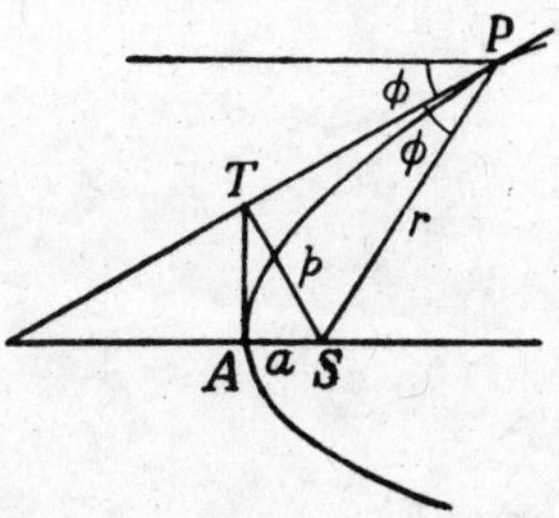

(2) *Parabola...pole at focus.*

If ST is perpendicular from focus S to the tangent at P, then T will lie on AT, the tangent at the vertex A.

Then $\angle SPT=\phi=\angle ATS$

$$\therefore \quad \frac{p}{r}=\sin\phi=\frac{a}{p}$$

or $$p^2=ar.$$

(3) *Ellipse* or *Hyperbola...pole at centre.*

If CD is semi-conjugate to CP,

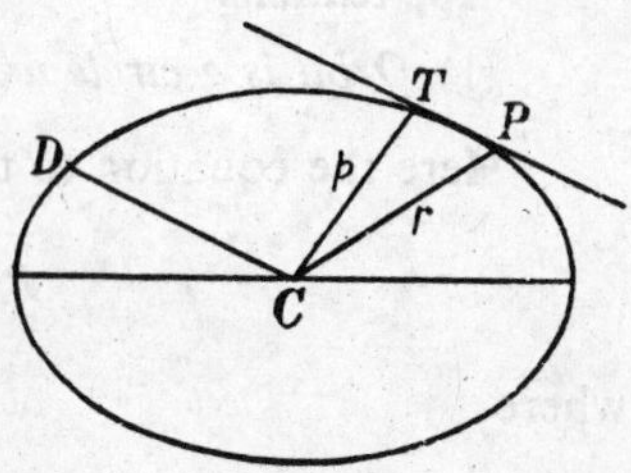

then $$CP^2+CD^2=a^2+b^2$$

and $$CD.CT=ab$$

i.e., $$CD^2=a^2+b^2-r^2$$

and $$CD=\frac{ab}{p}$$

$$\therefore \quad \frac{a^2b^2}{p^2}=a^2+b^2-r^2.$$

This is for ellipse.

For hyperbola, change the sign of b^2.

(4) *Ellipse or Hyperbola...pole at focus.*

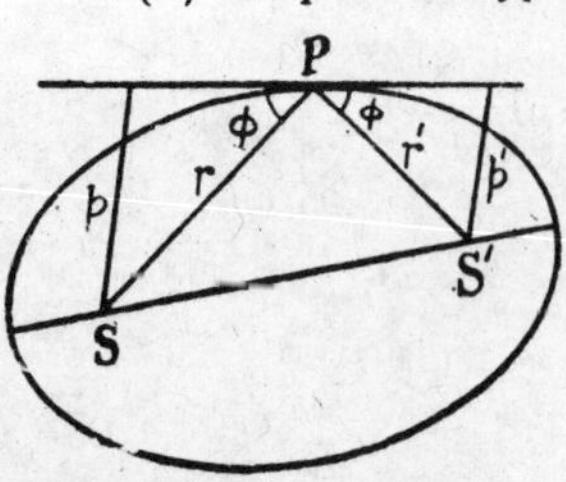

$$\sin\phi=\frac{p}{r}=\frac{p'}{r'}$$

or $$\frac{p'}{p}=\frac{r'}{r}$$

$$\therefore \quad \frac{pp'}{p^2}=\frac{r'}{r}=\frac{2a-r}{r}$$

because $$r+r'=2a$$

Aiso $$pp'=b^2$$

$$\therefore \quad \frac{b^2}{p^2}=\frac{2a}{r}-1$$

This is for ellipse.

For hyperbola, the equation is

$$\frac{b^2}{p^2}=\frac{2a}{r}+1 \text{ for the nearer branch}$$

and $$\frac{b^2}{p^2}=1-\frac{2a}{r} \text{ for the farther branch.}$$

(5) *Equiangular spiral.*

Here the angle ϕ, the angle between the radius vector and the tangent, is constant, equal to α say

$$\therefore \quad p=r\sin\alpha.$$

(6) $r^n=a^n \cos n\theta$.

Taking log. differential, we get

$$\frac{1}{r}\frac{dr}{d\theta}=-\tan n\theta$$

i.e., $$\cot\phi=-\tan n\theta$$

$$\therefore \quad \phi=90^\circ+\theta$$

Hence $$p=r\sin\phi=r\cos n\theta=\frac{r^{n+1}}{a^n}$$

Applications

(1) *Orbit is a circle under a force to a point on the circumference.*

Here the equation to the circle is $r^2=2ap$

$$\therefore \quad P=\frac{h^2}{p^3}\frac{dp}{dr}=\frac{h^2}{p^3}\cdot\frac{r}{a}=\frac{8a^2h^2}{r^5}=\frac{\mu}{r^5}$$

where $$\mu=8a^2h^2$$

Also $$v^2=\frac{h^2}{p^2}=\frac{4a^2h^2}{r^4}=\frac{\mu}{2r^4}$$

(2) *Orbit is an ellipse under a force to the centre.*

Ellipse is $$\frac{a^2b^2}{p^2}=a^2+b^2-r^2$$

$$P=\frac{h^2}{p^3}\frac{dp}{dr}=\frac{h^2r}{a^2b^2}=\mu r$$

where $$\mu=\frac{h^2}{a^2b^2}$$

Also $$v^2=\frac{h^2}{p^2}=\frac{b^2(a^2+b^2-r^2)}{a^2b^2}$$

$$=\mu(a^2+b^2-r^2)=\mu.CD^2$$

where CD is the semi-conjugate diameter to CP.

(3) *Orbit $r^n=a^n\cos n\theta$ under a force to the pole.*

Here $$p=\frac{r^{n+1}}{a^n}$$

$$\therefore \quad P=\frac{h^2}{p^3}\frac{dp}{dr}=\frac{(n+1)h^2a^{2n}}{r^{2n+3}}$$

(4) *Orbit is an ellipse under a force to the focus.*

Here ellipse is $$\frac{b^2}{p^2}=\frac{2a}{r}-1$$

$$\therefore \quad P=\frac{h^2}{p^3}\frac{dp}{dr}=\frac{ah^2}{b^2r^2}=\frac{\mu}{r^2}$$

where $h^2=\mu\frac{b^2}{a}=\mu l$, l being semi-latus rectum

Also $$v^2=\frac{h^2}{p^2}=\frac{\mu b^2}{ap^2}=\frac{\mu}{a}\left(\frac{2a}{r}-1\right)=\mu\left(\frac{2}{r}-\frac{1}{a}\right)$$

(5) *Orbit is a parabola under a force to the focus.*

Here parabola is $p^2=ar$

$$\therefore \quad P=\frac{h^2}{p^3}\frac{dp}{dr}=\frac{ah^2}{2p^4}=\frac{ah^2}{2a^2r^2}=\frac{h^2}{2ar^2}=\frac{\mu}{r^2}$$

where $h^2=2a\mu$

Also $$v^2=\frac{h^2}{p^2}=\frac{2a\mu}{ar}=\frac{2\mu}{r}.$$

EXAMPLES VIII (A)

1. *Prove the following cases of central orbits :*

(*i*) If orbit is Equiangular Spiral $r=ac^{\alpha\cot\theta}$ or $p=r\sin\alpha$,

then prove that $P=\dfrac{\mu}{r^3}$

(*ii*) If orbit is Bernoulli's Lemniscate $r^2=a^2\cos 2\theta$, then prove

that $p=\dfrac{\mu}{r^7}$

(*iii*) If orbit is $\dfrac{a}{r}=e^{n\theta}$, $n\theta$, cosh $n\theta$ or sinh $n\theta$, then prove that

$$P=\frac{\mu}{r^3}$$

(*iv*) If orbit is $r^n \cos n\theta = a^n$, then prove that $P = \mu r^{2n-3}$

(*v*) If orbit is $r = a \sin n\theta$, then prove that

$$P = h^2 \left(\frac{2a^2 n^2}{r^5} - \frac{n^2 - 1}{r^3} \right).$$

(*vi*) If orbit is a circle, pole within the circle, then prove that

$$P = \frac{\mu r}{p^3}$$

(*vii*) If orbit is a cardiode $r = a(1 + \cos \theta)$, then prove that

$$P = \frac{\mu}{r^4}$$

(*vii*) If the velocity at any point varies inversely as the distance of the point from the centre of force the orbit is an equiangular spiral.

2. A particle describes the equiangular spiral $r = ar^{\phi \cot \alpha}$ under a force F to the pole. Prove that $F \propto r^{-3}$.

8.3. Apses. An apse is a point on the path, the velocity at which is perpendicular to the radius vector.

Here $p = r$ $\therefore$ from $\dfrac{1}{p^2} = u^2 + \left(\dfrac{du}{d\theta}\right)^2$

we get
$$\frac{du}{d\theta} = 0$$

hence also
$$\frac{dr}{d\theta} = 0$$

Therefore at an apse, the radius vector is either maximum or minimum.

The distance of an apse from the centre of the force is called an *aspidal distance* and the angle between two aspidal distance is called an *apsidal angle.*

If the central force is a single valued function of the distance i.e., it depends only on the distance, $P = f(r)$ say

Now $\dfrac{d^2u}{d\theta^2} + u = \dfrac{P}{h^2 u^2}$.

Integrating with respect to u, we get

$$\left(\frac{du}{d\theta}\right)^2 + u^2 = \frac{2}{h^2} \int \frac{P}{u^2} du = \frac{2}{h^2} \int f(r)\, r^2 d\left(\frac{1}{r}\right) = -\frac{2}{h^2} \int f(r)\, dr$$

$$\therefore \quad \frac{1}{p^2} = u^2 + \left(\frac{du}{d\theta}\right)^2 = -\frac{2}{h^2} \int f(r)\, dr.$$

Hence p and $\therefore$ ϕ are functions of r only.

Thus if a point describes a central orbit, commencing from an apse, it will describe symmetrical orbits on either side of the apsidal distance. Therefore, the path is divided symmetrical by the apsidal distance. When it arrives at a second apse, the path on either side of this second apse is again symmetrical and, therefore, the next apsidal distance will be equal to the one before this second apse and the angle between the second and the third aspidal distances will be the same as the angle between the first and second.

Thus if a central force be a single-valued function of the distance there will be only two apsidal distances and one apsidal angle.

For example let $P=\mu u^n$, when n is an integer.

$$\frac{d^2u}{d\theta^2}+u=\frac{P}{h^2u^2}=\frac{\mu}{h^2}u^{n-2}.$$

Integrating we get $\dfrac{h^2}{2}\left[\left(\dfrac{du}{d\theta}\right)^2+u^2\right]=\dfrac{\mu}{n-1}u^{n-1}+\text{constant}.$

At an apse $\dfrac{du}{d\theta}=0 \quad \therefore \quad u^{n-1}-\dfrac{n-1}{2}\dfrac{h^2}{\mu}u^2+C=0.$

Whatever n or C may be, the equation cannot have more than two positive roots by Descarte's Rule. Hence there are at most two apsidal distances.

Velocity in a Circle. If v be the velocity in a circle at a distance r from the centre of force. then $P=\dfrac{v^2}{r}$ or $v^2=Pr$.

Velocity from Infinity. If a particle falls from infinity under an attraction P, to a point at distance r from the centre of force,

then $$\tfrac{1}{2}v^2=-\int_{\infty}^{r}P\,dr.$$

Velocity of fall to the point of projection. If a particle falls from the centre of repulsion under a force P to a point at a distance r,

then $$\tfrac{1}{2}v^2=\int_0^r P\,dr$$

8·4. Time in an orbit. The time of passing from one point A to another point P of a central orbit is given from the equation

$$h=r^2\dot{\theta}=x\dot{y}-\dot{x}y$$

where x, y are the co-ordinates of P with centre of force as origin.

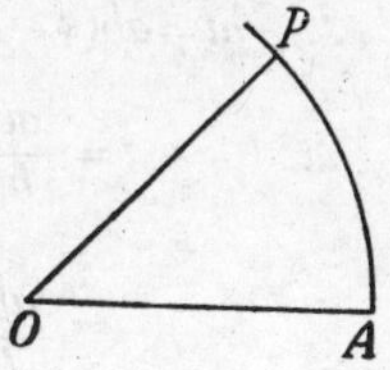

Also time for any arc AP is given by

$$t=\frac{\text{area } AOP}{\tfrac{1}{2}h}$$

Applications.

(1) *Elliptic orbit with centre of force at the centre.*

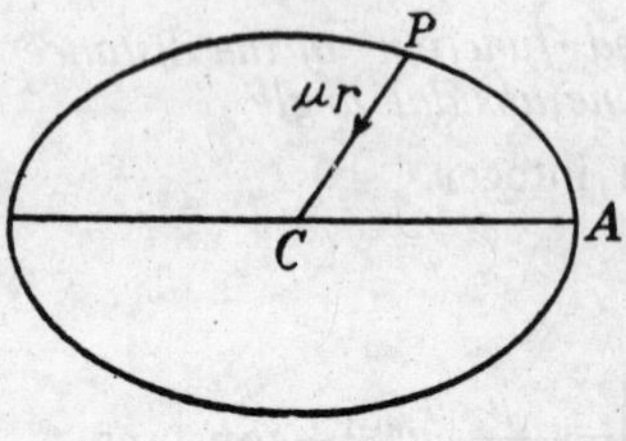

As in Application (2) of art. **8·2** the central force is $P=\mu r$

where $$\mu=\frac{h^2}{a^2b^2}$$

Here $x=a\cos\phi$, $y=b\sin\varphi$, φ being the accentric angle

$$\therefore \quad h=x\dot{y}-\dot{x}y=a\cos\phi\,.\,b\cos\varphi\dot{\phi}+a\sin\varphi\dot{\phi}\,b\sin\varphi=ab\dot{\phi}$$

hence $\qquad ht=ab\phi$ if the particle starts from $A(\phi=0)$.

$$\therefore \qquad t=\frac{ab}{h}\varphi=\frac{\varphi}{\sqrt{\mu}}$$

If T be the period,

then $$T=\frac{2\pi}{\sqrt{\mu}}$$

(2) *Elliptic orbit with centre of force at focus*

As in Application (4) of art. **8·2**

the central force is $P=\dfrac{\mu}{r^2}$

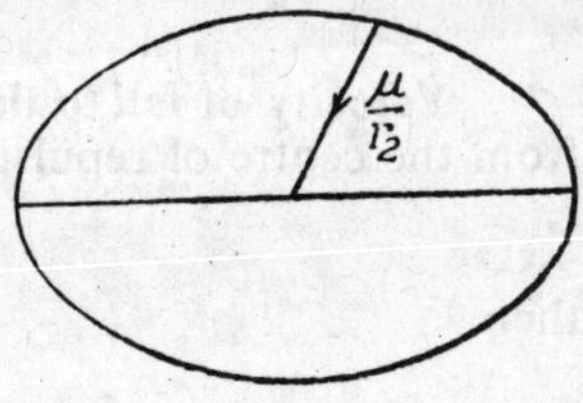

where $$\mu=\frac{h^2}{l}$$

Here $x=a\cos\phi-ae$, $y=b\sin\varphi$, φ being the eccentric angle

$$\therefore \qquad h=x\dot{y}-\dot{x}y=ab\dot{\phi}-abe\cos\varphi\dot{\phi}$$

$$\therefore \quad ht=ab(\varphi-e\sin\varphi) \quad \text{if } t=0 \quad \text{when } \varphi=0$$

$$\therefore \qquad t=\frac{ab}{h}(\varphi-e\sin\varphi)=\frac{ab}{\sqrt{\mu l}}(\varphi-e\sin\varphi)$$

$$=\frac{ab\,a^{\frac{1}{2}}}{\sqrt{\mu b^2}}(\varphi-e\sin\varphi)$$

$$=\sqrt{\frac{a^3}{\mu}}\,(\varphi-e\sin\varphi)$$

If T be the period

then $$T=\sqrt{\frac{a^3}{\mu}}.2\pi$$

(3) *Parabolic orbit with centre of force at the focus.*

As in Application (5) of art. **8·2**

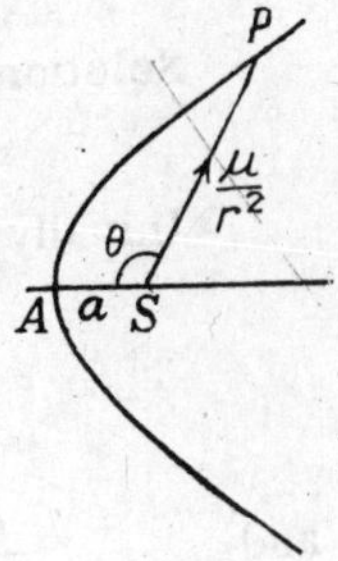

the central force is $P=\frac{\mu}{r^2}$

where $$\mu=\frac{h^2}{2a}$$

Here $x=am^2-a,\ y=2am$

$$\therefore\quad h=x\dot{y}-\dot{x}y=(am^2-a)2am$$

$$-2am\dot{m}\,2am$$

$$=-2a^2(1+m^2)m$$

$$\therefore\quad ht=-2a^2\left(m+\frac{m^3}{3}\right)_m^0=2a^2\left(m+\frac{m^3}{3}\right)$$

if it starts from a point $P(m)$.

$$\therefore\quad t=\frac{2a^2}{h}\left(m+\frac{m^3}{3}\right)=\sqrt{\frac{2a^3}{\mu}}\left(m+\frac{m^3}{3}\right)$$

This gives the time from P to A.

Ex. 1. *A particle moves with a central acceleration which varies inversely as the cube of the distance ; if it be projected from an apse at a distance a from the origin with a velocity which is $\sqrt{2}$ times the velocity for a circle of radius a, show that the equation to the path is*

$$r\cos\frac{\theta}{\sqrt{2}}=a.$$

Sol. The equation of motion is

$$\frac{d^2u}{d\theta^2}+u=\frac{\mu u^3}{h^2u^2}=\frac{\mu}{h^2}u$$

If V_1 be the velocity in a circle of radius a

then $$\frac{V_1^2}{a}=\text{force in the circle}=\frac{\mu}{a^3}$$

$$\therefore\quad V_1^2=\frac{\mu}{a^2}$$

Hence $\quad h^2=a^2\,.\,2V_1^2=2\mu$

∴ The equation of motion becomes

$$\frac{d^2u}{dv^2}+u=\tfrac{1}{2}u$$

or

$$\frac{d^2u}{d\theta^2}=-\tfrac{1}{2}u$$

Solution is $u=A\cos\dfrac{\theta}{\sqrt{2}}+B\sin\dfrac{\theta}{\sqrt{2}}$

Initially when $u=\dfrac{1}{a}$ at the apse, $\theta=0$, $\dfrac{du}{d\theta}=0$

$$\therefore \qquad \frac{1}{a}=A$$

and

$$\frac{du}{d\theta}=-\frac{A}{\sqrt{2}}\sin\frac{\theta}{\sqrt{2}}+\frac{B}{\sqrt{2}}\cos\frac{\theta}{\sqrt{2}}$$

$$\therefore \qquad 0=0+\frac{B}{\sqrt{2}} \quad \therefore \quad B=0$$

Hence

$$u=\frac{1}{a}\cos\frac{\theta}{\sqrt{2}}$$

or

$$r\cos\frac{\theta}{\sqrt{2}}=a$$

Ex. 2. *A particle moving under a constant force from the centre is projected in a direction perpendicular to the radius vector with the velocity acquired in falling to the point of projection from the centre. Show that its path is* $\left(\dfrac{a}{r}\right)^3=\cos^2\dfrac{3}{2}\theta$.

Sol. $\dfrac{d^2u}{d\theta^2}+u=-\dfrac{P}{h^2u^2}$ *where P is constant.*

If v_0 be the velocity of projection at a distance a from the centre

$$\tfrac{1}{2}v_0^2=\int_0^a P\,dr=Pa$$

$$\therefore \qquad h^2=v_0^2a^2=2Pa^3$$

Hence

$$\frac{d^2u}{d\theta^2}+u=-\frac{1}{2a^3u^2}$$

Integrating $\left(\dfrac{du}{d\theta}\right)^2+u^2=\dfrac{1}{a^3u}+C.$

Initially when $u=\frac{1}{a}$, $\frac{du}{d\theta}=0$, at the apse $\therefore$ $C=0$

Hence $\left(\frac{du}{d\theta}\right)^2+u^2=\frac{1}{a^3 u}$

or $$\left(\frac{du}{d\theta}\right)^2=\frac{1-a^3 u^3}{a^3 u}$$

or $$\sqrt{\frac{a^3 u}{1-a^3 u^3}}\, du=\pm d\theta \text{ [Put } a^3u^3=\cos^2 \phi$$

or $au=\cos^{2/3} \phi$

$\therefore$ $adu=-\frac{2}{3} \cos^{-1/3} \phi \sin \phi \, d\phi]$

$$\sqrt{\frac{a^2 \cos^{2/3} \phi}{\sin^2 \phi}} \cdot \frac{2.}{3a} \frac{\sin \phi}{\cos^{1/3} \phi}\, d\phi=\pm d\theta$$

or $$d\phi=\pm \tfrac{3}{2}\, d\theta$$

$\therefore$ $$\phi=\pm \tfrac{3}{2}\, \theta.$$

Hence $a^3u^3=\cos^2 \frac{3}{2} \theta$ i.e., $\left(\frac{a}{r}\right)^3=\cos^2 \frac{3}{2}\theta.$

Ex. 3. *A particle is projected at right angles at a distance a under a central force $\frac{\mu}{r^n}$ with the velocity from infinity ; show that the orbit is* $r^{\frac{n-3}{2}}=a^{\frac{n-3}{2}} \cos \frac{n-3}{2} \theta.$

Sol. $$\frac{d^2u}{d\theta^2}+u=\frac{\mu u^n}{h^2u^2}=\frac{\mu}{h^2} u^{n-2}$$

Integrating $\left(\frac{du}{d\theta}\right)^2+u^2=\frac{2\mu}{(n-1)\, h^2} u^{n-1}+C$

If v_0 be the velocity of projection $\frac{1}{2} v_0^2=-\int_\infty^a \frac{\mu}{r^n}\, dr$

$$=\frac{\mu}{(n-1)\, a^{n-1}}$$

Hence $h^2=a^2 v_0^2=\frac{2\mu a^2}{(n-1)\, a^{n-1}}=\frac{2\mu}{(n-1)\, a^{n-3}}$

$\therefore$ $$\left(\frac{du}{d\theta}\right)^2+u^2=a^{n-3} u^{n-1}+C$$

Initially when $\quad u=\frac{1}{a},\ \frac{du}{d\theta}=0 \quad \therefore \quad C=0.$

Hence $\quad \left(\frac{du}{d\theta}\right)^2=u^2\,(a^{n-3}\,u^{n-3}-1)$

$$\therefore \quad \frac{du}{u\sqrt{a^{n-3}\,u^{n-3}-1}}=\pm d\theta. \quad \text{Put } au=\sec^{\frac{2}{n-3}}\phi$$

$$\therefore \quad \frac{\frac{2}{n-3}\cdot\frac{1}{a}\sec^{\frac{2}{n-3}}\phi\tan\phi\,d\phi}{\frac{1}{a}\sec^{\frac{2}{n-3}}\phi\tan\phi}=\pm d\theta$$

or $$d\phi=\pm\frac{n-3}{2}\,d\theta$$

$$\therefore \quad \phi=\pm\frac{n-3}{2}\,\theta$$

Hence $$au=\sec^{\frac{2}{n-3}}\frac{n-3}{2}\,\theta$$

or $$\left(\frac{a}{r}\right)^{\frac{n-3}{2}}=\sec\frac{n-3}{2}\theta \text{ or } r^{\frac{n-3}{2}}=a^{\frac{n-3}{2}}\cos\frac{n-3}{2}\,\theta.$$

Ex. 4. *A particle is acted on by a central repulsive force which varies as the nth power of the distance ; if the velocity at any point of the path be equal to that which would be acquired in falling from the centre to the point, show that the path is*

$$r^{(n+3)/2}\cos\frac{n+3}{2}\theta=\text{const.}$$

Sol. Here $\quad \frac{1}{2}v^2=\mu\int_0^r r^n\,dr=\frac{\mu r^{n+1}}{n+1}$

Equation of motion is

$$\frac{d^2u}{d\theta^2}+u=-\frac{\mu}{h^2u^{n+2}}$$

Integrating we have

$$\left(\frac{du}{d\theta}\right)^2+u^2=\frac{2\mu}{(n+1)h^2\,u^{n+1}}+C$$

But $$\frac{v^3}{h^2}=\frac{1}{p^2}=\left(\frac{du}{d\theta}\right)^2+u^2$$

$$\therefore \quad \frac{2\mu}{h^2(n+1)u^{n+1}}=\frac{2\mu}{(n+1)h^2u^{n+1}}+C$$

$$\therefore \quad C=0$$

Hence $$\left(\frac{du}{d\theta}\right)^2=\frac{b^2}{u^{n+1}}-u^2$$

where $$b^2=\frac{2\mu}{(n+1)h^2}$$

i.e., $$d\theta=-\frac{u^{(n+1)/2}}{\sqrt{b^2-u^{n+3}}}\,du$$

Integrating $$\theta=\frac{2}{n+3}\cos^{-1}\frac{u^{(n+3)/2}}{b}+A \text{ hence.}$$

Ex. 5. *A particle, moving under a central acceleration* $\frac{\mu}{r^3}$ *is projected from an apse at a distance a with a velocity V ; show that the path is*

$$r\cosh\frac{\sqrt{\mu-a^2V^2}}{aV}\,\theta=a$$

or $$r\cos\frac{\sqrt{a^2V^2-\mu}}{aV}\,\theta=a$$

according as V is $<$ *or* $>$ *the velocity from infinity.*

Sol. Equation of motion is

$$\frac{d^2u}{d\theta^2}+u=\frac{\mu u}{h^2}; \text{ here } h^2=a^2V^2$$

$$\therefore \quad \frac{d^2u}{d\theta^2}+u=\frac{\mu}{a^2V^2}u$$

or $$\frac{d^2u}{d\theta^2}=\left(\frac{\mu}{a^2V^2}-1\right)u$$

If $\frac{\mu}{a^2V^2}>1$, then solution is

$$u=A\cosh\frac{\sqrt{\mu-a^2V^2}}{aV}\,\theta+B\sinh\frac{\sqrt{\mu-a^2V^2}}{aV}\,\theta$$

At the apse, $u=\frac{1}{a}$, $\theta=0$, $\frac{du}{d\theta}=0$

$$\therefore \qquad A=\frac{1}{a}, \quad B=0$$

Hence $$u=\frac{1}{a}\cosh\frac{\sqrt{\mu-a^2V^2}}{aV}\,\theta.$$

If $$\frac{\mu}{a^2V^2}<1$$

then the equation becomes

$$\frac{d^2u}{d\theta^2}=-\left(1-\frac{\mu}{a^2V^2}\right)\mu$$

Solution is $$u=A\cos\frac{\sqrt{a^2V^2-\mu}}{aV}\,\theta+B\sin\frac{\sqrt{a^2V^2-\mu}}{aV}\,\theta$$

Again when $$u=\frac{1}{a},\ \theta=0,\ \frac{du}{d\theta}=0$$

$$\therefore \qquad A=\frac{1}{a},\ B=0$$

Hence $$u=\frac{1}{a}\cos\frac{\sqrt{a^2V^2-\mu}}{aV}\,\theta$$

If V_0 be the velocity from infinity

$$\tfrac{1}{2}V_0^2=-\int_\infty^a\frac{\mu}{r^3}\,dr$$

$$=\frac{\mu}{2a^2}$$

$$\therefore \qquad V_0^2=\frac{\mu}{a^2}$$

$\therefore$ Conditions are $\dfrac{V_0^2}{V^2}\gtrless 1$

i.e., $$V\lesseqgtr V_0$$

Ex. 6. *A particle moves under a central repulsive force* $=\left(\dfrac{m\mu}{(distance)^3}\right)$ *and is projected from an apse at a distance a with velocity V. Show that the equation to path is* $r\cos p\theta=a$ *and the angle* θ *described in time t is* $\dfrac{1}{p}\tan^{-1}\left(\dfrac{pV}{a}t\right)$ *where* $p^2=\dfrac{a^2V^2+\mu}{a^2V^2}$.

Sol. $$\frac{d^2u}{d\theta^2}+u=-\frac{\mu u^3}{h^2u^2}=-\frac{\mu}{h^2}u$$

or $$\frac{d^2u}{d\theta^2}=-\left(1+\frac{\mu}{h^2}\right)u$$

Now $$h=vp=Va$$

$$\therefore \quad \frac{d^2u}{d\theta^2}=-\left(1+\frac{\mu}{a^2V^2}\right)u=-p^2\,u.$$

Solution is $$u=A\cos p\theta+B\sin p\theta$$

Initially $u=\dfrac{1}{a}$ where $\theta=0$ and $\dfrac{du}{d\theta}=0$ where $\theta=0$,

since the particle was projected from an apse.

$$\therefore \quad \frac{1}{a}=A$$

$$0=\frac{du}{d\theta}=\left(-Ap\sin p\theta+Bp\cos p\theta\right)_{\theta=0}=Bp$$

$$\therefore \quad B=0.$$

Hence $u=\dfrac{1}{a}\cos p\theta$ or $r\cos p\theta=a$.

To find t, $$r^2\dot\theta=h=aV$$

or $$\frac{d\theta}{dt}=\frac{aV}{a^2\sec^2 p\theta}=\frac{V}{a}\cos^2 p\theta$$

or $$\frac{V}{a}t=\int_{\theta=0}^{\theta}\sec^2 p\theta\,d\theta=\frac{1}{p}\tan p\theta.$$

$$\therefore \quad \tan p\theta=\frac{pV}{a}t \quad \therefore \quad \theta=\frac{1}{p}\tan^{-1}\left(\frac{pV}{a}t\right).$$

Ex. 7. *The eccentricity of the earth's orbit round the sun is $\frac{1}{60}$; show that the earth's distance from the sun exceeds the length of the semi-major axis of the orbit about two days more than half the year.*

Sol. AA', BB' are the major and minor axes, $SB=a$

$\therefore$ earth's distance is greater than a when E lies in the arc $BA'B'$

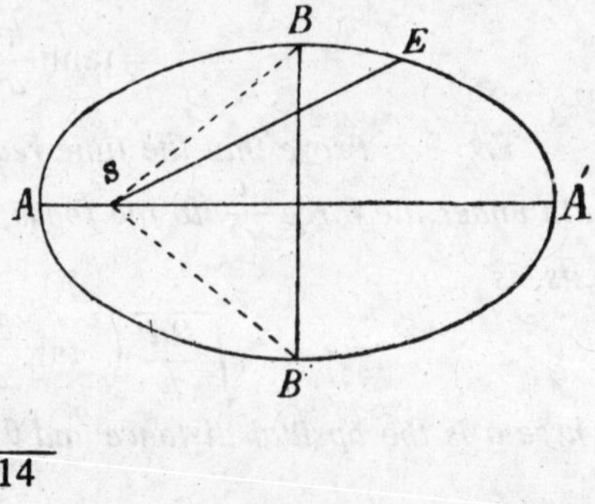

$$\therefore \quad \frac{\text{time of describing arc } B'AB}{\text{periodic time}}$$

$$=\frac{\text{area } SB'\,A'BS}{\text{whole area}}$$

$$=\frac{\frac{1}{2}\pi\,ab+\frac{1}{2}\,2b.ae}{\pi ab}=\frac{1}{2}+\frac{e}{\pi}$$

$$=\frac{1}{2}+\frac{1}{60\times 3{\cdot}14}$$

$\therefore$ Required time $=\left(\dfrac{1}{2}+\dfrac{1}{60\times3{\cdot}14}\right)$ year

$=\frac{1}{2}$ year $+2$ days approx.

Ex. 8. *Show that a particle can describe a rectangular hyperbola under a force from a fixed centre varying as the distance and show that the time the radius vector to the particle from the centre takes in sweeping out an angle θ from the vertex is given by $\tan\theta=\tanh(t\sqrt{\mu})$, μ being acceleration at unit distance.*

Sol. Rectangular hyperbola is $x^2-y^2=a^2$

or $$r^2\cos 2\theta=a^2$$

Taking log differential, $\dfrac{2}{r}\dfrac{dr}{d\theta}=2\tan 2\theta$

$$\therefore \quad \cot\phi=\tan 2\theta$$

$\therefore \quad \varphi=\dfrac{\pi}{2}-2\theta$ hence $p=r\sin\varphi=r\cos 2\theta=\dfrac{a^2}{r}$

$$\therefore \quad P=-\frac{h^2}{p^3}\frac{dp}{dr}=\frac{h^2}{p^3}\frac{a^2}{r^2}=\frac{h^2 r}{a^4}=\mu r$$

where $$h^2=\mu a^4$$

$$r^2\dot{\theta}=h \text{ gives } 2\frac{h}{a^2}t=2\int\frac{d\theta}{\cos 2\theta}$$

$$=\log\tan\left(\frac{\pi}{4}+\theta\right)+A$$

Since $t=0$ when $\theta=0$ $\quad\therefore\quad A=0$

$$\therefore \quad e^{2ht/a^2}=\tan\left(\frac{\pi}{4}+\theta\right)=\frac{1+\tan\theta}{1-\tan\theta}$$

or $$\tan\theta=\frac{e^{2ht/a^2}-1}{e^{2ht/a^2}+1}=\frac{e^{ht/a^2}-e^{-ht/a^2}}{e^{ht/a^2}+e^{-ht/a^2}}$$

$$=\tanh\frac{ht}{a^2}=\tanh(t\sqrt{\mu}).$$

Ex. 9. *Prove that the time required to describe an arc of a parabola under the force $\dfrac{\mu}{r^2}$ to the focus, starting from the nearer end of the axis, is*

$$\sqrt{\frac{2a^3}{\mu}}\left(\tan\frac{\theta}{2}+\frac{1}{3}\tan^3\frac{\theta}{2}\right)$$

where a is the apsidal distance and θ is measured from the axis.

Sol Parabola is $\frac{l}{r}=1+\cos\theta$

or $$\frac{2a}{r}=2\cos^2\frac{\theta}{2}$$

or $$r=a\sec^2\frac{\theta}{2}$$

$$r^2\dot{\theta}=h$$

$\therefore$ $$ht=\int_0^\theta r^2\,d\theta=a^2\int_0^\theta \sec^4\frac{\theta}{2}\,d\theta$$

$$=2a^2\int_0^\theta\left(1+\tan^2\frac{\theta}{2}\right)d\left(\tan\frac{\theta}{2}\right)$$

i.e., $$ht=2a^2\left[\tan\frac{\theta}{2}+\tfrac{1}{3}\tan^3\frac{\theta}{2}\right]$$

But $$h^2=2a\mu$$

Hence $$t=\sqrt{\frac{2a^3}{\mu}}\left[\tan\frac{\theta}{2}+\tfrac{1}{3}\tan^3\frac{\theta}{2}\right].$$

Ex. 10. *If the period of a planet be 365 days and $e=1/60$, show that the times of describing the two portions of the orbit bounded by the latus rectum passing through the centre of force are*

$$\tfrac{1}{2}(365)\left[1+\frac{1}{15\pi}\right] \text{ days very nearly.}$$

Sol. Ellipse is $\frac{l}{r}=1+e\cos\theta$

From $$r^2\dot{\theta}=h$$

we get $$ht=\int_0^\theta\frac{l^2}{(1+e\cos\theta)^2}\,d\theta$$

For the small arc, time required is

$$2t=2\frac{l^2}{h}\int_0^{\pi/2}\frac{d\theta}{(1+e\cos\theta)^2}$$

$$=\frac{2l^{3/2}}{\sqrt{\mu}}\int_0^{\pi/2}(1-2e\cos\theta+\ldots)\,d\theta,\ e<<1$$

$$=\frac{2a^{3/2}(1-e^2)^{3/2}}{\sqrt{\mu}}\left[\left(\frac{\pi}{2}-2e\sin\theta\right)_0^{\pi/2}\right]$$

$$=\frac{2a^{3/2}(1-e^2)^{3/2}}{\sqrt{\mu}}\left(\frac{\pi}{2}-2e\right)$$

$$=\frac{2a^{3/2}}{\sqrt{\mu}}\left(\frac{\pi}{2}-2e\right), \text{ as first approx. } 365=T=\frac{2\pi}{\sqrt{\mu}}a^{3/2}$$

$$=\frac{365}{\pi}\left(\frac{\pi}{2}-2e\right)$$

$$=\frac{365}{2}\left(1-\frac{1}{15\pi}\right)$$

Time for the other arc

$$=365-\frac{365}{2}\left(1-\frac{1}{15\pi}\right)=\text{etc.}$$

Ex. 11. *Show that the only law for a central attraction, for which the velocity in a circle at any distance is equal to the velocity acquired in falling from infinity to that distance, is that of the inverse cube.*

Sol. If v_0 be the velocity for the circular orbit, P the central attraction, then $v_0^3=Pr$.

If v be the velocity in falling from infinity,

then $$\tfrac{1}{2}v^2=-\int_\infty^r P\,dr$$

Since $v=v_0$, we get $$Pr=-2\int_\infty^r P\,dr$$

Differentiating with respect to r, we get

$$P+r\frac{dP}{dr}=-2P$$

or $$r\frac{dP}{dr}=-3P \quad \text{or} \quad \frac{dP}{P}=-3\frac{dr}{r}$$

Integrating $\log P=-3\log r+C$ $\therefore$ $P=\frac{\mu}{r^3}$.

Ex. 12. *A particle moves in a curve under a central attraction so that its velocity at any point is equal to that in a circle at the same distance and under the same attraction; show that the law of force is that of the inverse cube and that the path is an equiangular spiral.*

Sol. Given $v^2=Pr$ or $$\frac{v^2}{r}=P=\frac{h^2}{p^3}\frac{dp}{dr}$$

i.e., $$\frac{h^2}{p^2r}=\frac{h^2dp}{p^3dr}$$

or $$\frac{dr}{r}=\frac{dp}{p} \;\therefore\; p=Ar \text{ which is an equiangular spiral.}$$

and $$P=\frac{v^2}{r}=\frac{h^2}{p^2r}=\frac{h^2}{A^2\,r^3}\propto\frac{1}{r^3}.$$

EXAMPLES VIII (B)

1. The law of force is μu^3; a particle is projected from an apse at distance a with velocity $\frac{2}{a}\sqrt{\frac{\mu}{3}}$. Show that the orbit is $r\cos\frac{\theta}{2}=a$.

2. A particle is projected from an apse at a distance a with a velocity from infinity under the action of a central acceleration μu^{2n+3}; prove that the path is $r^n=a^n\cos n\theta$.

3. The velocity at any point of a central orbit is $1/n$th of what it would be for a circular orbit at the same distance; show that the central force varies nversely as r^{2n^2+1} and the orbit is $r^{n^2-1}=a^{n^2-1}\cos(n^2-1)\theta$.

4. A particle moves with a central acceleration $\frac{\mu}{r^5}$ and is projected from an apse at a distance a with a velocity equal to n times that which would be acquired from infinity; show that the other apsidal distance is $\frac{a}{\sqrt{n^2-1}}$.

5. A particle moves with a central acceleration $\mu(r^5-c^4r)$ being projected from an apse at distance c with velocity $c^3\sqrt{2\mu/3}$. Show that its path is
$$x^4+y^4=c^4.$$

6. A particle acted on by a central attractive force $\frac{\mu}{r^3}$ is projected with a velocity $\frac{\sqrt{\mu}}{a}$ at an angle $\frac{\pi}{4}$ with its initial distance a, from the centre of force; prove that orbit is $r=ae^{-\theta}$.

7. Show that in elliptic motion about a focus under attraction $\frac{\mu}{r^2}$ the radial velocity is given by the equation
$$r^2\dot{r}^2=\frac{\mu}{a}\left\{a(e-1)+r\right\}\left\{a(1+e)-r\right\}.$$

8. If the central acceleration is $\frac{\mu}{r^5}$, the velocities at the two aspidal distances satisfy the relation $v_1{}^2+v_2{}^2=\frac{2h^4}{\mu}$.

9. A particle is projected in a plane with velocity $\sqrt{(\mu/3a^6)}$ at a distance a from a centre of force attracting according to the law μ/r^7, in a direction inclined at 30° to the radius vector; show that the orbit is $r^2=2a^2\cos 2\theta$.

10. If the law of force be $\mu(u^4-\frac{10}{9}au^5)$ and the particle be projected from an apse at a distance $5a$ with velocity equal to $\sqrt{\frac{5}{7}}$ of that in a circle at the same distance; show that the orbit is the limacon $r=a(3+2\cos\theta)$.

11. If the acceleration at a distance r is μr^{-5} and the particle is projected at a distance a from the centre of force with velocity $\sqrt{\mu/2a^4}$, prove that the orbit is a circle through O of a diameter a cosec α where α is the inclination of the direction of projection to the radius vector.

12. A particle subject to central attractive acceleration μ/r^3+f is projected from an apex at a distance a with a velocity $\sqrt{\mu}/a$. Prove that at any subsequent time t, $r=a-\frac{1}{2}ft^2$.

13. Assuming that the eccentricity e of a planet's orbit is a small fraction, show that the ratio of the time taken by the planet to travel over the two halves of its orbit separated by the minor axis is nearly $1+\dfrac{4e}{\pi}$.

14. Prove that the two parts into which the earth's orbit is divided by the latus rectum are described in 178·63 and 186·37 days respectively

$$(e=\tfrac{1}{60}).$$

15. A particle is describing a parabola, given by the equation $x+c=am^2$, $y=2am$, under a central force to the origin. Prove that the time in passing from the point m_1 to the point m_2 is given by

$$ht=2a(m_1-m_2)\left\{c+\frac{a}{3}(m_1^2+m_1m_2+m_2^2)\right\}.$$

8·5. *If the central acceleration is equal to* $\dfrac{\mu}{r^2}$, *to find the orbit.*

Here $$\frac{h^2}{p^3}\frac{dp}{dr}=\frac{\mu}{r^2}.$$

Integrating we have $$\frac{h^2}{p^2}=\frac{2\mu}{r}+C.$$

Comparing with $\dfrac{b^2}{p^2}=\dfrac{2a}{r}\mp 1$, which is an ellipse or hyperbola referred to the focus, we get

$$\frac{h^2}{b^2}=\frac{\mu}{a}=\frac{C}{\mp 1}$$

$\therefore$ $$h=\sqrt{\frac{\mu b^2}{a}}=\sqrt{\mu l}$$ where l is the semi-latus rectum

and

$$C=\mp\frac{\mu}{a}.$$

Hence the orbit is an ellipse, parabola ($p^2=ar$) or hyperbola according as C is negative, zero or positive.

Also $$v^2=\frac{h^2}{p^2}=\frac{2\mu}{r}+C=\mu\left(\frac{2}{r}\mp\frac{1}{a}\right).$$

Hence for elliptic orbit, $$v^2=\mu\left(\frac{2}{r}-\frac{1}{a}\right)$$

" hyperbolic " " $$v^2=\mu\left(\frac{2}{r}+\frac{1}{a}\right)$$

for parabolic orbit $$v^2=\frac{2\mu}{r}$$

Also since $\frac{1}{2}h$ is the areal velocity

$$\tfrac{1}{2}h.\ T=\text{area of the ellipse}=\pi ab$$

i.e., periodic time $T=\dfrac{2\pi ab}{h}=\dfrac{2\pi ab}{\sqrt{\mu\dfrac{b^2}{a}}}=\dfrac{2\pi}{\sqrt{\mu}}\,a^{3/2}$

When the central acceleration is from the centre, we have

$$\frac{h^2}{p^3}\frac{dp}{dr}=-\frac{\mu}{r^2}$$

$$\therefore \quad \frac{h^2}{p^2}=-\frac{2\mu}{r}+C.$$

Comparing with $\dfrac{b^2}{p^2}=1-\dfrac{2a}{r}$ which is the farther branch of the hyperbola,

we get $$\frac{h^2}{b^2}=\frac{\mu}{a}=C$$

so that $$h=\sqrt{\mu l} \text{ and } v^2=\frac{h^2}{p^2}=\mu\left(\frac{1}{a}-\frac{2}{r}\right).$$

Thus when the central force is repulsive varying inversely as the square of the distance, the farther branch of the hyperbola is described.

8·6. Kepler's Laws of Planetary Motion.

1. *Each planet describes an ellipse having the sun in one of its foci.*

2. *The radius vector, drawn from the sun to the planet, describes equal areas in equal times.*

3. *The squares of the periodic times of the various planets are proportional to the cubes of the major axes of their orbits,*

Deductions. The second law says that the areal velocity is constant or $r^2\dot{\theta}$=constant ; hence the force on planet is entirely directed to the sun.

From the first law it follows that the force on the planet is inversely proportional to the square of its distance from the sun.

From the third law it follows, since $T=\dfrac{2\pi}{\sqrt{\mu}}\,a^{\frac{3}{2}}$

that μ is constant *i.e.*, acceleration at unit distance from the sun is the same for all planets.

Meaning of μ.

From *Newton's law of gravitation*, we know that if S, P be the masses of the sun and of the planet respectively, the mutual attraction between them is

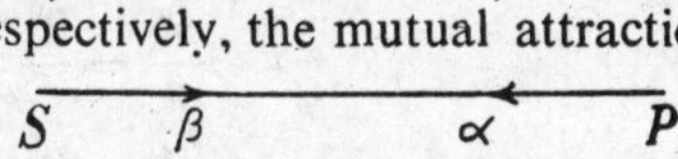

$$F=\gamma\,\frac{S\,.\,P}{r^2}$$

where r is the distance between, γ the constant of gravitation.

Thus the acceleration of the planet is

$$\alpha=\frac{F}{P}=\gamma\frac{S}{r^2}$$

and that of the sun is $\beta=\frac{F}{S}=\gamma\frac{P}{r^2}$

Hence the acceleration of planet relative to the sun is

$$\alpha+\beta=\gamma\frac{(S+P)}{r^2}=\frac{\mu}{r^2}$$

$$\therefore \quad \mu=\gamma(S+P).$$

Thus the periodic time of the planet is given by

$$T=\frac{2\pi a^{3/2}}{\sqrt{\gamma(S+P)}}$$

For two planets P, P_1 having semi-major axis a, a_1 in their orbits, the periodic times T, T_1 are related as

$$\frac{T}{T_1}=\sqrt{\frac{S+P_1}{S+P}}\left(\frac{a}{a_1}\right)^{3/2}$$

or $$\frac{S+P}{S+P_1}\cdot\frac{T^2}{T_1^2}=\frac{a^3}{a_1^3}.$$

If the planet P has a satellite p, d being its mean distance from P, its periodic time is

$$t=\frac{2\pi}{\sqrt{\gamma(P+p)}}d^{3/2}.$$

$$\therefore \quad \frac{S+P}{P+p}\cdot\frac{T^2}{t^2}=\frac{a^3}{d^3}.$$

Ex. 13. *A particle describes an ellipse under a force* $\frac{\mu}{r^2}$ *and has a velocity* v *at a distance* r *from the centre of force, show that its periodic time is*

$$\frac{2\pi}{\sqrt{\mu}}\left(\frac{2}{r}-\frac{v^2}{\mu}\right)^{-3/2}.$$

Sol. $$v^2=\mu\left(\frac{2}{r}-\frac{1}{a}\right)$$

$$\therefore \quad \frac{v^2}{\mu}=\frac{2}{r}-\frac{1}{a}$$

or $$\frac{1}{a}=\frac{2}{r}-\frac{v^2}{\mu}$$

$$\therefore \qquad T=\frac{2\pi}{\sqrt{\mu}}\, a^{3/2}=\text{etc.}$$

Ex. 14. *If V_1 and V_2 are the velocities of a planet when it is respectively nearest and farthest from the sun, prove that*

$$(1-e)V_1=(1+e)\,V_2.$$

Sol.
$$V^2=\mu\left(\frac{2}{r}-\frac{1}{a}\right)$$

$$\therefore \qquad V_1^2=\mu\left\{\frac{2}{a(1-e)}-\frac{1}{a}\right\}=\frac{\mu}{a}\,\frac{1+e}{1-e}$$

$$V_2^2=\mu\left\{\frac{2}{a(1+e)}-\frac{1}{a}\right\}=\frac{\mu}{a}\,\frac{1-e}{1+e}$$

Hence
$$\frac{V_1^2}{V_2^2}=\text{etc.}$$

Ex. 15. *If ω be the angular velocity of a planet at the nearer end of the major axis, prove that its period is*

$$\frac{2\pi}{\omega}\sqrt{\frac{1+e}{(1-e)^3}}$$

Sol.
$$r^2\dot{\theta}=h$$

$$\therefore \qquad a^2(1-e)^2\,\omega=h, \qquad h^2=\mu l=\mu a(1-e^2)$$

$$\therefore \qquad a^4(1-e)^4\,\omega^2=h^2=\mu a(1-e^2)$$

$$\therefore \qquad \mu=\frac{a^3(1-e)^3\,\omega^2}{1+e}$$

Hence
$$T=\frac{2\pi}{\sqrt{\mu}}\,a^{3/2}=\text{etc.}$$

Ex. 16. *If the velocity in a given elliptic orbit (major axis $2a$) is the same at a certain point P whether the orbit is being described in periodic time T about one focus S or in periodic time T' about the other focus S', prove that*

$$SP=\frac{2a\,T'}{T+T'}, \qquad S'P=\frac{2a\,T}{T+T'}.$$

Sol.
$$T=\frac{2\pi}{\sqrt{\mu}}\,a^{3/2}, \quad T'=\frac{2\pi}{\sqrt{\mu'}}\,a^{3/2},$$

$$V^2=\mu\left(\frac{2}{r}-\frac{1}{a}\right),$$

$$V'^2=\mu'\left(\frac{2}{r'}-\frac{1}{a}\right)$$

$$\therefore \quad \mu\left(\frac{2}{r}-\frac{1}{a}\right)=\mu'\left(\frac{2}{r'}-\frac{1}{a}\right)$$

or $$\mu\,\frac{2a-r}{ar}=\mu'\frac{a-r'}{ar'}$$

or $$\mu\,\frac{r'}{r}=\mu'\frac{r}{r'}$$

or $$\left(\frac{r'}{r}\right)^2=\frac{\mu'}{\mu}$$

Also $$\frac{T}{T'}=\sqrt{\frac{\mu'}{\mu}}=\frac{r'}{r}$$

$$\therefore \quad \frac{T+T'}{T'}=\frac{r+r'}{r}=\frac{2a}{r}$$

and $$\frac{T'}{T}=\frac{r}{r'}$$

$$\therefore \quad \frac{T+T'}{T}=\frac{r+r'}{r'}=\frac{2a}{r'}.$$

Ex. 17. *A particle describes an ellipse under a force to the focus S. When the particle is at one extremity of the minor axis, its kinetic energy is doubled without any change in the direction of motion. Prove that the particle proceeds to describe a parabola.*

Sol. At the end of minor axis, $r=a$

$$V^2=\mu\left(\frac{2}{r}-\frac{1}{a}\right)=\mu\left(\frac{2}{a}-\frac{1}{a}\right)=\frac{\mu}{a}$$

If V_1 be the velocity when the kinetic energy is doubled,

$$\tfrac{1}{2}MV_1^2=2\,.\,\tfrac{1}{2}MV^2$$

i.e., $$V_1^2=2V^2=\frac{2\mu}{a}$$

Hence it moves in a parabola as for a parabola, $v^2=\dfrac{2\mu}{r}$.

Ex. 18. *If the velocity of the earth at any point of its orbit, assumed circular, were increased by about one-half, prove that it would describe a parabola about the sun as focus. Show also that if a body were projected from the earth with a velocity exceeding 7 miles per second, it will not return to the earth.*

Sol. For a circular orbit,

$$V^2=\mu\left(\frac{2}{a}-\frac{1}{a}\right)=\frac{\mu}{a}$$

For a parabolic orbit,

$$V_1^2=\frac{2\mu}{a}$$

$$\therefore \qquad V_1=\sqrt{2}V$$

If R be earth's radius, $\frac{\mu_1}{r^2}$ be its attraction then on earth's surface $\frac{\mu_1}{R^2}=g$.

The velocity for a parabolic path about the earth's centre

$$=\sqrt{\frac{2\mu_1}{R}}=\sqrt{2gR}$$

$$=(2\times 32\times 4000\times 8760\times 3)^{\frac{1}{2}} \text{ ft./sec}$$

$$=7 \text{ miles/sec approx.}$$

Now the earth has a velocity 18·5 miles/second in its orbit round the sun and the body had the same velocity before projection hence in order that it may move in a parabola, its velocity must be $18{\cdot}5\times\sqrt{2}=26$ miles/sec. approx. *i.e.*, must be increased by about 7·5 miles/sec.

Ex. 19. *The time of revolution of Jupiter about the sun is 11·86 years and his mean distance 483 millions of miles ; the distance of his first satellite is 261,000 miles and time of revolution 1 day 18½ hours : show that the mass of Jupiter is little less than one-thousandth of that of the sun.*

Sol. $$\frac{S+P}{P+p}=\frac{a^3\,t^2}{d^3\,T^2}=\left(\frac{483\times 10^6}{261\times 10^3}\right)^3\left(\frac{1\frac{3}{4}}{11{\cdot}86\times 365}\right)^2$$

$$=1000 \text{ about.}$$

Ex. 20. *Show that the velocity of a particle moving in an ellipse about a centre of force in the focus is compounded of two constant velocities, $\frac{\mu}{h}$ perpendicular to the radius and $e\frac{\mu}{h}$ perpendicular to the major axis.*

Sol. S is the centre of force, S' the empty focus. Let p, p' be perpendiculars from S, S' on the tangent at P.

Now $vp=h$ also $pp'=b^2$

$$\therefore \qquad v=\frac{h}{p}=\frac{h}{b^2}\,p'$$

$$=\frac{h}{b^2}\cdot S'Z$$

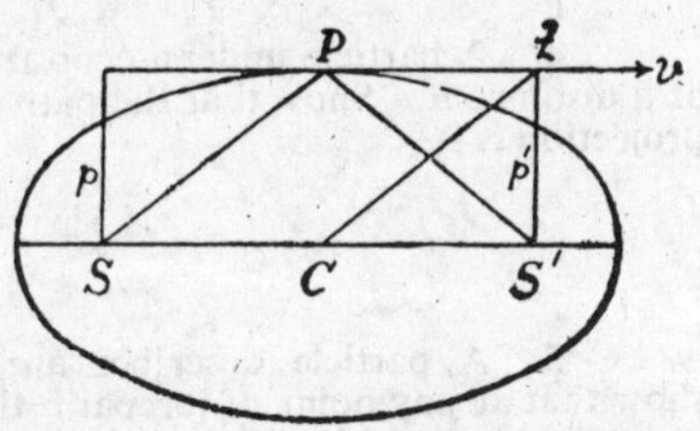

Also $h^2=\mu l=\mu\frac{b^2}{a}$

$$\therefore \qquad b^2 = \frac{ah^2}{\mu}$$

Hence
$$v = \frac{\mu}{ah}.S'Z$$

Thus the velocity is $\frac{\mu}{ah}$ times $S'Z$ and perpendicular to $S'Z$.

But
$$\vec{S'Z} = \vec{S'C} + \vec{CZ}$$

$\therefore$ the velocity is $\frac{\mu}{ah}$ times $S'C = (ae)$ and perp. to $S'C$

and $\frac{\mu}{ah}$ times $CZ\ (=a)$ and perp. to CZ

which is parallel to SP

Thus the velocity is $e\ \frac{\mu}{h}$ perp. to major axis

and $\frac{\mu}{h}$ perp. to radius vector.

EXAMPLES VIII (C)

1. A particle describes an ellipse about one focus. Show that the greatest and least angular velocities occur at the ends of the major axis and also that if α, β be these angular velocities the mean angular velocity is

$$\frac{2(\alpha\beta)^{\frac{3}{4}}}{\sqrt{\alpha}+\sqrt{\beta}}.$$

[Hint. $\alpha = \frac{h}{r_1^2}$, $\beta = \frac{h}{r_2^2}$, $r_1 = a(1-e)$, $r_2 = a(1+e)$

mean angular velocity $= \frac{2\pi}{T}$.]

2. If a particle describes an ellipse as a central orbit about a focus, prove that the velocity at the end of the minor axis is a geometric mean between the velocities at the ends of any diameter.

3. Prove that if when the particle is at a distance r from the focus, its velocity is v in a direction making an angle ϕ with the radius vector then

$$e^2\mu^2 = (v^2 r - \mu)^2 \sin^2\phi + \mu^2 \cos^2\phi.$$

[Hint. $v^2 = \mu\left(\frac{2}{r} - \frac{1}{a}\right)$, $h^2 = \mu a(1-e^2)$, $h = vp = vr \sin\phi$.]

4. A particle under a central acceleration μ/r^2 is projected with velocity V at a distance R. Show that the path is a rectangular hyperbola if the angle of projection is

$$\sin^{-1}\frac{\mu}{VR\left(V^2 - \frac{2\mu}{R}\right)^{\frac{1}{2}}}.$$

5. A particle describes an ellipse about a centre of forces at the focus, show that at any point of its path the angular velocity about the other focus varies inversely as the square of the normal at the point.

9

Resisting Medium

9·1. In a resisting medium there is some resistance to the motion of a particle due to the presence of air or any other fluid in the medium. This resistance usually varies either with the velocity or with some power of the velocity.

9·2. Resistance varying as the velocity

(*a*) **Descent**. *A particle is falling under gravity in a medium whose resistance varies as the velocity—to find the motion.*

If at time t, the particle has fallen through a distance x when v is the velocity then the equation of motion is

$$\ddot{x}=g-k\dot{x}$$

resistance being $\quad kv=k\dot{x}$

i.e.,
$$\ddot{x}=g\left(1-\frac{\dot{x}}{V}\right)$$

where
$$V=\frac{g}{k}.$$

If $\dot{x}$ becomes equal to V then the acceleration would be zero, then the motion will be unresisted and the particle will fall with the constant velocity V, This velocity $V=\frac{g}{k}$ is called the *terminal velocity*.

Writing the above equation as

$$\ddot{x}+\frac{g}{V}\dot{x}=g \qquad \text{...(1)}$$

and integrating with rest with t we get

$$\dot{x}+\frac{g}{V}\,x=gt+C$$

Initially $\quad t=0=x=\dot{x}$

$\therefore \quad C=0$

Integrating again, we get

$$x\,e^{\frac{g}{V}t}=g\int t\,e^{\frac{g}{V}t}\,dt+C_1$$

$$=g\left[\frac{V}{g}\,t\,e^{\frac{gt}{V}}-\frac{V^2}{g^2}\,e^{\frac{gt}{V}}\right]+C_1$$

Initially $\quad t=0=x$

$$\therefore \quad C_1=\frac{V^2}{g}$$

Hence $$x=Vt-\frac{V^2}{g}+\frac{V^2}{g}\,e^{-\frac{gt}{V}} \qquad ...(2)$$

Also taking (1) as a linear equation in $\dot{x}$ and integrating we get

$$\dot{x}\,e^{\frac{gt}{V}}=g\int e^{\frac{gt}{V}}\,dt+C_2$$

$$=V\,e^{\frac{gt}{V}}+C_2$$

Initially $\quad t=0=\dot{x}$

$\therefore \quad C_2=-V$

Hence $$\dot{x}=V-V\,e^{-\frac{gt}{V}} \qquad ...(3)$$

The equation (3) shows that as t becomes very very large, $\dot{x}$ tends to V *i.e.*, the particle ultimately attains the velocity V. That is why V is called the terminal velocity.

Integrating (3) we get

$$x=Vt+\frac{V^2}{g}\,e^{-\frac{gt}{V}}+C_3$$

$$=Vt+\frac{V^2}{g}\,e^{-\frac{gt}{V}}-\frac{V^2}{g}$$

which is the same as equation (2) obtained earlier.

Also the equation of motion can be written as

$$\frac{dv}{dt}=g\left(1-\frac{v}{V}\right)$$

we get
$$\frac{V\,dv}{V-v}=g\,dt$$

Integrating we get

$$-V\log(V-v)=gt+C_4$$

or
$$V-v=C_5\,e^{-\frac{gt}{V}}$$

Initially $t=0, \quad v=0$

$\therefore \quad C_5=V$

Hence
$$V-v=V\,e^{-\frac{gt}{V}}$$

or
$$v=V\left(1-e^{-\frac{gt}{V}}\right) \quad ...(4)$$

Also we can take the equation of motion in the form

$$v\frac{dv}{dx}=g\left(1-\frac{v}{V}\right)$$

or
$$\frac{v\,dv}{V-v}=\frac{g}{V}\,dx$$

or
$$\left(\frac{V}{V-v}-1\right)dv=\frac{g}{V}\,dx$$

Integrating we get

$$-V\log(V-v)-v=\frac{gx}{V}+C_6$$

$$=\frac{gx}{V}-V\log V$$

Hence
$$\frac{gx}{V}=V\log\frac{V}{V-v}-v$$

or
$$x=\frac{V^2}{g}\log\frac{V}{V-v}-\frac{V}{g}\,v \quad ...(5)$$

(*b*) **Ascent.** *A particle is projected upwards with a velocity U in a medium whose resistance varies as the velocity—to find the motion.*

The equation of motion is

$$\ddot{x}=-g-k\dot{x}=-g\left(1+\frac{\dot{x}}{V}\right)$$

where $V=\dfrac{g}{k}$,

the terminal velocity.

or $$\ddot{x}+\frac{g}{V}\dot{x}=-g$$

Integrating we get

$$\dot{x}\,e^{\frac{gt}{V}}=-Ve^{\frac{gt}{V}}+C$$

Initially $t=0,\quad \dot{x}=U$

$\therefore$ $C=U+V$

Hence $$\dot{x}=(U+V)e^{-\frac{gt}{V}}-V \qquad ...(1)$$

Integrating again we get

$$x=-\frac{V}{g}(U+V)\,e^{-\frac{gt}{V}}-Vt+C_1$$

Initially $t=0=x$

$\therefore$ $$C_1=\frac{V}{g}(U+V)$$

Hence $$x=-\frac{V}{g}(U+V)\,e^{-\frac{gt}{V}}-Vt+\frac{V}{g}(U+V) \qquad ...(2)$$

At the highest point

$$\dot{x}=0\quad\therefore\quad \frac{gt}{V}=\log\frac{U+V}{V} \qquad ...(3)$$

Therefore the greatest height

$$=-\frac{V^2}{g}-\frac{V^2}{g}\log\frac{U+V}{V}+\frac{V}{g}(U+V)$$

$$=\frac{UV}{g}-\frac{V^2}{g}\log\frac{U+V}{V} \qquad ...(4)$$

Also we can write the equation of motion as

$$v\frac{dv}{dx}=-\frac{g}{V}(V+v)$$

or

$$\frac{v\,dv}{V+v}=-\frac{g}{V}\,dx$$

or

$$\left(1-\frac{V}{V+v}\right)dv=-\frac{g}{V}\,dx$$

Integrating we get

$$v-V\log(V+v)=-\frac{g}{V}x+C_2$$

Initially $v=U,\ x=0$

$\therefore$ $U-V\log(V+U)=C_2$

Hence $v-U-V\log\frac{V+v}{V+U}=-\frac{g}{V}x$

or

$$x=\frac{V^2}{g}\log\frac{V+v}{V+U}+\frac{V}{g}(U-v) \qquad ...(5)$$

At the highest point $v=0$, therefore the greatest height

$$=\frac{V^2}{g}\log\frac{V}{V+U}+\frac{VU}{g} \qquad ...(5)$$

which is the same as (4).

If $U=\lambda V$

then greatest height

$$=\frac{V^2}{g}\log\frac{V}{V+\lambda V}+\frac{\lambda V^2}{g}$$

$$=\frac{V^2}{g}\left[\lambda-\log(1+\lambda)\right].$$

Ex. 1. *A heavy particle is projected in a resistance medium, the resistance varying as the velocity. If v_1 and v_2 are its velocities at any point in its upward and downward paths and t the interval between its passage through this point, prove that*

$$v_1+v_2=gt,\ V-v_2=(V+v_1)\,e^{-\frac{gt}{V}}$$

where V is the terminal velocity.

Sol. Since any point may be tpken as the point of projection let at a point P_1 the upward velocity be v_1, then the time to the highest point above P and the greatest height reached above P are given by

$$t_1=\frac{V}{g}\log\frac{V+v_1}{V}$$

and
$$h=\frac{V\,v_1}{g}-\frac{V^2}{g}\log\frac{V+v_1}{V}.$$

In its downward motion when the particle comes to P_1 if t_2 be the time and v_2 the velocity at P,

$$v_2=V\left(1-e^{-\frac{gt_2}{V}}\right)$$

or
$$t_2=\frac{V}{g}\log\frac{V}{V-v_2}$$

Thus we get

$$V-v_2=V\,e^{-\frac{gt_2}{V}}$$

and
$$V+v_1=V\,e^{\frac{gt_1}{V}}$$

$\therefore$
$$V-v_2=(V+v_1)\,e^{-\frac{gt}{V}}$$

where
$$t=\text{total time}=t_1+t_2.$$

Also
$$t=t_1+t_2=\frac{V}{g}\log\frac{V+v_1}{V-v_2}$$

Further since the distance fallen in its downward motion is equal to the greatest height in its upward motion, we have

$$\frac{V^2}{g}\log\frac{V}{V-v_2}-\frac{V}{g}v_2=\frac{V\,v_1}{g}-\frac{V^2}{g}\log\frac{V+v_1}{V}$$

Hence
$$V\log\frac{V+v_1}{V-v_2}=v_1+v_2$$

i.e.,
$$gt=v_1+v_2.$$

9·3. Resistance varying as the square of the velocity.

(*a*) **Descent.** *A particle is falling under gravity in a medium whose resistance varies as the square of the velocity—to find the motion.*

The equation of motion for vertical fall is

$$\ddot{x}=g-kv^2=g\left(1-\frac{v^2}{V^2}\right)$$

where
$$V=\sqrt{\frac{g}{k}}$$

the terminal velocity.

Writing this as

$$v\frac{dv}{dx}=g\left(1-\frac{v^2}{V^2}\right)$$

so that $$\frac{2g}{V^2}x=\int\frac{2v\,dv}{V^2-v^2}+C=C-\log(V^2-v^2)$$

Initially $x=0=v$

$\therefore$ $C=\log V^2$

Hence $$v^2=V^2\left(1-e^{-\frac{2gx}{V^2}}\right)\quad\ldots(1)$$

Also we can write

$$\frac{dv}{dt}=g\left(1-\frac{v^2}{V^2}\right)$$

$$\therefore\quad\frac{gt}{V^2}=\int\frac{dv}{V^2-v^2}+C_1$$

$$=C_1+\frac{1}{2V}\log\frac{V+v}{V-v}$$

but $C_1=0$
from initial conditions

$$\therefore\quad\frac{V+v}{V-v}=e^{\frac{2gt}{V}}$$

$$\therefore\quad\frac{v}{V}=\frac{e^{\frac{2gt}{V}}-1}{e^{\frac{2gt}{V}}+1}=\frac{1-e^{-\frac{2gt}{V}}}{1+e^{-\frac{2gt}{V}}}$$

Thus when t becomes very large, v tends to V.

Also $$\frac{v}{V}=\frac{e^{\frac{gt}{V}}-e^{-\frac{gt}{V}}}{e^{\frac{gt}{V}}+e^{-\frac{gt}{V}}}=\tanh\left(\frac{gt}{V}\right)$$

$$v=V\tanh\left(\frac{gt}{V}\right)\quad\ldots(2)$$

This can be written as

$$\frac{dx}{dt}=V\tanh\left(\frac{gt}{V}\right)$$

Integrating we get

$$x=\frac{V^2}{g}\log\cosh\frac{gt}{V}+C_2$$

but C_2 is zero from initial conditions hence

$$x=\frac{V^2}{g}\log\cosh\left(\frac{gt}{V}\right) \qquad ...(3)$$

(*b*) **Ascent.** *A particle is projected upwards with a velocity U in a medium whose resistance varies as the square of the velocity—to find the motion.*

The equation of motion is

$$\ddot{x}=-g-kv^2=-g\left(1+\frac{v^2}{V^2}\right)$$

where $$V=\sqrt{\frac{g}{k}}$$

or $$v\frac{dv}{dx}=-g\left(1+\frac{v^2}{V^2}\right)$$

$\therefore$ $$\frac{2gx}{V^2}=-\int\frac{2v\,dv}{V^2+v^2}+C=C-\log(V^2+v^2)$$

Initially $$x=0,\ v=U$$

$\therefore$ $$C=\log(V^2+U^2)$$

Hence $$\frac{2gx}{V^2}=\log\frac{V^2+U^2}{V^2+v^2} \qquad ...(1)$$

The equation of motion can also be written as

$$\frac{dv}{dt}=-g\left(1+\frac{v^2}{V^2}\right)$$

$\therefore$ $$-\frac{gt}{V^2}=\int\frac{dv}{V^2+v^2}+C_1=\frac{1}{V}\tan^{-1}\frac{v}{V}+C_1$$

From initial conditions

$$t=0,\ v=U$$

we have $$C_1=-\frac{1}{V}\tan^{-1}\frac{U}{V}$$

Hence $$\frac{gt}{V}=\tan^{-1}\frac{U}{V}-\tan^{-1}\frac{v}{V} \qquad ...(2)$$

At the highest point, $v=0$, therefore the greatest height is given by

$$h=\frac{V^2}{2g}\log\frac{V^2+U^2}{V^2} \qquad ...(3)$$

and the time to that point is

$$t=\frac{V}{g}\tan^{-1}\frac{U}{V} \qquad \ldots(4)$$

From (3) and (4) it follows that

$$\frac{U}{V}=\tan\frac{gt}{V}$$

and

$$h=\frac{V^2}{2g}\log\left(1+\frac{U^2}{V^2}\right)$$

$$=\frac{V^2}{2g}\log\left(1+\tan^2\frac{gt}{V}\right)$$

$$=\frac{V^2}{2g}\log\sec^2\frac{gt}{V}=\frac{V^2}{g}\log\sec\frac{gt}{V}$$

giving the greatest height reached where t is the time to reach it.

Ex. 2. *A particle falls from rest under gravity in a medium whose resistance varies as the square of the velocity ; if v be its velocity and v_0 be the velocity which would be acquired if there was no resistance and V the terminal velocity, show that*

$$\frac{v^2}{v_0^2}=1-\frac{1}{2}\frac{v_0^2}{V^2}+\frac{1}{2\cdot 3}\frac{v_0^4}{V^4}$$

Sol. The velocity v at height x is given by

$$v^2=V^2\left(1-e^{\frac{-2gx}{V^2}}\right)$$

$$\therefore \quad e^{\frac{-2gx}{V^2}}=1-\frac{v^2}{V^2}$$

and

$$v_0^2=2gx$$

$$\therefore \quad \frac{v^2}{V^2}=1-e^{\frac{-v_0^2}{V^2}}$$

$$=1-\left[1-\frac{v_0^2}{V^2}+\frac{1}{1.2}\frac{v_0^4}{V^4}-\frac{1}{1.2.3}\frac{v_0^6}{V^6}+\ldots\right]$$

$$=\text{etc.}$$

Ex. 3. *A particle, projected upwards with a velocity U in a medium whose resistance varies as the square of the velocity, will return to the point of projection with velocity* $v=\frac{UV}{\sqrt{U^2+V^2}}$ *after a time*

$$\frac{V}{g}\left[\tan^{-1}\left(\frac{U}{V}\right)+\tanh^{-1}\left(\frac{v}{V}\right)\right].$$

Sol. If h be the greatest height

$$h=\frac{V^2}{2g}\log\frac{V^2+U^2}{V^2}$$

or

$$\frac{V^2+U^2}{V^2}=e^{\frac{2gh}{V^2}}$$

When falling, $$v^2=V^2\left(1-e^{\frac{-2gh}{V^2}}\right)$$

$$=V^2\left(1-\frac{V^2}{V^2+U^2}\right)$$

$$=\frac{U^2V^2}{V^2+U^2}.$$

If t be the time to the highest point

$$t=\frac{V}{g}\tan^{-1}\frac{U}{V}$$

When falling, time $=\frac{V}{g}\tanh^{-1}\frac{v}{V}$

$\therefore$ Total time $=\frac{V}{g}\left[\tan^{-1}\frac{U}{V}+\tanh^{-1}\frac{v}{V}\right].$

Ex. 4. *A particle moving in a straight line is subject to a resistance kv^3 when v is the velocity. Show that if v is the velocity at time t when the distance is s*

$$v=\frac{u}{1+kus},\quad t=\frac{s}{u}+\tfrac{1}{2}ks^2$$

where u is the inital velocity.

Also show that

$$s=\frac{1}{ku}\left[\sqrt{1+2ku^2t}-1\right]$$

and

$$v=\frac{u}{\sqrt{1+2ku^2t}}.$$

Sol. The equation of motion is

$$v\frac{dv}{ds}=-kv^3$$

$\therefore$ $$\frac{dv}{ds}=-kv^2$$

or

$$\frac{dv}{v^2}=-kds$$

Integrating $\frac{1}{v}=ks+C$

Initially $\frac{1}{u}=C$

$\therefore$ $\frac{1}{v}=\frac{1}{u}+ks=\frac{1+kus}{u}$

$\therefore$ $v=\frac{u}{1+kus}$

i.e., $\dot{s}=\frac{u}{1+kus}$

or $(1+kus)ds=udt$

Integrating $ut=s+ku\frac{s^2}{2}+C$

but $C=0$ as $t=0=s$

Hence, $t=\frac{s}{u}+\frac{1}{2}ks^2.$

Also $\ddot{s}=-k\dot{s}^3$

$\therefore$ $\frac{\ddot{s}}{\dot{s}^3}=-k$

Integrating $\frac{1}{2\dot{s}^2}=kt+C$

Initially $\frac{1}{2u^2}=C$

$\therefore$ $\frac{1}{2\dot{s}^2}=\frac{1}{2u^2}+kt$

or $\frac{1}{v^2}=\frac{1}{u^2}+2kt$

or $v=\frac{u}{\sqrt{1+2ku^2t}}$

i.e., $\dot{s}=\frac{u}{\sqrt{1+2ku^2t}}$

$\therefore$ Integrating $s=\frac{u}{ku^2}\sqrt{1+2ku^2t}+C$

Initially $0=\frac{1}{ku}+C$

hence
$$s=\frac{1}{ku}\left[\sqrt{1+2ku^2t}-1\right].$$

EXAMPLES IX (A)

1. A particle is projected with velocity V along a smooth horizontal plane in a medium whose resistance per unit mass is k (velocity). Show that the velocity v after a time t and the distance s in that time are given by

$$v=Ve^{-kt}, \quad s=\frac{V}{k}(1-e^{-kt}).$$

2. Prove that with a resistance k times the square of the velocity, a particle placed at rest at distance a from the origin and attracted to the origin by a force μ (distance) will first come to rest again at a distance b where

$$(1+2ak)e^{-2ak}=(1-2bk)e^{2kb}.$$

3. A heavy particle is projected vertically upwards in a medium the resistance of which varies as the square of the velocity. It has a kinetic energy K in its upward path at a given position. When it passes the same point on the way down, show that its loss of kinetic energy is $K^2/(K+K')$ when K' is the limit to which the energy approaches in its downward course.

4. A particle is projected upwards in a medium whose resistance is $\frac{gv}{V}$ where v is the velocity. If V be large compared to U, the velocity of projection, prove that the vertical height reached, the time of ascent and the time of descent are decreased respectively by the fractions

$$\frac{2}{3}\frac{U}{V}, \quad \frac{1}{2}\frac{U}{V} \text{ and } \frac{1}{6}\frac{U}{V}$$

of their values when there is no resistance and that the particle returns to the point of projection with velocity $U\left(1-\frac{2}{3}\frac{U}{V}\right)$.

9·4. Motion of a projectile in a resisting medium, resistance varying as the velocity.

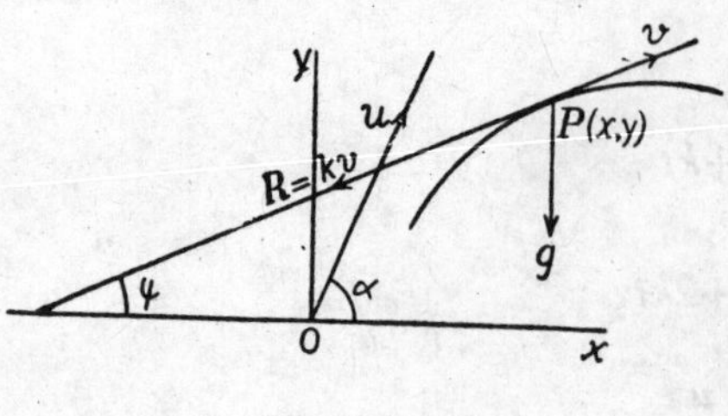

Take the point of projection, as origin, the horizontal line and the vertical line as axes of x and y, U the velocity of projection at angle α with x-axis.

P be the position of the particle at time t, its co-ordinates be (x, y), v be the velocity at P so that resistance per unit mass is kv.

The equations of motion are

$$\ddot{x}=-k\dot{x}, \quad \ddot{y}=-g-k\dot{y}, \quad (\dot{x}, \dot{y}) \text{ being components of } v.$$

$$\therefore \quad \frac{d\dot{x}}{dt}=-k\dot{x}$$

or
$$\frac{d\dot{x}}{\dot{x}}=-k\,dt$$

Integrating $\log \dot{x} = -kt + \text{const.}$

or $\dot{x} = Ce^{-kt}$

Initially $t=0,\ \dot{x} = U \cos \alpha$

$\therefore$ $C = U \cos \alpha$

Hence $\dot{x} = U \cos \alpha\ e^{-kt}$

Integrating again $x = -\dfrac{U}{k} \cos \alpha\ e^{-kt} + \dfrac{C_1}{k}.$

Initially $t=0=x,$

$\therefore$ $C_1 = U \cos \alpha$

Hence $$x = \frac{U}{k} \cos \alpha\ (1 - e^{-kt}) \qquad \ldots(1)$$

Also $$\frac{d\dot{y}}{dt} = -g - k\dot{y}$$

or $$\frac{d\dot{y}}{g + k\dot{y}} = -dt$$

Integrating $g + k\dot{y} = C_2 e^{-kt}$

Initially $t=0,\quad \dot{y} = U \sin \alpha$

$\therefore$ $C_2 = g + k\ U \sin \alpha$

Hence $$\dot{y} = -\frac{g}{k} + \left(U \sin \alpha + \frac{g}{k} \right) e^{-kt}$$

Integrating again

$$y = -\frac{gt}{k} - \frac{1}{k} \left(U \sin \alpha + \frac{g}{k} \right) e^{-kt} + C_3$$

Initially $t=0=y$

$\therefore$ $$C_3 = \frac{1}{k} \left(U \sin \alpha + \frac{g}{k} \right)$$

Hence $$y = -\frac{gt}{k} + \frac{1}{k} \left(U \sin \alpha + \frac{g}{k} \right) \left(1 - e^{-kt} \right) \qquad \ldots(2)$$

To eliminate t between (1) and (2) we proceed as follows :

From (1) $$1 - e^{-kt} = \frac{k}{U \cos \alpha} x$$

or $$e^{-kt} = 1 - \frac{kx}{U \cos \alpha}$$

$\therefore \qquad -kt = \log\left(1 - \dfrac{kx}{U \cos \alpha}\right)$

$\therefore$ from (2),

$$y = \frac{g}{k^2} \log\left(1 - \frac{kx}{U \cos \alpha}\right) + \frac{1}{k}\left(U \sin \alpha + \frac{g}{k}\right)\frac{kx}{U \cos \alpha}$$

i.e.,

$$y = x\left(\tan \alpha + \frac{g}{kU \cos \alpha}\right) + \frac{g}{k^2} \log\left(1 - \frac{kx}{U \cos \alpha}\right) \qquad \text{...(3)}$$

which is the equation to the path.

We see that when $t \to \infty$, $\dot{x} \to 0$, $\dot{y} \to -\dfrac{g}{k} = -V$

where $\qquad V = \dfrac{g}{k}$ is called the terminal velocity.

For the greatest height,

$$\dot{y} = 0$$

$\therefore$

$$e^{-kt} = \frac{g}{kU \sin \alpha + g}$$

or

$$-kt = \log \frac{g}{kU \sin \alpha + g} = -\log\left(1 + \frac{kU \sin \alpha}{g}\right)$$

i.e.,

$$t = \frac{1}{k} \log\left(1 + \frac{kU \sin \alpha}{g}\right)$$

$=$ time to greatest height,

Then greatest height is

$$y = -\frac{g}{k^2} \log\left(1 + \frac{kU \sin \alpha}{g}\right) + \frac{1}{k^2}\left(kU \sin \alpha + g\right)\left(1 - \frac{g}{kU \sin \alpha + g}\right)$$

$$= -\frac{g}{k^2} \log\left(1 + \frac{kU \sin \alpha}{g}\right) + \frac{U \sin \alpha}{k}.$$

Ex. 5. ***Prove that the path can be put in the form***

$$y + ax = b \log x$$

by a proper choice of ***axes.***

Sol. From (1) we have $x=\dfrac{U\cos\alpha}{k}\left(1-e^{-kt}\right)$

or $$e^{-kt}=1-\frac{kx}{U\cos\alpha}=\frac{kX}{U\cos\alpha}$$

From (2) we have $y=-\dfrac{gt}{k}+\left(\dfrac{U\sin\alpha}{k}+\dfrac{g}{k^2}\right)\left(1-e^{-kt}\right)$

$$=\frac{g}{k^2}\log\frac{kX}{U\cos\alpha}+\frac{kU\sin\alpha+g}{k^2}\left(1-\frac{kX}{U\cos\alpha}\right)$$

or $$\frac{g}{k^2}\log X-\frac{kU\sin\alpha+g}{kU\cos\alpha}X=y-\frac{g}{k^2}\log\frac{k}{U\cos\alpha}-\frac{kU\sin\alpha+g}{k^2}$$

$$=Y \text{ say}$$

This equation can be put as

$$y+ax=b\log x.$$

Ex 6. *If the resistance varies as the velocity and the range on the horizontal plane through the point of projection is a maximum, show that the angle α which the direction of projection makes with the vertical is given by*

$$\frac{\lambda(1+\lambda\cos\alpha)}{\cos\alpha+\lambda}=\log(1+\lambda\sec\alpha)$$

when λ is the ratio of the velocity of projection to the terminal velocity.

Sol. To find the range on the horizohtal plane. we put $y=0$.

We note that in this case α is the angle with the vertical.

$\therefore$ from (2), $$gt=\left(U\cos\alpha+\frac{g}{k}\right)\left(1-e^{-kt}\right) \quad ...(a)$$

and the range is $$R=x=\frac{U\sin\alpha}{k}(1-e^{-kt})$$

R is the maximum when

$$\frac{dR}{d\alpha}=0$$

i.e., $$\cos\alpha(1-e^{-kt})+k\sin\alpha\, e^{-kt}\frac{dt}{d\alpha}=0$$

subject to the condition

$$g\frac{dt}{d\alpha}=-U\sin\alpha(1-e^{-kt})+\left(U\cos\alpha+\frac{g}{k}\right)ke^{-kt}\frac{dt}{d\alpha}$$

or $$\left\{(kU\cos\alpha+g)e^{-kt}-g\right\}\frac{dt}{d\alpha}=U\sin\alpha(1-e^{-kt})$$

Eliminating $\frac{dt}{d\alpha}$ we get

$$\frac{(kU\cos\alpha+g)\,e^{-kt}-g}{k\sin\alpha\;e^{-kt}}=-U\tan\alpha$$

or $$(kU\cos\alpha+g)\,e^{-kt}=g-kU\frac{\sin^2\alpha}{\cos\alpha}\;e^{-kt}$$

or $$e^{-kt}=\frac{g\cos\alpha}{kU+g\cos\alpha}$$

Hence from (*a*)

$$\frac{g}{k}\log\frac{kU+g\cos\alpha}{g\cos\alpha}=\left(U\cos\alpha+\frac{g}{k}\right)\left(1-\frac{g\cos\alpha}{kU+g\cos\alpha}\right)$$

$$=U\frac{kU\cos\alpha+g}{kU+g\cos\alpha}$$

Now by putting $V=\frac{g}{k}$, $\lambda=\frac{U}{V}$, we get

$$V\log\frac{U+V\cos\alpha}{V\cos\alpha}=U\frac{U\cos\alpha+V}{U+V\cos\alpha}$$

or $$\log(1+\lambda\sec\alpha)=\frac{\lambda(1+\lambda\cos\alpha)}{\cos\alpha+\lambda}.$$

Ex. 7. *Prove that in the motion of a projectile in a resisting medium the equation* $\frac{d^2y}{dx^2}=-\frac{g}{u^2}$ *is satisfied whatever the law of resistance, u being the horizontal component of velocity, the axes of x and y being horizontal and vertically upwards.*

Sol. $$\ddot{x}=-R\cos\psi,\qquad \ddot{y}=-g-R\sin\psi$$

$$\frac{dy}{dx}=\frac{\dot{y}}{\dot{x}}=\tan\psi$$

$$\frac{d^2y}{dx^2}=\frac{\dot{x}\ddot{y}-\dot{y}\ddot{x}}{\dot{x}^3}=-\frac{\dot{x}(g+R\sin\psi)-\dot{y}R\cos\psi}{\dot{x}^3}$$

$$=-\frac{g\dot{x}}{\dot{x}^3}\text{ since }\dot{x}\sin\psi=\dot{y}\cos\psi$$

$$=-\frac{g}{\dot{x}^2}=-\frac{g}{u^2}$$

9·5. Resistance varying as the square of the velocity.

u=horizontal ccmponent of velocity at P

$$=v \cos \psi$$

Resolving horizontally and normally we get

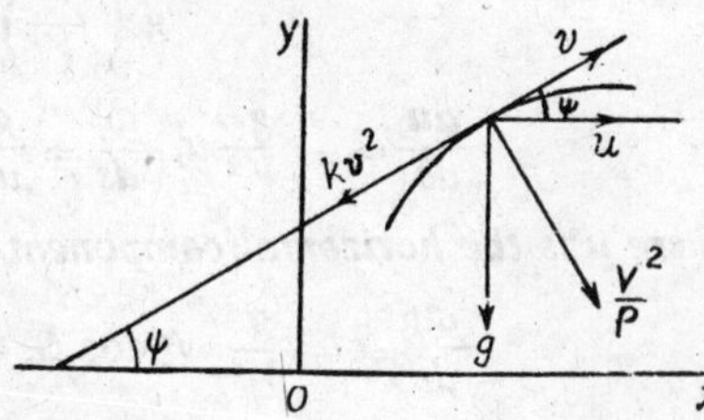

$$\frac{du}{dt}=-kv^2 \cos \psi$$

$$\frac{v^2}{\rho}=g \cos \psi$$

But $\rho=-\frac{ds}{d\psi}$ as ψ is decreasing

$\therefore$ $$g \cos \psi=-v^2 \frac{d\psi}{ds}=-v\frac{d\psi}{dt}$$

$\therefore$ dividing by the first equation we get

$$\frac{v\, d\psi}{du}=\frac{g}{kv^2}$$

or $$\frac{du}{d\psi}=\frac{kv^3}{g}=\frac{ku^3 \sec^3 \psi}{g}$$

or $$\frac{du}{u^3}=\frac{k}{g}\sec^3\psi\, d\psi$$

Integrating

$$-\frac{1}{2u^2}=\frac{k}{2g}\left[\tan \psi \sec \psi + \log(\tan \psi+\sec \psi)\right]+\text{const.}$$

Also $$\frac{du}{dt}=-kv^2 \cos \psi=-kvu=-ku\frac{ds}{dt}$$

$\therefore$ $$\frac{du}{u}=-kds$$

$\therefore$ $$u=u_0 e^{-ks}$$

if when $s=0, u=u_0$

$\therefore$ $$\frac{e^{2ks}}{u_0^2}=-\frac{k}{g}$$

$$\left[\tan \psi \sec \psi+\log (\tan \psi+\sec \psi)\right]+\text{const.}$$

Ex. 8. *If the resistance per unit mass is*

$$g\left(\frac{v}{V}\right)^2, \textit{ prove that}$$

$$\frac{du}{ds}=-\frac{g}{V^2}u,\ \frac{d\psi}{ds}=\frac{g}{u^2}\cos^3\psi,$$

where u is the horizontal component of velocity.

$$\frac{du}{dt}=-\frac{g}{V^2}\,v^2\cos\psi,\ \frac{v^2}{\rho}=g\cos\psi,\ u=v\cos\psi$$

$$\therefore\quad \frac{du}{dt}=-\frac{g}{V^2}uv=-\frac{gu}{V^2}\frac{ds}{dt}$$

or $$\frac{du}{ds}=-\frac{gu}{V^2}$$

$$\frac{d\psi}{ds}=\frac{1}{\rho}=\frac{g\cos\psi}{v^2}=\frac{g\cos^3\psi}{u^2}.$$

EXAMPLES (IX B)

1. Prove that in the motion of a projectile in a resisting medium, resistance being k times the velocity, projected with velocity U at an angle α with the horizon,

(*i*) the direction will again make an angle α after time

$$\frac{1}{k}\log\left(1+\frac{2kU}{g}\sin\alpha\right).$$

(*ii*) time to the greatest height is less than half the time of flight.

(*iii*) the acceleration retains a fixed direction and diminishes without limit to zero.

2. A heavy particle is projected in a resisting medium. If v be the velocity at any time, ϕ the inclination to the vertical of the direction of motion and f the retardation, prove that

$$\frac{1}{v}\frac{dv}{d\phi}+\cot\phi+\frac{f}{g\sin\phi}=0.$$

3. If the resistance to a projectile's motion is constant and is equal to kg, show that the velocity v at any point is given by

$$v(1-\sin\psi)^k=u_0\cos^{k-1}\psi,$$

ψ being the slope and u_0 the velocity at the highest point.

10

Constrained Motion

10·1. Equation of Energy. *A particle is moving on a smooth plane curve under the action of given forces; to find the motion.*

Let P be the position of the particle at time t, its co-ordinates with respect to fixed axes be (x, y) and its distance measured along the curve from some fixed point A on it be s. Let (X, Y) be the components of the external force acting on it and R be the reaction of the curve normally to the curve.

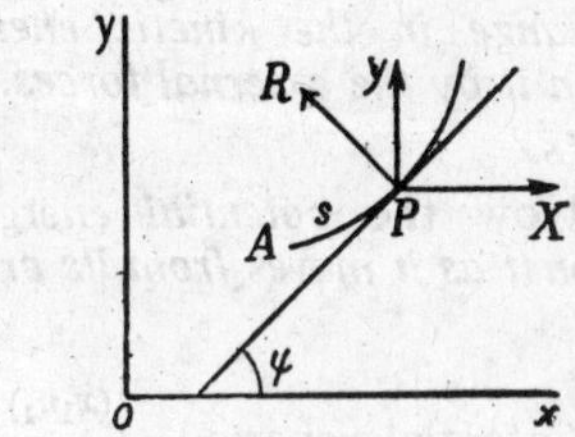

Resolving tangentially and normally, we have

$$mv\frac{dv}{ds}=X\cos\psi+Y\sin\psi \qquad \ldots(1)$$

$$m\frac{v^2}{\rho}=R-X\sin\psi+Y\cos\psi \qquad \ldots(2)$$

From (1), $\quad mv\dfrac{dv}{ds}=X\dfrac{dx}{ds}+Y\dfrac{dy}{ds}$

Therefore by integrating

$\frac{1}{2}mv^2=\int(X\,dx+Y\,dy)+C$ where C is some constant.

If (X, Y) be derived from some potential function $\phi\,(x, y)$

so that $X=\frac{\partial \phi}{\partial x}$, $Y=\frac{\partial \phi}{\partial y}$, we get

$$\tfrac{1}{2}mv^2=\int\left(\frac{\partial \phi}{\partial x}\,dx+\frac{\partial \phi}{\partial y}\,dy\right)+C$$

$$=\int d\phi+C=\phi\,(x, y)+C$$

If the particle be projected from a point $(x_0 y_0)$ with velocity v_0,

then $\tfrac{1}{2}mv_0^2=\phi\;(x_0, y)_0+C$

$\therefore$ $\tfrac{1}{2}mv^2-\tfrac{1}{2}mv_0^2=\phi\,(x, y)-\phi(x_0, y_0)$

This is independent of the path followed in moving from the point (x_0, y_0) to the present position (x, y) but depends only on the initial and final positions of the particle.

Now the function $\phi(x, y)=\int(X\,dx+Y\,dy)$ is the work done on the particle by the external forces and is called the *work-function.*

Such forces which have a potential function are called *conservative forces*. Thus we get that *when the force system is conservative, the change in the kinetic energy of the particle is equal to the work done on it by the external forces. This relation is called the equation of energy.*

Now the *potential energy of a particle is defined as the work done on it as it moves from its present position to some standard position.* Thus

$$\text{Potential energy}=\int_{(xy)}^{(x_1y_1)}(X\,dx+Y+dy)$$

where (x_1y_1) is the standard position

$$=\phi(x_1, y_1)-\phi(x, y)$$

$\therefore$ Kinetic energy + Potential energy

$$=\tfrac{1}{2}mv_0^2+\phi(x, y)-\phi(x_0, y_0)+\phi(x_1, y_1)-\phi(x, y)$$

$$=\tfrac{1}{2}mv_0^2-\phi(x_0, y_0)+\phi(x_1\, y_1)$$

$$=\text{constant.}$$

Thus again *when the force system is conservative, the sum of the kinetic energy and the potential energy is constant throughout the motion. This is called the principle of conservation of energy.*

Ex. 1. *A particle slides down the smooth curve $y=a\sinh\frac{x}{a}$ the axis of x being horizontal and the axis of y downwards, starting from rest at the point where the tangent is inclined at α to the horizon; show that it will leave the curve when it has fallen through a vertical distance $a\sec\alpha$.*

Sol. Let P be the position of the particle at time t, whose depth below x-axis is y.

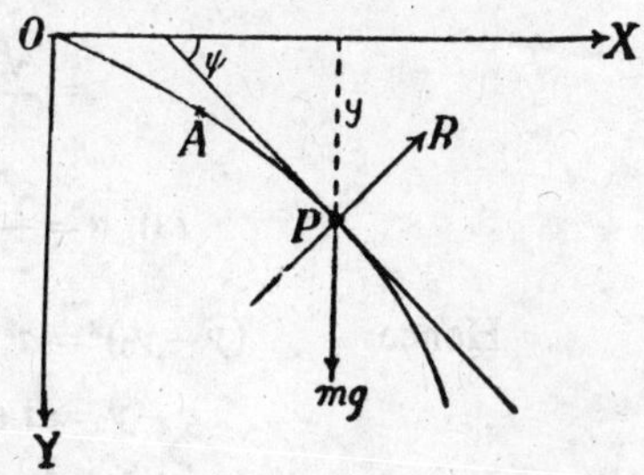

If v be the velocity there, the equation of energy gives $v^2=2g(y-y_0)$ where y_0 is the depth of A, the starting point.

The normal equation is

$$\frac{mv^2}{\rho}=mg\cos\psi-R.$$

The particle will leave the curve when $R=0$

$$v^2=g\rho\cos\psi.$$

Equation to the curve is $y=a\sinh\dfrac{x}{a}$

$$\therefore \qquad \tan\psi=\frac{dy}{dx}=\cosh\frac{x}{a}$$

$$\frac{d^2y}{dx^2}=\frac{1}{a}\sinh\frac{x}{a}$$

$$\therefore \qquad \rho=\frac{\left(1+\cosh^2\frac{x}{a}\right)^{\frac{3}{2}}}{\frac{1}{a}\sinh\frac{x}{a}}$$

$$\therefore \qquad \rho\cos\psi=\frac{\left(1+\cosh^2\frac{x}{a}\right)^{\frac{3}{2}}}{\frac{1}{a}\sinh\frac{x}{a}}\times\frac{1}{\left(1+\cosh^2\frac{x}{a}\right)^{\frac{1}{2}}}$$

$$=\frac{1+\cosh^2\frac{x}{a}}{\frac{1}{a}\sinh\frac{x}{a}}=\frac{2+\sinh^2\frac{x}{a}}{\frac{1}{a}\sinh\frac{x}{a}}$$

$$=\frac{2+\frac{y^2}{a^2}}{\frac{y}{a^2}}=\frac{2a^2+y^2}{y}$$

$$\therefore \qquad 2g(y-y_0)=g\rho\cos\psi$$

i.e., $$2(y-y_0)=\frac{2a^2+y^2}{y}$$

or $$2y^2-2yy_0=2a^2+y^2 \quad \text{or } y^2-2yy_0=2a^2$$

or $(y-y_0)^2=2a^2+y_0^2$. But $\tan\psi=\cosh\frac{x}{a}=\sqrt{1+\sinh^2\frac{x}{a}}$

$$=\frac{\sqrt{a^2+y^2}}{a}$$

$$\therefore \qquad \tan\alpha=\frac{\sqrt{a^2+y_0^2}}{a} \text{ or } a^2+y_0^2=a^2\tan^2\alpha$$

Hence $\qquad (y-y_0)^2=a^2+a^2\tan^2\alpha=a^2\sec^2\alpha$

$$\therefore \qquad y-y_0=a\sec\alpha.$$

Ex. 2. *A particle describes a smooth curve under the action of gravity, describing equal vertical distances in equal times, and starting in a vertical direction. Show that the curve is a semi-cubical parabola the tangent at the cusp of which is vertical.*

Sol. Take x-axis horizontal and y-axis vertical.

Given $\qquad \dot{y}=\text{constant}=V$ say.

From equation of energy $v^2-V^2=2g.y$

i.e., $\qquad \dot{x}^2+\dot{y}^2-V^2=2gy$

i.e., $\qquad \dot{x}^2=2gy.$

But $\qquad \dot{x}=\frac{dx}{dy}\dot{y}=V\frac{dx}{dy}$

$$\therefore \qquad V^2\left(\frac{dx}{dy}\right)^2=2gy \text{ or } \frac{dx}{dy}=\frac{\sqrt{2gy}}{V}$$

$\therefore \qquad x=\frac{2}{3}\frac{\sqrt{2gy^3}}{V}$ the constant of integration

is zero as $\qquad x=y=0$ initially.

i.e., the curve is of the form

$$ax^2=y^3 \text{ where } a \text{ is a constant.}$$

This is a semi-cubical parabola, whose cusp is the origin and tangent there is the y-axis.

Ex. 3. *A particle of mass m moves in a smooth circular tube of radius a, under the action of a force, equal to $m\mu\times$distance, to a point inside the tube at a distance c from its centre ; if the particle be placed very nearly at its greatest distance from the centre of force, show that it will describe the quadrant ending at its least distance in time*

$$\sqrt{\frac{a}{\mu c}}\log(\sqrt{2}+1).$$

Sol. C is the centre of force, A the starting point, P the position of the particle at any instant, where $\angle AOP=\theta$.

The equation of energy gives

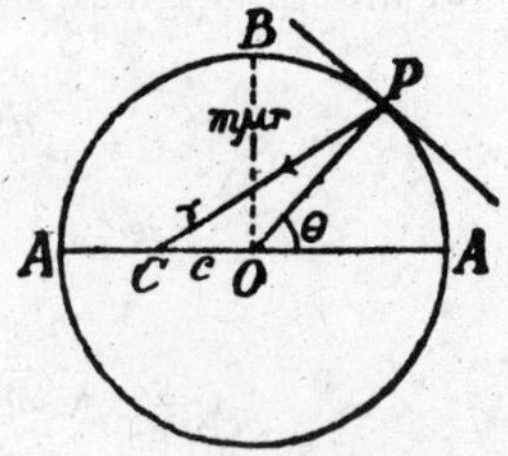

$$\tfrac{1}{2}mv^2=-\int_A^P m\mu r\,dr$$

i.e., $$\tfrac{1}{2}a^2\dot{\theta}^2=-\int_{a+c}^{r}\mu r\,dr=-\frac{\mu}{2}r^2\Big]_{a+c}^{r}$$

$$=-\frac{\mu}{2}\Big[r^2-(a+c)^2\Big]$$

$$=-\frac{\mu}{2}\Big[a^2+c^2-2ac\cos\theta-(a+c)^2\Big]$$

$$=2\mu\, ac\sin^2\theta/2$$

$$\therefore\qquad \dot{\theta}=2\sqrt{\frac{\mu c}{a}}\ \sin\frac{\theta}{2}.$$

If t be the time B to A'

$$2\sqrt{\frac{\mu c}{a}}t=\int_{\pi/2}^{\pi}\frac{d\theta}{\sin\theta/2}=2\log\tan\frac{\theta}{4}\Big]_{\pi/2}^{\pi}$$

$$=2\log\frac{\tan\pi/4}{\tan\pi/8}=-2\log\tan\frac{\pi}{8}=2\log\frac{\cos\pi/8}{\sin\pi/8}$$

$$=\log\frac{\cos^2\pi/8}{\sin^2\pi/8}=\log\frac{1+\cos\pi/4}{1-\cos\pi/4}=\log\frac{\sqrt{2}+1}{\sqrt{2}-1}$$

$$=\log(\sqrt{2}+1)^2$$

$$=2\log(\sqrt{2}+1)$$

$$t=\sqrt{\frac{a}{\mu c}}\log(\sqrt{2}+1).$$

Ex. 4. *A bead is constrained to move on a smooth wire in the form of an equiangular spiral. It is attracted to the pole of the spiral by a force* $=\frac{m\mu}{r^2}$ *and starts from rest at a distance b from the pole. Show that, if the equation of the spiral be* $r=ae^{\theta\cot\alpha}$, *the time of arriving at the pole is* $\frac{\pi}{2}\sqrt{\frac{b^3}{2\mu}}\sec\alpha$.

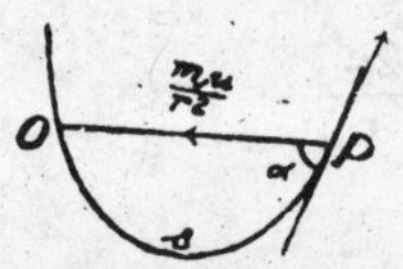

Sol. The equation of er

$$\tfrac{1}{2}m\dot{s}^2=-\int_b^r\frac{m\mu}{r^2}\,dr=\frac{r.}{r}\Big]_b$$

$$=m\mu\left(\frac{1}{r}-\frac{1}{b}\right).$$

But

for an equiangular spiral of angle $\alpha, \dfrac{dr}{ds}=\cos\alpha$

$$\therefore \qquad ds=dr\sec\alpha \quad \therefore \quad \dot{s}=\dot{r}\sec\alpha.$$

$$\dot{r}^2\sec^2\alpha=2\mu\frac{b-r}{br}$$

$$\therefore \qquad \dot{r}=-\sqrt{2\mu}.\cos\alpha\sqrt{\frac{b-r}{br}}, \text{ minus because } r \text{ is decreasing}$$

$$\text{or} \quad \sqrt{2\mu}\cos\alpha.t=-\int_b^0\sqrt{\frac{br}{b-r}}\,dr=\int_0^b\sqrt{\frac{br}{b-r}}\,dr$$

Put $\qquad r=b\sin^2\theta.$

$$\therefore \quad \sqrt{2\mu}\cos\alpha\,.t=\int_0^{\pi/2}\sqrt{\frac{b^2\sin^2\theta}{b\cos^2\theta}}\,2b\sin\theta\cos\theta\,d\theta$$

$$=2b^{3/2}\int_0^{\pi/2}\sin^2\theta\,d\theta$$

$$=2b^{3/2}\frac{\pi}{4}=\frac{\pi}{2}b^{3/2}$$

$$t=\frac{\pi}{2}\sqrt{\frac{b^3}{2\mu}}\sec\alpha.$$

Ex. 5. *A small bead of mass m moves in a smooth circular wire being acted upon by a force $m\mu/r^2$ to a point within the circle at a distance b from the centre ; show that in order that the bead may move completely round the circle, its velocity at the point of the wire nearest the centre of force must not be less than*

$$\sqrt{\frac{4\mu b}{(a^2-b^2)}}.$$

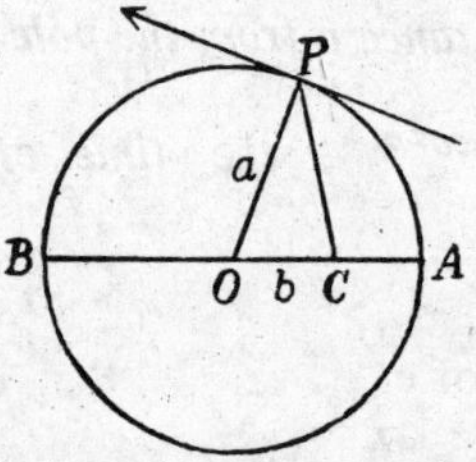

Sol. The equation of energy is

$$\tfrac{1}{2}m(v^2-V_A^2)=-\int_{a-b}^{r}\frac{m\mu}{r^2}\,dr$$

$$=m\mu\left(\frac{1}{r}-\frac{1}{a-b}\right)$$

$$\therefore \quad v^2=V^2_A-\frac{2\mu}{a-b}+\frac{2\mu}{r}$$

$$\therefore \qquad V^2_B=V^2_A-\frac{2\mu}{a-b}+\frac{2\mu}{a+b}>0$$

in order it may move completely round hence

$$V^2_A > 2\mu\left(\frac{1}{a-b} - \frac{1}{a+b}\right) = \frac{2\mu.2b}{a^2-b^2} = \frac{4\mu b}{a^2-b^2}.$$

Ex. 6. *A smooth parabolic curve is fixed with its plane vertical and axis horizontal. show that a heavy particle placed on the arc at a height above the axis equal to the latus rectum, will run off the curve on arriving at the end of the latus rectum and will then describe a parabola having the same latus rectum as the given arc.*

Sol. A is the starting point, P the position at time t.

If v be the velocity at P, then by equation of enegry

$$v^2 = 2g\,(4a-y).$$

The normal equation is

$$m\frac{v^2}{\rho} = mg\cos\psi - R.$$

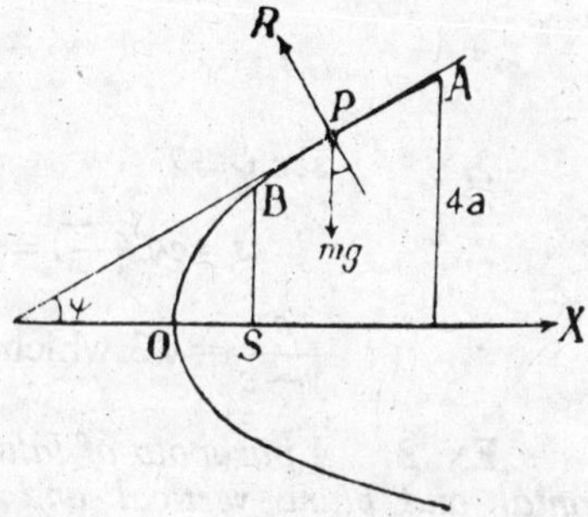

The particle will leave the curve when $R=0$

$$\therefore \quad v^2 = g\rho\cos\psi$$

Hence $g\rho\cos\psi = 2g\,(4a-y)$.

Equation to parabola is $y^2 = 4ax$

$$\therefore \quad -\tan\psi = \frac{dy}{dx} = \frac{2a}{y}, \quad \frac{d^2y}{dx^2} = -\frac{4a^2}{y^3}$$

$$\therefore \quad \rho = \frac{(1+4a^2/y^2)^{3/2}}{-4a^2/y^3} = -\frac{(4a^2+y^2)^{3/2}}{4a^2}$$

Hence $\rho\cos\psi = 2(4a-y)$ gives

$$-\frac{(4a^2+y^2)^{3/2}}{4a^2} \times -\frac{y}{\sqrt{4a^2+y^2}} = 2\,(4a-y)$$

or $\quad y(4a^2+y^2) = 8a^2\,(4a-y)$

or $\quad y^3 + 12a^2y = 32a^3$ which is satisfied by $y=2a$.

Then $\quad v^2 = 2g\,(4a-2a) \quad = 4ag.$

$$\text{Latus rectum of the free path} = \frac{2v^2\sin^2\psi}{g}$$

$$= \frac{8ag}{g} \cdot \frac{4a^2}{4a^2+4a^2} = 4a.$$

Ex. 7. *A particle falls from rest at the vertex of an inverted catenary; prove that the particle will leave the curve when the path described is $\sqrt{3}$ times the vertical distance through which the particle has fallen.*

Sol. Equation of energy gives $v^2=2g(y-c)$

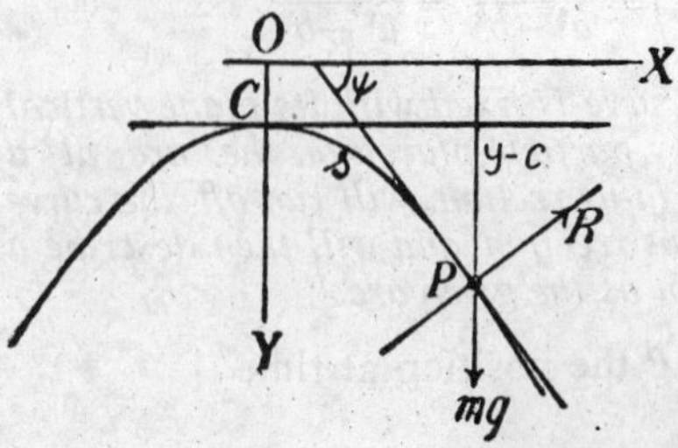

The normal equation is

$$m\frac{v^2}{\rho}=mg\cos\psi-R.$$

Hence the particle will leave when $R=0$ *i.e.*, when

$$g\rho\cos\psi=2g(y-c).$$

Equation to a catenary is $s=c\tan\psi,\ y=c\sec\psi$

$$\therefore\quad \rho=c\sec^2\psi$$

Hence $c\sec^2\psi\cos\psi$

$$=2(c\sec\psi-c)$$

$$\therefore\qquad \sec\psi=2$$

$$\therefore\qquad s=c\sqrt{4-1}=\sqrt{3}c \text{ and } y-c=c\sec\psi-c=c$$

$$\therefore\qquad \frac{s}{y-c}=\sqrt{3} \text{ which is the answer.}$$

Ex. 8. *A parabola of latus rectum 4a, is placed with axis horizontal and plane vertical and a heavy smooth particle is projected from the vertex with a velocity due to a height h and moves on the concave side of the arc. Prove that the vertical height attained before leaving the arc is two-thirds of the greatest height and if θ be the angle described about the focus before leaving the curve*

$$h=a\left(\tan^2\frac{\theta}{2}+3\tan\frac{\theta}{2}\right).$$

Sol. If u be the velocity of projection. $u^2=2gh$

The equation of energy gives

$$v^2=u^2-2gy.$$

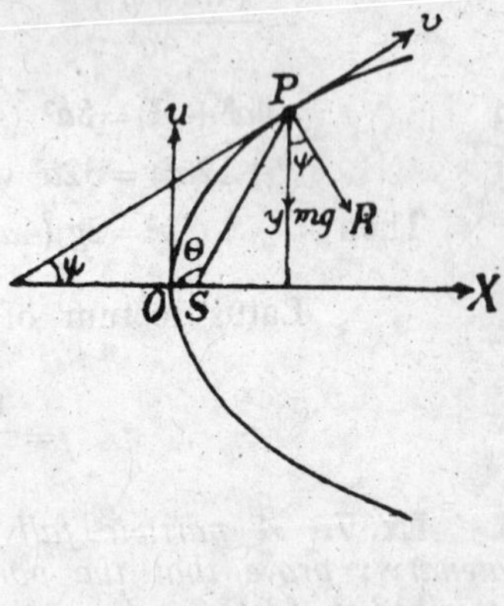

The normal equation is

$$m\frac{v^2}{\rho}=R+mg\cos\psi$$

The particle will leave the arc when $R=0$

$$\therefore\quad g\rho\cos\psi=u^2-2gy=2g(h-y)$$

i.e., $\rho\cos\psi=2(h-y).$

Equation to the parabola is

$$y^2=4ax$$

$$\therefore\qquad \rho=-\frac{(4a^2+y^2)^{3/2}}{4a^2}$$

$$\tan\psi = -\frac{2a}{y} \text{ as in Ex. 6., worked out}$$

$$\therefore \quad -\frac{(4a^2+y^2)^{\frac{3}{2}}}{4a^2} \times -\frac{y}{\sqrt{4a^2+y^2}} = 2(h-y)$$

or $\quad y(4a^2+y^2) = 8a^2(h-y)$

or $\quad y^3 + 12a^2 y = 8a^2 h.$

If $\quad \angle OSP = \theta, \ \frac{2a}{r} = 1 + \cos\theta = 2\cos^2\theta/2$

$$\therefore \quad r = \frac{a}{\cos^2\theta/2}$$

Hence $\quad y = r\sin\theta = 2a\tan\theta/2$

$$\therefore \quad 8a^3\tan^3\theta/2 + 12a^2\, 2a\tan\theta/2 = 8a^2 h$$

or $\quad a(\tan^3\theta/2 + 3\tan\theta/2) = h.$

After leaving the arc, height attained

$$= \frac{v^2\sin^2\psi}{2g} = \frac{g\rho\cos\psi.\sin^2\psi}{2g} = \frac{1}{2}\cdot\frac{y(4a^2+y^2)}{4a^2}\,\frac{4a^2}{4a^2+y^2}$$

$$= \frac{y}{2} = H \text{ say.}$$

$\therefore$ Greatest height attained above the axis $= y + H$

$$= y + \frac{y}{2} = \frac{3}{2}y$$

$\therefore \quad y = \frac{2}{3}$ (greatest height).

Ex. 9. *A particle moves on the outside of an ellipse of eccentricity e where major axis is vertical, starting from rest at the highest point. Show that it will leave the curve at a point whose eccentric angle is ϕ where $e^2\cos^3\phi = 3\cos\phi - 2$.*

Sol. Let P be the position at time t.

$$P \to a\cos\phi,\ b\sin\phi.$$

Equation of energy gives $v^2 = 2g.AN = 2g(a - a\cos\phi)$

$$= 2ga(1-\cos\phi).$$

Normal equation is $\frac{mv^2}{\rho} = mg\cos(\psi - 90) - R = mg\sin\psi - R.$

It will have the curve when $R=0$

$\therefore \quad \rho \sin \psi = 2a(1-\cos \phi).$

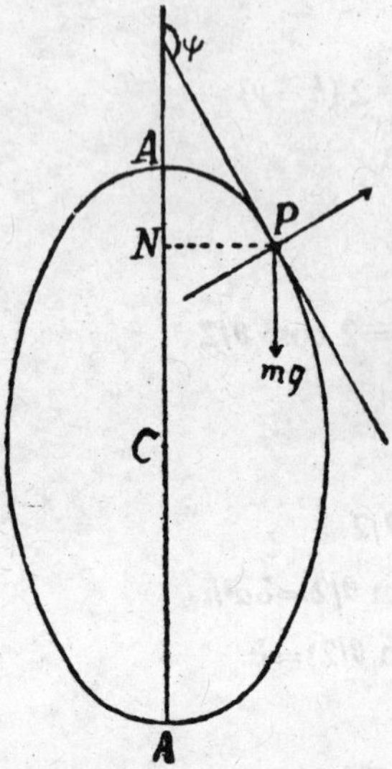

Equation to the tangent at **P** is

$$\frac{x}{a} \cos \phi + \frac{y}{b} \sin \phi = 1.$$

$$\therefore \quad \tan \psi = -\frac{b}{a} \cot \phi$$

$\rho = \dfrac{CD^3}{ab}$ when C is the centre of the ellipse, CD is the semi-diameter conjugate to CP

$$= \frac{(a^2 \sin^2 \phi + b^2 \cos^2 \phi)^{\frac{3}{2}}}{ab}$$

$$\therefore \quad \frac{(a^2 \sin^2 \phi + b^2 \cos^2 \phi)^{\frac{3}{2}}}{ab}.$$

$$\times \frac{b \cos \phi}{\sqrt{a^2 \sin^2 \phi + b^2 \cos^2 \phi}}$$

$$= 2a(1-\cos \phi)$$

i.e., $\quad \cos \phi (a^2 \sin^2 \phi + b^2 \cos^2 \phi) = 2a^2(1-\cos \phi)$

or $\quad \cos \phi [a^2 \sin^2 \phi + a^2(1-e^2) \cos^2 \phi] = 2a^2(1-\cos \phi)$

or $\quad \cos \phi (1-e^2 \cos^2 \phi) = 2(1-\cos \phi)$

or $\quad e^2 \cos^3 \phi = 3 \cos \phi - 2.$

Ex. 10. *From the lowest point of a smooth hollow elliptic cylinder of axes 2a, 2b, with minor axis vertical, particle is projected in a vertical plane perpendicular to the axis of the cylinder ; show that it will leave the cylinder if the velocity of projection lies between*

$$\sqrt{2gb} \text{ and } \sqrt{\frac{g(a^2+b^2)}{b}}.$$

Sol. If v be the velocity at $P(x, y)$

$(x = a \cos \phi, \; y = b \sin \phi)$, the equation of energy is

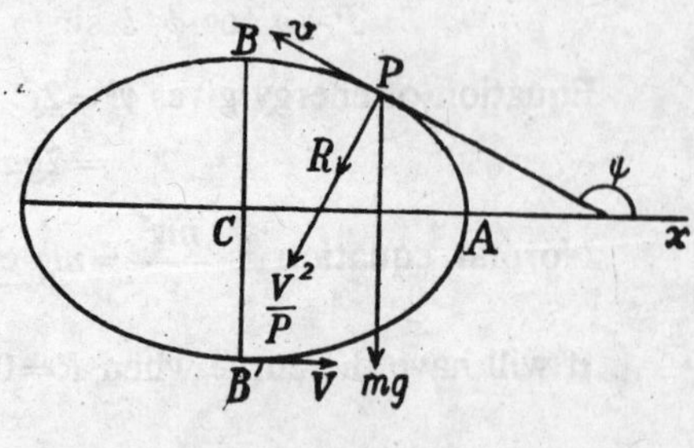

$$\tfrac{1}{2} m (v^2 - V^2) = -mg(y + b)$$

where V is the velocity of projection from B'.

The normal equation is

$$\frac{v^2}{\rho}=R+mg\cos(\psi-180)=R-mg\cos\psi$$

The particle will leave when $R=0$

then $$V^2-2gb\,(1+\sin\varphi)=-g\,\rho\cos\psi$$

i.e., $$V^2-2gb\,(1+\sin\varphi)-g\rho\cos\psi$$

Equation to tangent at P is

$$\frac{x}{a}\cos\varphi+\frac{y}{b}\sin\varphi=1$$

$$\therefore \qquad \tan\psi=-\frac{b}{a}\cot\varphi$$

Since $\psi>90°$, $$\cos\psi=-\frac{a\sin\varphi}{\sqrt{a^2\sin^2\phi+b^2\cos^2\varphi}}$$

Also $\rho=\dfrac{CD^3}{ab}$ where CD is conjugate semi-diameter to CP

$$=\frac{(a^2\sin^2\varphi+b^2\cos^2\phi)^{3/2}}{ab}$$

Hence $$V^2=2gb(1+\sin\phi)+\frac{g}{b}\sin\phi(a^2\sin^2\phi+b^2\cos^2\phi)$$

If V_A be the velocity of projection in order that the particle may leave at A and V_B in order that it may leave at B,

we have $$V_A^2=(V^2)_{\phi=0}=2gb$$

$$V_B^2=(V^2)_{\varphi=90}=4gb+\frac{ga^2}{b}$$

Now the particle will leave somewhere in the arc AB

$$\therefore \qquad V_A^2<V^2<V_B^2 \quad \text{hence.}$$

Ex. 11. *From the lowest point of a smooth hollow cylinder whose cross-section is one-half of the lemniscate $r^2=a^2\cos 2\theta$, with axis vertical and node downwards, a particle is projected with velocity V along the inner surface in the plane of the cross-section; show that it will make a complete revolution if $3V^2>7ag$.*

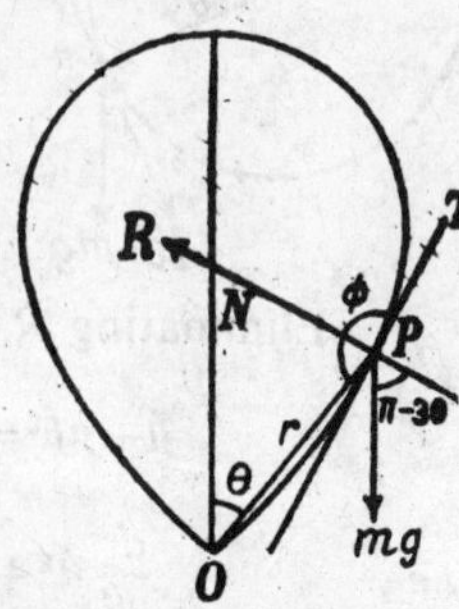

$$r^2=a^2\cos 2\theta$$

Taking log differential

$$\frac{1}{r}\frac{dr}{d\theta}=-\tan 2\theta$$

$$\therefore \quad \cot\phi=-\tan 2\theta$$

$$\therefore \qquad \varphi=90+2\theta$$

$\therefore \qquad p=r\sin\varphi=r\cos 2\theta=\dfrac{r^3}{a^2}$

Also $\qquad \angle NPT=90^\circ$

$\therefore \qquad \angle OPN=2\theta$

$\therefore \qquad \angle ONP=\pi-3\theta$

Equation of energy is $\frac{1}{2}m(v^2-V^2)=-gy,\ y=r\cos\theta$

$$v^2=V^2-2gr\cos\theta$$

normal equation is

$$m\frac{v^2}{\rho}=R-mg\cos(\pi-3\theta)=R+mg\cos 3\theta$$

$$\rho=r\frac{dr}{dp}=\frac{a^2}{3r}$$

$$\therefore \qquad R=m(V^2-2gr\cos\theta).\frac{3r}{a^2}-mg\cos 3\theta$$

Hence $\quad R$ is zero at the highest point if

$$O=(V^2-2ga)\frac{3}{a}-g$$

i.e., $\qquad 3V^2=7ag$

Hence it will make complete revolution if $3V^2>7ag$.

Ex. 12. *A bead slides down a rough circular wire, which is in a vertical plane, starting from rest at the end of the horizontal diameter. When it has described an angle θ about the centre, show that the square of the angular velocity is*

$$\frac{2g}{a(1+4\mu^2)}\left[(1-2\mu^2)\sin\theta+3\mu(\cos\theta-e^{-2\mu\theta})\right].$$

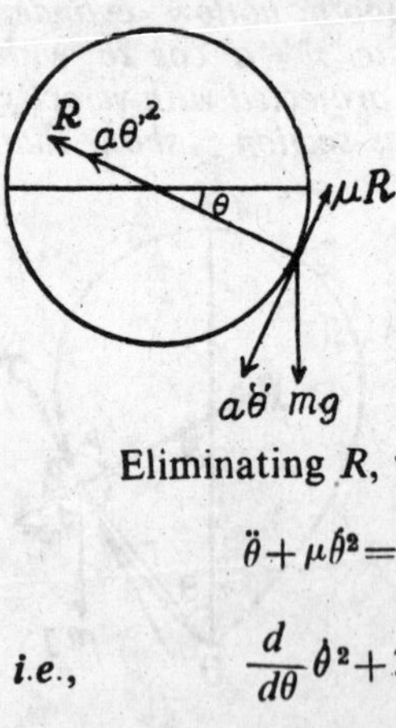

Sol. The forces are reaction R normally towards centre, friction μR tangentially, opposite to the direction of motion and mg vertically downwards.

The equations of motion are

$$ma\ddot{\theta}=mg\cos\theta-\mu R$$

$$ma\dot{\theta}^2=R-mg\sin\theta$$

Eliminating R, we get

$$\ddot{\theta}+\mu\dot{\theta}^2=\frac{g}{a}(\cos\theta-\mu\sin\theta)$$

i.e., $$\frac{d}{d\theta}\dot{\theta}^2+2\mu\dot{\theta}^2=\frac{2g}{a}(\cos\theta-\mu\sin\theta)$$

Integrating $\dot{\theta}^2 e^{2\mu\theta}=\dfrac{2g}{a}\displaystyle\int e^{2\mu\theta}(\cos\theta-\mu\sin\theta)d\theta+C$

$$=\frac{2g}{a}\left[\frac{2\mu\, e^{2\mu\theta}\cos\theta+e^{2\mu\theta}\sin\theta}{1+4\mu^2}\right.$$
$$\left.-\mu\,\frac{2\mu\, e^{2\mu\theta}\sin\theta-e^{2\mu\theta}\cos\theta}{1+4\mu^2}\right]+C$$

$$=\frac{2g}{a}\cdot\frac{e^{2\mu\theta}}{1+4u^2}\left\{3\mu\cos\theta+(1-2\mu^2)\sin\theta\right\}+C$$

Initially $\theta=0=\dot{\theta}$

$\therefore$ $\dfrac{2g}{a}\cdot\dfrac{3\mu}{1+4\mu^2}+C=0$ hence.

Ex. 13. *A particle is projected along the inner surface of a rough sphere and is acted on by no forces ; show that it will return to the point of projection after time* $\dfrac{a}{\mu V}(e^{2\pi\mu}-1)$ *where a is the radius of the sphere, V the velocity of projection.*

Sol. Since there are no forces, the particle will continue to move in the plane section *i.e.*, in a circle.

Only forces are R and μR.

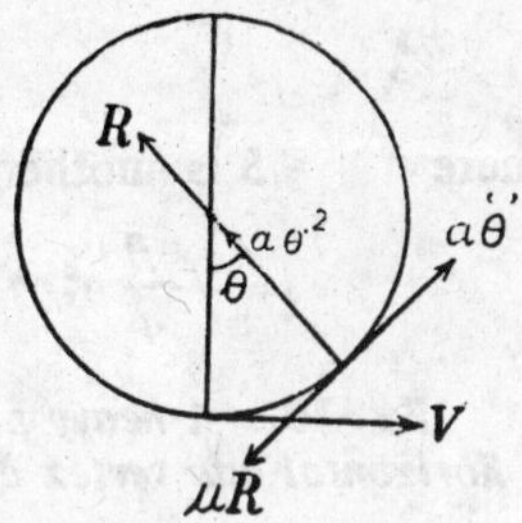

Equations of motion are

$ma\ddot{\theta}=-\mu R$ and $ma\dot{\theta}^2=R$

$\therefore$ $\ddot{\theta}+\mu\dot{\theta}^2=0$

or $\dfrac{\ddot{\theta}}{\dot{\theta}}+\mu\dot{\theta}=0$

Integrating $\log\dot{\theta}+\mu\theta=C$

Initially $\theta=0,\ a\dot{\theta}=V$

$\therefore$ $\log\dfrac{V}{a}=C$

Hence $\log\dfrac{a\dot{\theta}}{V}+\mu\theta=0$

or $a\dot{\theta}=Ve^{-\mu\theta}$

$\therefore$ $t=\dfrac{a}{V}\displaystyle\int_0^{2\pi}e^{\mu\theta}\,d\theta=\dfrac{a}{\mu V}(e^{2\pi\mu}-1).$

Ex. 14. *A bead moves along a rough curved wire which is such that it changes its direction of motion with constant angular velocity. Show that a possible form of the wire is an equiangular spiral.*

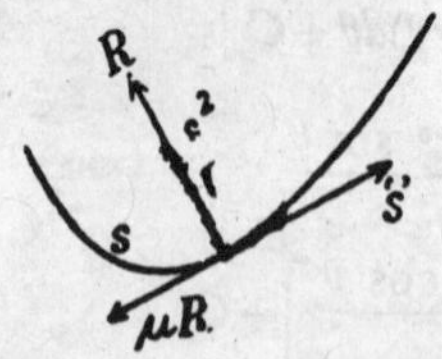

Sol. Equations of motion are

$$m\ddot{s}=-\mu R$$

$$m\frac{\dot{s}^2}{\rho}=R$$

Eliminating R, $\ddot{s}+\mu\dfrac{\dot{s}^2}{\rho}=0$

or $$\frac{d}{ds}\dot{s}^2+2\mu\frac{\dot{s}^2}{\rho}=0$$

or $$\frac{d\psi}{ds}\frac{d}{d\psi}\dot{s}^2+2\mu\frac{\dot{s}^2}{\rho}=0$$

or $$\frac{d}{d\psi}\dot{s}^2+2\mu\dot{s}^2=0$$

Integrating $\dot{s}^2e^{2\mu\psi}=\text{const.}$

or $\dot{s}\,e^{\mu\psi}=C$

Given that $\dot{\psi}=\text{const.}$

$\therefore$ $$\frac{ds}{d\psi}\dot{\psi}\,e^{\mu\psi}=C$$

$\therefore$ $$\frac{ds}{d\psi}=B\,e^{-\mu\psi}$$

where B is another constant

$$s=-\frac{B}{\mu}e^{-\mu\psi}+\text{const.}=a+b\,e^{-\mu\psi}\text{—equiangular spiral.}$$

Ex. 15. *A heavy particle slides down a rough cycloid whose base is horizontal and vertex downwards ; show that*

(*a*) *if it starts from rest at the cusp and comes to rest at the vertex then* $\mu^2e^{\mu\pi}=1.$

(*b*) *if it starts from rest at a point where the tangent makes an angle* θ *with the horizontal and comes to rest at the vertex then* $\mu e^{\mu\theta}=\sin\theta-\mu\cos\theta.$

Sol When the motion is down the arc, the friction μR acts up the tangent.

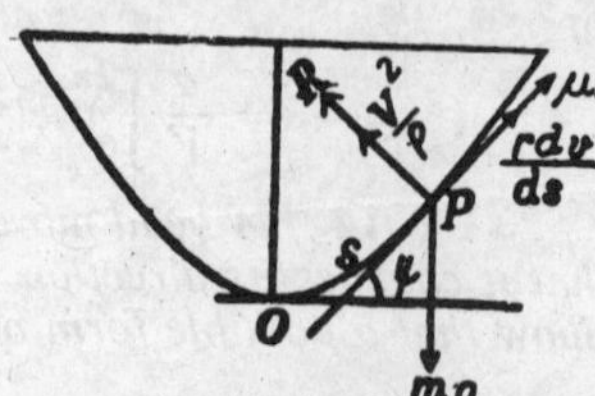

Equations of motion are

$$mv\frac{dv}{ds}=\mu R-mg\sin\psi$$

$$m\frac{v^2}{\rho}=R-mg\cos\psi$$

Eliminating R,

$$v\frac{dv}{ds}-\mu\frac{v^2}{\rho}=g(\mu\cos\psi-\sin\psi)$$

i.e.,
$$\frac{1}{2}\frac{d}{ds}v^2-\mu\frac{v^2}{\rho}=g(\mu\cos\psi-\sin\psi)$$

or
$$\frac{d}{d\psi}v^2-2\mu v^2=2g\rho(\mu\cos\psi-\sin\psi).$$

Integrating

$$v^2e^{-2\mu\psi}=2g\int\rho e^{-2\mu\psi}(\mu\cos\psi-\sin\psi)d\psi+C$$

for a cycloid, $s=4a\sin\psi,\ \rho=4a\cos\psi$

$$\therefore\quad v^2e^{-2\mu\psi}=8ag\int e^{-2\mu\psi}\cos\psi(\mu\cos\psi-\sin\psi)d\psi+C$$

Put $z=e^{-\mu\psi}(\mu\cos\psi-\sin\psi)$

$$\therefore\quad dz=-e^{-\mu\psi}(1+\mu^2)\cos\psi d\psi$$

Hence we get

$$v^2e^{-2\mu\psi}=-\frac{8ag}{1+\mu^2}\int z\,dz+C=-\frac{4ag}{1+\mu^2}z^2+C$$

$$=-\frac{4ag}{1+\mu^2}(\mu\cos\psi-\sin\psi)^2e^{-2\mu\psi}+C$$

(*a*) Conditions are

$v=0$ when $\psi=\pi/2$ and when $\psi=0$

$$\therefore\quad C=\frac{4ag}{1+\mu^2}e^{-\pi\mu}=\frac{4ag}{1+\mu^2}\mu^2$$

or $\mu^2=e^{-\pi\mu}$

or $\mu^2e^{\pi\mu}=1$

(*b*) Conditions are

$v=0$ when $\psi=\theta$ and when $\psi=0$

$$C=\frac{4ag}{1+\mu^2}(\mu\cos-\theta\sin\theta)^2e^{-2\mu\theta}=\frac{4ag}{1+\mu^2}\mu^2$$

giving $\mu e^{\mu\theta}=\sin\theta-\mu\cos\theta.$

EXAMPLES X

1. A particle slides on the curve $x=2\sqrt{a(y-a)}$ with a velocity due to a fall from the x-axis which is horizontal, the y-axis being downwards. Prove that the pressure on the curve at any time is zero so that the curve is a free path under gravity.

2. A particle slides down a catenary whose plane is vertical and vertex upwards the velocity at any point being due to the level of the directrix. Prove that the pressure at any point varies inversely as the distance of that point from the directrix.

3. A particle moves in a smooth tube in the form of a catenary being attracted to the directrix with a force proportional to the distance from the directrix. Prove that the period of oscillation is independent of the amplitude.

4. A small bead is projected with any velocity along the smooth circular wire under the action of a force varying inversely as the fifth power of the distance from a centre of force situated on the circumference. Prove that the pressure on the wire is constant.

5. A particle is projected from the vertex of a smooth parabolic tube of latus rectum $4a$ along the tube and is acted upon by a repulsive force mgr/c from the focus; if the velocity of projection is that which would be acquired in moving from focus to the vertex, prove that time of describing an angle θ about the focus is $2\sqrt{\dfrac{c}{g}}\log\tan\frac{1}{4}(\pi+\theta)$.

6. A particle is projected horizontally with velocity V along the inside of a rough vertical circle from the lowest point. Prove that if it completes the circuit it will return to the lowest point with a velocity v given by

$$v^2=V^2e^{-4\pi\mu}-2ag.\frac{2\mu^2-1}{1+4\mu^2}\left(1-e^{-4\pi\mu}\right).$$

7. A ring which can slide on a rough circular wire in a vertical plane is projected from the lowest point with such a velocity as will just take it to the horizontal diameter; if the ring returns to the lowest point, show that its velocity on arrival is to the velocity of projection as

$$\sqrt{(1-2\mu^2-3\mu e^{-\pi\mu})}:\sqrt{(1-2\mu^2+3\mu e^{\pi\mu})}.$$

8. A heavy particle slides down a rough cycloid where base is horizontal and vertex downwards. Show that

(*a*) if it starts from rest at the cusp and comes to rest on the other side at a point where the tangent is at $\angle 45°$ to vertical, then

$$3\pi\mu+4\log(1+\mu)=2\log 2$$

(*b*) if it starts from rest at the cusp, then its velocity at the vertex is to its velocity at the same point when the cycloid is smooth as

$$\sqrt{(e^{-\pi\mu}-\mu^2)}:\sqrt{1+\mu^2}$$

(*c*) if it starts from rest at the cusp and comes to rest before reacting the vertex then

$$\mu e^{\frac{1}{2}\pi\mu}>1.$$

11

Moment of Inertia

11·1. Definitions. Consider a body of mass M and take a particle of it of mass m and at a distance r from a given line. Form the product mr^2. Corresponding to every particle of the body we get a product of this type so that the total sum of all these products i Σmr^2. This quantity Σmr^2 is called the *moment of inertia* of the body with respect to the given line. If this sum can be put in the form MK^2 then K is called the *radius of gyration* of the body about the line. Similarly we can define the moment of inertia with respect to a plane or a point.

If the body be a lamina and (x, y) be the co-ordinates of the element of mass m referred to rectangular axes ox, oy in the plane of the lamina, the moments of inertia of the lamina about the axes ox, oy are respectively Σmy^2 and Σmx^2. The quantity Σmxy is called the *product* of inertia with respect to the axes ox, oy.

In three dimensions we take three axes ox, oy, oz mutually at right angles. If (x, y, z) be the co-ordinates of the element m, the moments of inertia of the body about the axes ox, oy, oz are respectively

$$A=\Sigma m(y^2+z^2), \quad B=\Sigma m(z^2+x^2), \quad C=\Sigma m(x^2+y^2).$$

The products of inertia with respect to the axes of y and z, of z and x and of x and y are respectively

$$D=\Sigma myz, \ E=\Sigma mzx, \ F=mxy.$$

The moment of inertia with respect to the origin is

$$H=\Sigma m(x^2+y^2+z^2)=\Sigma mr^2$$

where r is the distance of particle m from origin.

In the case of a plane lamina, taking the plane of the lamina as (xy) plane so that co-ordinates of any element m of the lamina are (x, y, o), we have

$$A=\Sigma my^2,\ B=\Sigma mx^2,\ C=\Sigma m(x^2+y^2)$$

so that $\quad C=A+B$

This proves the important result :

The sum of the moment of inertia of a lamina about two rectangular axes ox, oy in its plane is equal to the moment of inertia of the lamina about an axis through O perpendicular to the plane.

11·2. *Moments of inertia in some simple cases.*

(a) *Rod of length 2a and mass M.*

Let O be the middle point. Take an element dx of the rod at distance x from O, its mass $=\dfrac{M}{2a}dx$.

Its moment of inertia about a perp. line through $O=\dfrac{M}{2a}dx.x^2$.

Therefore, $M.I.$ of the rod about this line

$$=\int_{-a}^{a}\frac{M}{2a}x^2dx=M.\frac{a^2}{3}.$$

Similarly $M.I.$ about a perp. line through one end

$$=\int_{0}^{2a}\frac{M}{2a}\ x^2dx=M.\ \frac{4a^2}{3}.$$

(b) *Uniform rectangular lamina of sides 2a, 2b and mass M.*

Let O be the $C.G.$ of the lamina, OX, OY, parallel to the sides $2a$, $2b$.

To find $M.I.$ about OY, take a strip parallel to OY of breadth dx at a distance x from O. Its mass $=\dfrac{M}{4ab}2b\ dx=\dfrac{M}{2a}dx$ and its $M.I.$ about $OY=\dfrac{M}{2a}dx.x^2$.

Therefore, $M.I.$ of the lamina about OY

$$=\int_{-a}^{a}\frac{M}{2a}\ x^2dx=M.\ \frac{a^2}{3}.$$

Similarly $M.J.$ about $OX=M.\ \dfrac{b^2}{3}.$

Therefore, *M.I.* about a perpendicular through O to the plane of the lamina $= M\dfrac{a^2}{3} + M\dfrac{b^2}{3} = M\dfrac{a^2+b^2}{3}$.

(c) *Uniform rectangular parallelopiped of edges 2a, 2b, 2c.*

Let OX, OY, OZ be the axes through the centre O perp. to the faces. To find *M.I.* about OX, divide the parallelopiped into rectangular laminas perp. to OX. Each lamina has sides $2b$ and $2c$ and *M.I.* of each about OX = its mass multiplied by $\dfrac{b^2+c^2}{3}$.

Hence *M.I.* of the whole $= M.\dfrac{b^2+c^2}{3}$.

Similarly about the other two axes.

For a cube, *M.I.* about any of these axes is $M.\dfrac{2a^2}{3}$

(d) *Uniform circular ring of radius a.*

Take $X'OX$ any diameter and P any point on the ring where $\angle XOP = \theta$.

Take an element at $P = a d\theta$, its mass$= \dfrac{M}{2\pi a} a d\theta$, its *M.I.* about $OX = \dfrac{M}{2\pi} d\theta . a^2 \sin^2 \theta$

$\therefore$ *M.I.* of the ring about $X'OX$

$$= \int_0^{2\pi} \frac{M}{2\pi} a^2 \sin^2 \theta d\theta = M\frac{a^2}{2}.$$

But $X'OX$ is any diameter, hence *M.I.* about any diameter is $M\dfrac{a^2}{2}$.

Another Method. Since every point of the ring is at distance a from the centre O, *M.I.* about $O = Ma^2$. Also due to symmetry *M.I.* about any diameter is the same $\therefore$ if A, B be the moment of inertia about x and y-axes, then $A = B$.

so that $$2A = A + B = \Sigma m(y^2+x^2) = \Sigma ma^2 = Ma^2$$

$$\therefore \quad A = M\frac{a^2}{2}.$$

Therefore *M.I.* about a perp. through O to the plane

$$= M\frac{a^2}{2} + M\frac{a^2}{2} = Ma^2$$

(e) *Uniform circular disc of radius a.*

Take a ring of radius r and breadth dr, its mass

$$=\frac{M.2\pi rdr}{\pi a^2}=\frac{2Mrdr}{a^2},$$

its $M.I.$ about any diameter

$$=\frac{2M.rdr}{a^2}\cdot\frac{r^2}{2}$$

$\therefore$ $M.I.$ of the disc $=\int_0^a \frac{2Mrdr}{a^2}\cdot\frac{r^2}{2}=\frac{M.a^2}{4}.$

$\therefore$ $M.I.$ about a perp. through O to the plane$=\frac{M.a^2}{2}.$

(f) *Uniform triangular lamina about one side.*

To find $M.I.$ about BC, divide the lamina by strips parallel to BC. Let PQ be one such strip of breadth dx at distance x from A and let p be the perp. AN.

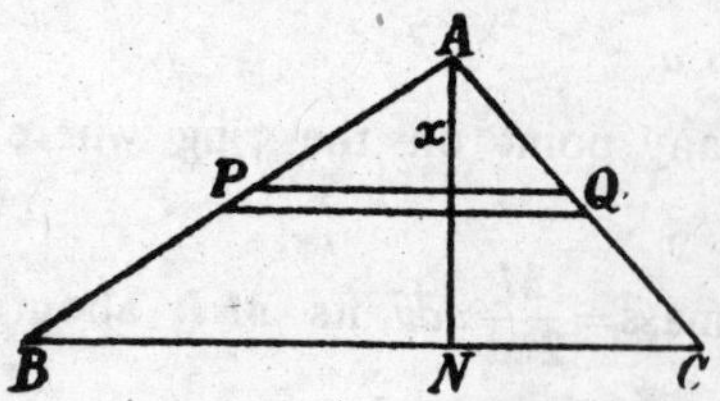

Then $\frac{PQ}{a}=\frac{x}{p}$

$\therefore$ mass of the strip$=\frac{M}{\frac{1}{2}ap}\cdot PQ.\,dx$

$$=\frac{2M}{p^2}xdx.$$

Its $M.I.$ about BC

$$=\frac{2M}{p^2}xdx.\,(p-x)^2.$$

$\therefore$ $M.I.$ of the triangle about $BC=\frac{2M}{p^2}\int_0^p (p-x)^2xdx=\frac{1}{6}M.p^2.$

(g) *Uniform elliptic disc of semi-axes a, b.*

Let OX, OY be the semi-axes. To find $M.I.$ about OX, take a strip parallel to it, of length $2x$ and breadth dy ; its mass

$$=\frac{M}{\pi ab}\cdot 2xdy.$$

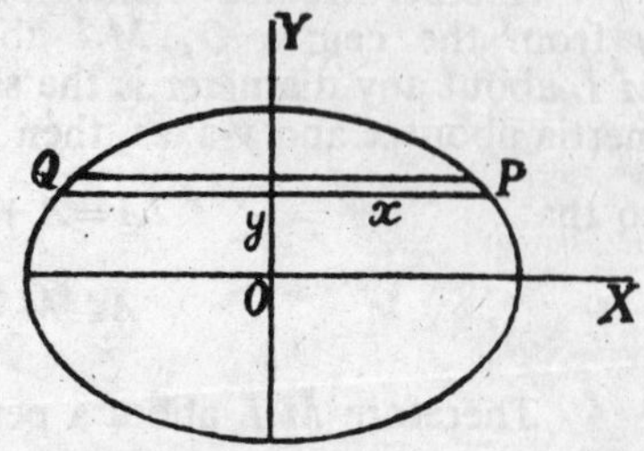

its $M.I.$ about $OX=\frac{M}{\pi ab}\cdot 2xdy.y^2$

$\therefore$ $M.I.$ of the disc about OX

$$=\frac{2M}{\pi ab}\int_{-b}^{b} xy^2dy$$

Put $x=a\cos\phi$, $y=b\sin\phi$ for an ellipse $\frac{x^2}{a^2}+\frac{y^2}{b^2}=1$

so that $dy=b\cos\phi\, d\phi$.

Hence $M.I.=\dfrac{2M}{\pi ab}\cdot 2\int_0^b xy^2dy=\dfrac{4M}{\pi ab}\int_0^{\pi/2} a\cos\phi . b^2\sin^2\phi .$

$b\cos\phi d\phi .$

$$=\frac{4Mb^2}{\pi}\int_0^{\pi/2}\cos^2\phi\sin^2\phi d\phi=\frac{4M.b^2}{\pi}\frac{\pi}{16}$$

$$=M.\frac{b^2}{4}.$$

Similarly $M.I.$ about $OY=M.\dfrac{a^2}{4}$.

Therefore, $M.I.$ about perp. through O to the plane $=M\dfrac{a^2+b^2}{4}$.

(*h*) *Hollow sphere about a diameter.*

Due to symmetry, $M.I.$ about any diameter is the same. Take any three perp. diameters as axes of x, y, z then if A, B, C be the moments of inertia about these axes, we have

$$A=B=C$$

$$\therefore \quad 3A=A+B+C=\Sigma m(y^2+z^2+z^2+x^2+x^2+y^2)$$

$$=2\Sigma m(x^2+y^2+z^2)=2\Sigma ma^2=2Ma^2$$

$$\therefore \quad A=M\frac{2a^2}{3}.$$

This is the $M.I.$ about any diameter.

(*i*) *Solid sphere about a diameter.*

Take a shell of radius r and breadth dr.

Its mass $=\dfrac{4\pi r^2dr}{\frac{4}{3}\pi a^3}.M$; its $M.I.$ about any diameter

$$=\frac{3Mr^2dr}{a^3}\cdot\frac{2r^2}{3}.$$

$\therefore$ $M.I.$ of the solid

$$=\frac{2M}{a^3}\int_0^a r^4dr=M.\frac{2}{5}a^2.$$

(*j*) *Solid ellipsoid about a principal diameter.*

Let the equation of the ellipsoid be $x^2/a^2+y^2/b^2+z^2/c^2=1$.

$$A=8\int_{x=0}^{a}\int_{y=0}^{b\sqrt{1-x^2/a^2}}\int_{z=0}^{c\sqrt{1-x^2/a^2-y^2/b^2}}\rho(y^2+z^2)dxdydz$$

Putting $x=ax^{1/2}$, $y=by^{1/2}$, $z=cz^{1/2}$, we see that the equation of ellipsoid becomes $x+y+z=1$ and $dx=\frac{a}{2}X^{-1/2}dX$, $dy=\frac{b}{2}Y^{-1/2}dY$, $dz=\frac{c}{2}Z^{-1/2}dZ$ so that

$$A=\int_{X=0}^{1}\int_{Y=0}^{1-X}\int_{Z=0}^{1-X-Y}\rho abc(b^2Y+c^2Z)X^{-1/2}Y^{-1/2}Z^{-1/2}dXdYdZ$$

$$=\rho abc[b^2\iiint X^{-1/2}\,Y^{1/2}\,Z^{-1/2}\,dXdYdZ+c^2\iiint X^{-1/2}\,Y^{-1/2}\,Z^{1/2}\,dXdYdZ]$$

$$=\rho abc[b^2\iiint X^{1/2-1}\,Y^{3/2-1}\,Z^{1/2-1}dXdYdZ+c^2\iiint X^{1/2-1}Y^{1/2-1}Z^{3/2-1}\,dXdYdZ]$$

where $X+Y+Z\leqslant 1$

$$=\rho abc\left[b^2\frac{\Gamma 1/2\,\Gamma 3/2\,\Gamma 1/2}{\Gamma 7/2}+c^2\frac{\Gamma 1/2\,\Gamma 1/2\,\Gamma 3/2}{\Gamma\,7/2}\right]$$

(By Dirichlet's integral)

$$=\frac{1}{5}M(b^2+c^2),\text{ since } M=\frac{4}{3}\pi\rho abc$$

Routh's Rule. *M.I.* about an axis of symmetry

$$=\frac{M(\text{sum of squares of perp. semi-axis})}{D}$$

where D is 3, 4, 5 according as the body is rectangular (including rod), circular (including elliptical) or spherical (including ellipsoidal).

11.3. *Products of Inertia in some simple cases*

(i) *Product of Inertia of an elliptic quadrant about its axes.*

$$\text{P.I.}=\int_0^a\int_0^y \rho dxdy.xy=\frac{1}{2}\rho\int_0^a y^2x.dx$$

$$=\frac{\rho}{2}\cdot\frac{b^2}{a^2}\int_0^a(a^2-x^2)xdx=\frac{\rho b^2}{2a^2}\,\frac{a^2}{4}=\frac{1}{4}\pi ab\rho\frac{ab}{2\pi}=\frac{Mab}{2\pi}.$$

Note. If the circular quadrant is considered then $b=a$ and $\text{P.I.}=\frac{Ma^2}{2\pi}$.

(ii) *Products of Inertia of a semi-circular disc about the diameter and tangent at its end.*

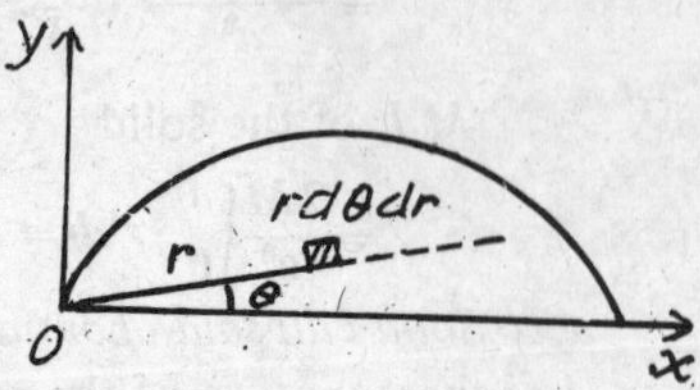

Let the circle be $r=2a\cos\theta$. Then,

$$F=\int_{\theta=0}^{\pi/2}\int_{\theta=0}^{2a\cos\theta}\rho\, r\cos\theta\, r\sin\theta.\, rd\theta dr$$

$$=4\rho a^4\int_0^{\pi/2}\sin\theta.\cos^5\theta\;d\theta=\frac{2}{3}\rho a^4=\frac{4}{3}Ma^2/\pi$$

11·4. Theorem of Parallel Axes. *Given the moments and products of inertia about all axes through the C.G. of a body to find those about all parallel axes.*

Let m be the mass of a particle of the body, its co-ordinates with any origin O be x, y, z and x', y', z' with $C.G.$ as origin and axes parallel and let the co-ordinates of $C.G.$ be $\bar{x}, \bar{y}, \bar{z}$, so that

$$x=\bar{x}+x', y=\bar{y}+y', z=\bar{z}+z'.$$

Now $M.I.$ about z-axis $=\Sigma m(x^2+y^2)$

$=M(\bar{y}^2+\bar{y}^2)+\Sigma m(x'^2+y'^2)$, for $\Sigma mx'$, $\Sigma my'$ are zero since $C.G.$ is the origin of the system $(x'y'z')$.

$\therefore$ $M.I.$ about z-axis $=M.I.$ of the whole mass collected at $C.G.+M.I.$ of the body about a line through $C.G.$ parallel to z-axis.

Since z-axis is any line, we have that the *M.I. of a body about any line is equal to that about a parallel line through the C.G. plus M.I. of the whole mass collected at the C.G. about the original line.* It follows, therefore, that, *of all lines parallel to a given straight line, that one has the least M.I. which passes through the C.G.*

Similarly product of inertia about x and y-axes $=\Sigma mxy$

$$=M\bar{x}\bar{y}+\Sigma mx'y'.$$

This gives that the *product of inertia about any two perp. lines is equal to that about two parallel lines through the C.G. plus that of the whole mass collected at C.G. about the original lines.*

11·5 *Given the moments and products of inertia of a lamina about rectangular axes Ox, Oy in its plane, to find the moment of inertia about any axis through O in its plane.*

Let OA be any line making angle θ with OX, P be the point (x, y) PL perp. to OA.

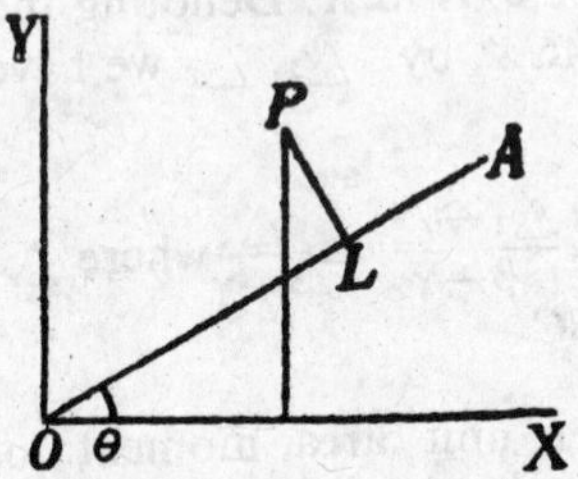

$$\therefore \quad PL=y\cos\theta-x\sin\theta.$$

$M.I.$ about $OA=\Sigma mPL^2$

$$=\Sigma m(y\cos\theta-x\sin\theta)^2$$

$$=\cos^2\theta\,\Sigma my^2-2\sin\theta\cos\theta\,\Sigma mxy+\sin^2\theta\,\Sigma mx^2$$

$$=A\cos^2\theta-2F\sin\theta\cos\theta+B\sin^2\theta.$$

Ex. 1. *Find the moment of inertia of an arc of a circle about (1) the diameter bisecting the arc, (2) an axis through the centre perp. to the plane, (3) axis through the middle point perp. to the plane.*

Sol. Let O be the centre, A the middle point of the arc. Take an element $ad\theta$ at P, its mass $=\dfrac{M}{2a\alpha}\cdot ad\theta$ where 2α is the angle sub, tended at O by the arc.

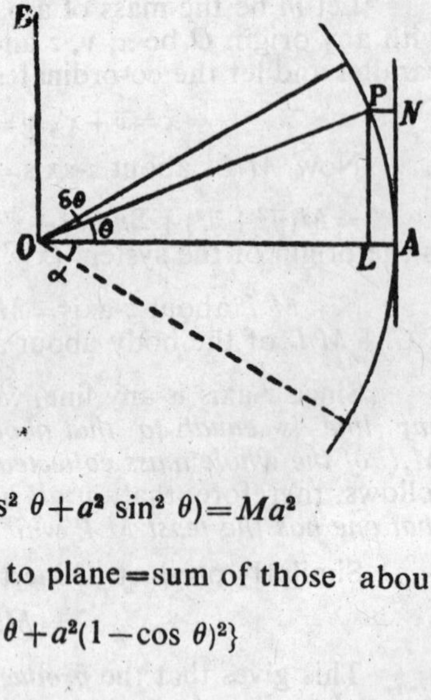

M.I. about OA

$$=\int_{-\alpha}^{\alpha}\frac{M}{2a\alpha}\cdot ad\theta\ a^2\sin^2\theta$$

$$=\frac{Ma^2}{2}\left(1-\frac{\sin 2\alpha}{2\alpha}\right).$$

M.I. about axis through O perp. to plane = sum of those about OA and OB

$$=\int_{-\alpha}^{\alpha}\frac{M}{2a\alpha}\ ad\theta.\ \ (a^2\cos^2\theta+a^2\sin^2\theta)=Ma^2$$

M.I. about axis through A perp. to plane = sum of those about AO and $AN=\int_{-\alpha}^{\alpha}\frac{M}{2a\alpha}\ ad\theta.\ \{a^2\sin^2\theta+a^2(1-\cos\theta)^2\}$

$$=2Ma^2\left(1-\frac{\sin\alpha}{\alpha}\right).$$

Ex. 2. *Prove that the moment of inertia of a triangular lamina ABC about any axis through A in its plane is* $\frac{M}{6}(\beta^2+\beta\gamma+\gamma^2)$ *where M is the mass of the lamina ; β, γ are perpendiculars from B, C on the axis.*

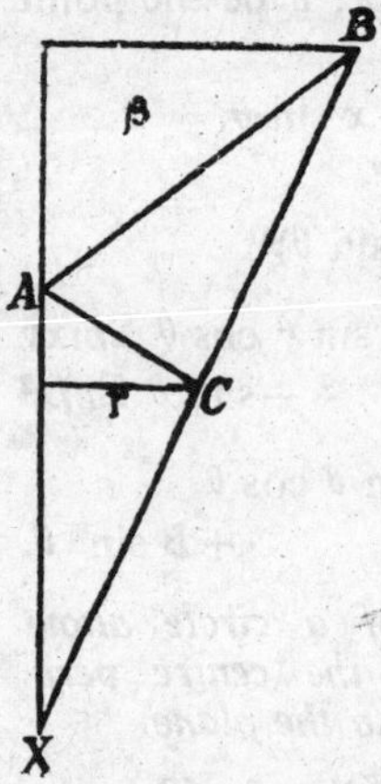

Sol. Let BC cut the axis in X. Denoting the areas of triangles ABX, ACX, by $\triangle_1$, $\triangle_2$ we have $\triangle_1=\frac{1}{2}AX.\beta$, $\triangle_2=\frac{1}{2}AX.\gamma$

$$\therefore\ \frac{\triangle_1}{\beta}=\frac{\triangle_2}{\gamma}=\frac{\triangle_1-\triangle_2}{\beta-\gamma}=\frac{\triangle}{\beta-\gamma}\text{ where }\triangle$$

is the area of triangle ABC

If m be the mass per unit area, moment of inertia about $AX=\frac{1}{6}m(\triangle_1\beta^2-\triangle_2\gamma^2)$

$$=\frac{1}{6}\ m\triangle\ \frac{\beta^3-\gamma^3}{\beta-\gamma}=\frac{1}{6}M.\ (\beta^2+\beta\gamma+\gamma^2).$$

Ex. 3. *Show that the moment of inertia of a right solid cone, whose height is h and radius of base is a, is* $\frac{3Ma^2}{20}\cdot\frac{6h^2+a^2}{h^2+a^2}$ *about a slant side and* $\frac{3M}{80}(h^2+4a^2)$ *about a line through its C.G. perp. to the axis.*

Sol. Take the element between two circular sections at distances x and $x+dx$ from O

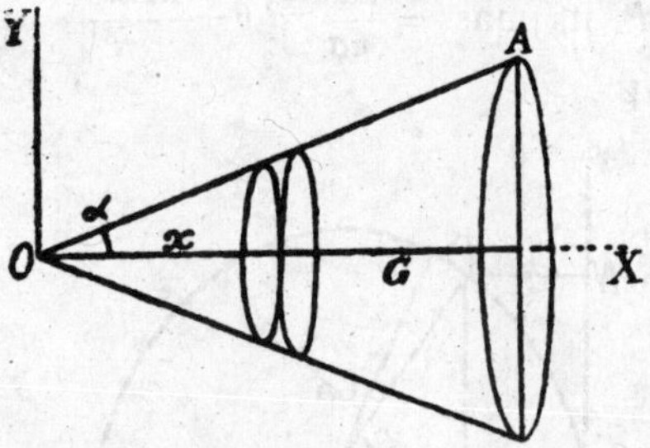

If ρ be the density, its mass $=\pi x^2 \tan^2 \alpha dx.\rho$ and its moment of inertia about OX

$$=\pi x^2 \tan^2 \alpha\, dx.\, \rho \frac{x^2 \tan^2 \alpha}{2}$$

$\therefore$ *M.I.* about OX

$$=\int_0^h \frac{\pi}{2} \rho \tan^4 \alpha\, x^4\, dx$$

$$=\frac{\pi}{2} \rho \tan^4 \alpha \frac{h^5}{5}=\frac{\pi}{10} \rho a^4 h$$

$$=\tfrac{3}{10} Ma^2=A;$$
$$M=\tfrac{1}{3}\pi a^2 h\rho$$

Moment of inertia of the circular disc about OY

$$=\pi x^3 \tan^1 \alpha\, dx\rho \times \left(\frac{x^2 \tan^2 \alpha}{4}+x^2\right)$$

$$\therefore \text{ M.I. about } OY=\int_0^h \pi x^2 \tan^2 \alpha\rho \left(\frac{x^2 \tan^2 \alpha}{4}+x^2\right)dx$$

$$=\frac{\pi}{4}\rho \tan^2 \alpha(\tan^2 \alpha+4)\frac{h^5}{5}=\frac{\pi}{20}\rho\, a^2(a^2+4h^2)h$$

$$=\tfrac{3}{20} M(a^2+4h^2)=B$$

M.I. about $OA=A \cos^2 \alpha+B \sin^2 a$

$$=\tfrac{3}{10}Ma^2 \cos^2 \alpha+\tfrac{3}{20}M(a^2+4h^2)\sin^2 \alpha$$

$$=\frac{3}{20} Ma^2\left(2 \cos^2 \alpha+\sin^2\alpha+4 \frac{h^2}{a^2} \sin^2 \alpha\right)$$

$$=\frac{3}{20} Ma^2 \cos^2 \alpha\left(2+\tan^2 \alpha+\frac{4h^2}{a^2} \tan^2 \alpha\right)$$

$$=\frac{3}{20} Ma^2 \frac{2+\frac{a^2}{h^2}+\frac{4h^2}{a^2}.\frac{a^2}{h^2}}{1+\frac{a^2}{h^2}}=\frac{3}{20}Ma^2 \frac{6h^2+a^2}{h^2+a^2}$$

Also $B=$ *M.I.* about a parallel line through $G+M\left(\frac{3h}{4}\right)^2$

$\therefore$ *M.I.* about the parallel line through G

$$=\tfrac{3}{20} M(a^2+4h^2)-\tfrac{3}{16}Mh^2$$
$$=\frac{3M}{80}(h^2+4a^2).$$

Ex. 4. *Prove that the product of inertia of a semi-circular wire* *about diameter and tangent at its extremity is* $2M\dfrac{a^2}{\pi}$.

Sol. Take an element $ad\theta$ at P, its mass$=\dfrac{M}{\pi a}\cdot d\theta=\dfrac{Md\theta}{\pi}$

Product of inertia about OX, OY

$$\int \frac{M}{\pi}\, d\theta .\, PN\, .\, PL$$

$$=\int_0^\pi \frac{M}{\pi}\, d\theta . a(\ \ -\cos\theta)\ a\sin\theta$$

$$=\int_0^\pi \frac{Ma^2}{\pi}(\sin\theta-\tfrac{1}{2}\sin 2\theta)\,d\theta$$

$$=\frac{2Ma^2}{\pi}.$$

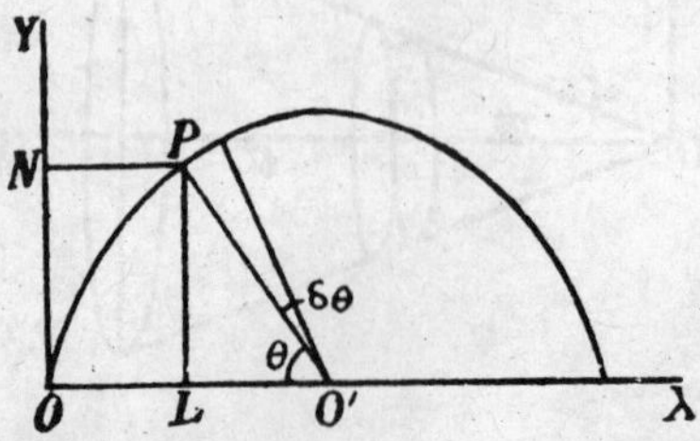

Ex. 5. *Find the moment of inertia of a circular disc about a line in its own plane whose perpendicular distance from its centre is c.*

Sol. M.I. about a diameter parallel to the given line$=\dfrac{1}{4}Ma^2$

$\therefore$ M.I. about the given line$=\dfrac{1}{4}Ma^2+Mc^2=M(c^2+\dfrac{1}{4}a^2)$.

Ex. 6. *Find the moment of inertia of a uniform circular cylinder about an axis through its centre of gravity, perpendicular to its axis.*

Sol. Let a be the radius of the cylinder and let $2l$ be its length.

Therefore the elementary mass of a circular slice of thickness δx at a distance x from $G=\dfrac{M}{2l}\delta x$

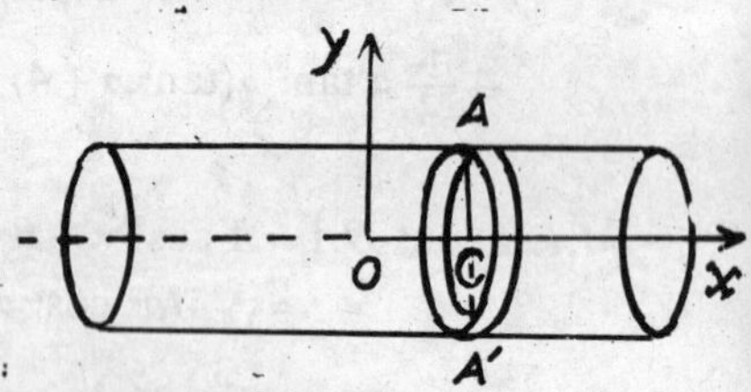

$\therefore$ its M.I. about $ACA'=\left(\dfrac{M}{2l}\delta x\right)\left(\dfrac{a^2}{4}\right)$

$\therefore$ Using theorem of parallel axes, M.I. about GY

$$=\frac{M\delta x}{2l}\left(\frac{a^2}{4}+x^2\right)$$

$\therefore$ for the whole cylinder, M.I. about GY

$$=\int_{-l}^{l}\frac{M}{2l}\left(\frac{a^2}{4}+x^2\right)dx=M\left(\frac{1}{4}a^2+\frac{1}{3}l^2\right).$$

EXAMPLES XI (A)

1. *Find the moment of inertia of the following :*

(*a*) An isosceles triangle ABC about an axis through the vertex A perp. t (*i*) the opposite side, (*ii*) the plane of the triangle

(*i*) $\dfrac{Ma^2}{24}$.   (*ii*) $\dfrac{Ma^2}{8}\left(\dfrac{1}{3}+\cot^2\dfrac{A}{2}\right)$.

(*b*) Solid right circular cylinder of length l and radius a about a diameter at one end

$$M\left(\frac{a^2}{4}+\frac{l^2}{3}\right).$$

(*c*) Area bounded by $r^2=a^2 \cos 2\theta$ about its axis

$$\frac{Ma^2}{16}\left(\pi-\frac{8}{3}\right).$$

(*d*) Triangular area ABC about a perp. to the plane through A

$$\frac{M}{12}(3b^2+3c^2-a^2).$$

(*e*) Truncated cone about its axis, radii of ends being a, b

$$\frac{3M}{10}\ \frac{a^5-b^5}{a^3-b^3}.$$

(*f*) Hollow sphere about a diameter, a, b being external and internal radii

$$\frac{2M}{5}\ \frac{a^5-b^5}{a^3-b^3}.$$

(*g*) Hollow cylinder about a tangent in the plane passing through $C.G$ and perp. to axis

$$M\left(\frac{3a^2}{2}+\frac{h^2}{3}\right).$$

(*h*) Solid cylinder as in (*g*)

$$M\left(\frac{5a^2}{4}+\frac{h^2}{3}\right).$$

(*i*) Parabolic area (latus rectum $4a$) cut off by an ordinate at distance h from the vertex (*i*) about tangent at vertex (*ii*) about axis

$$(i)\ \tfrac{3}{7}Mh^2\ ;\ (ii)\ \tfrac{4}{5}\ Mah\ .$$

(*j*) Semi-circular lamina about a tangent parallel to bounding diameter

$$Ma^2\left(\frac{5}{8}-\frac{8}{3\pi}\right).$$

(*k*) Elliptic lamina about (*i*) diameter of length r, (*ii*) tangent at distance p from centre

$$(i)\ \frac{Ma^2b^2}{4r^2} \qquad (ii)\ \tfrac{4}{5}\ Mp^2.$$

2. *If k_1, k_2 be the radii of gyration of an elliptic lamina about two conjugate diameters prove thet* $\dfrac{1}{k_1^2}+\dfrac{1}{k_2^2}=4\left(\dfrac{1}{a^2}+\dfrac{1}{b^2}\right)$.

3. Prove that the sum of the moments of inertia of an elliptic lamina about any two perp. tangents is constant.

4. Find the product of inertia in the following cases ;

(*a*) Right-angled triangle about side $2a$, $2b$ containing the right angle

$$\frac{Mab}{3}.$$

(*b*) Disc in the form of a quadrant of a circle about the bounding radii

$$\frac{Ma^2}{2\pi}.$$

5. Show that the M.I. of a semi-circular lamina about a tangent paralle to the bounding diameter is $Ma^2\left(\dfrac{5}{4}-\dfrac{8}{3\pi}\right)$.

6. Prove the formula $I=A \cos^2 \alpha-2F\sin\alpha \cos \alpha+B \sin^2 \alpha$ for the moment of inertia of a lamina about the line $y=x \tan \alpha$, stating the meanings of A, B, F.

7. Show that the *M.I.* of a rectangle of sides $2a$, $2b$, about a line through the centre, making an angle θ with side $2a$, is $\frac{1}{3}M(b^2\cos^2\theta+a^2\sin^2\theta)$.

Hence deduce that the *M.I.* of the rectangle about a diagonal is

$$\frac{2M}{3}\frac{a^2b^2}{a^2+b^2}.$$

8 Show that the sum of the moment of inertia of an elliptic lamina about any two perpendicular tangents is constant.

11.6 Momental Ellipse. If A, B and F be the moments and product of inertia of a lamina about the rect. axes Ox, Oy in its plane, the moment of inertia about a line at angle θ with Ox is given by

$$I=\Sigma\, m.\, PN^2=\Sigma\, m\,(y\cos\theta-x\sin\theta)^2$$

$$I=A\cos^2\theta-2F\sin\theta\cos\theta+B\sin^2\theta.$$

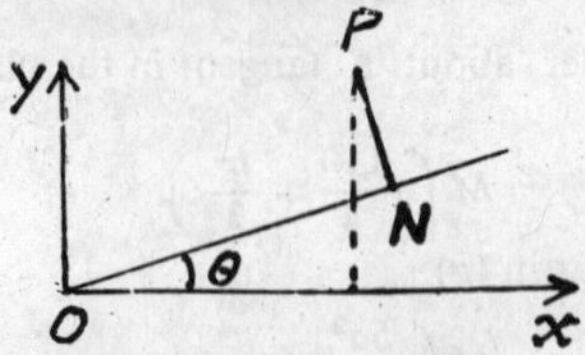

Let the line revolve about O in the plane of the lamina and be of such length OP that the moment of inertia of the lamina about OP be inversely proportional to the square of its length, that is, if $OP=r$, then $I=\frac{K}{r^2}$ when K is a constant. Thus

$$\frac{K}{r^2}=A\cos^2\theta-2F\sin\theta\cos\theta+B\sin^2\theta.$$

If P be the point x, y so that $x=r\cos\theta$, $y=r\sin\theta$, the locus of the point P is

$$Ax^2-2Fxy+By^2=K,$$

which is a conic having O as centre.

This conic *must be an ellipse*, for I, the moment of inertia about any axis through O is clearly always finite and so r is never infinite. For different constant values of K, the equation represents a family of concentric similar and similarly situated ellipses. An ellipse of this family is called a *momental ellipse of the lamina at the point O*

Now an ellipse is a figure having double symmetry about its major and minor axes, the semi-axes are the maximum and minimum radii vector from the centre ; also when the ellipse is referred to these axes as co-ordinate axes, its equation has no term in xy.

Hence we conclude that *at every point O of a lamina*

(*i*) *there is one pair of rectangular axes for which F=0 i.e., product of inertia about these axes vanishes.*

(*ii*) *the major and minor axes of the momental ellipse are the axes of least and greatest moments of inertia.*

(*iii*) *moments of inertia about axes equally inclined to the major and minor axes are equal.*

11.7 Principal axes. Definition. *A pair of axes about which the product of inertia vanishes is called Principal Axes.*

From Article 11·5, we conclude that

(*i*) *at each point of a lamina there exists one pair of principal axes and these are the axes of the momental ellipse at that point.*

(*ii*) *the principal axes at a point are axes of greatest and least moments of inertia at that point.*

If A, B, F be the moments and product of inertia about Ox, Oy, then the product of inertia about two axes making angles $\theta, \frac{\pi}{2}+\theta$ with x-axis

$$=\Sigma m(y\cos\theta-x\sin\theta)(x\cos\theta+y\sin\theta)$$
$$=(A-B)\cos\theta\sin\theta+F\cos 2\theta$$
$$=\frac{A-B}{2}\sin 2\theta+F\cos 2\theta.$$

This vanishes when $\tan\ 2\theta=\frac{2F}{B-A}$.

This gives the inclinations of the principal axes to the axes of co-ordinates.

We further note that if a lamina has an axis of symmetry the principal axes at any point of this axis are the axis itself and the perp. Also if the principal moments of inertia be equal, momental ellipse becomes a circle and moment of inertia about any line through that point is the same.

11.8 Foci of Inertia

Let GX, GY be the principal axes at the C.G. of the lamina; A, B the moments of inertia about them. On Gx take points S and H such that $GS=GH=\sqrt{\frac{A-B}{M}}$.

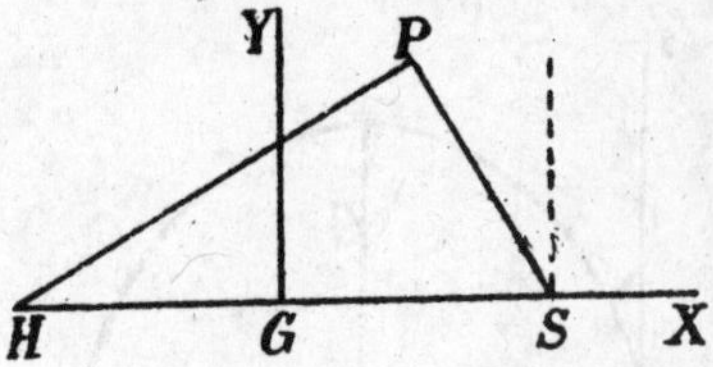

If Sy' be perp. to Sx, moment of inertia about Sy' $=$ $B+M\,GS^2$ $=B+A-B=A=$ that about Sx. Product of inertia about Sx, Sy'

$$\Sigma m(x-GS)y$$

$=\Sigma mxy-GS\Sigma my=0$ since Gx, Gy are principal axes so that $\Sigma mxy=0$ and G is the centroid so that $\Sigma my=0$. Thus the momental ellipse at S is a circle and moments of inertia about all axes through

S are equal to A. Similarly the moment of inertia about any line through H is also A. Thus if we take a point P in the lamina, moments of inertia about PS and PH are equal. Thus if we construct the momental ellipse at P, radii vectors in the directions PS and PH must be equal so that the principal axes of the momental ellipse are the bisectors of the angles between PS and PH. These are the principal axes of the lamina at P. If then we draw with S and H as foci, an ellipse to pass through P, the principal axes of the lamina at P are the tangent and normal to this ellipse at P. The points S and H are called the foci of inertia.

11·8. Equimomental Particles. If a lamina and a set of particles have (*i*) the same total mass, (*ii*) same C.G., (*iii*) same principal axes and (*iv*) same principal moments, then the lamina and the particles have the same momental ellipse and hence the moments and products of inertia about all axes in the plane of the lamina are identical. The particles are then said to be equimomental with the lamina.

Ex. 7. *If A and B be the moments of inertia of a uniform lamina about two perp. axes Ox, Oy in its plane, F the product of inertia about these lines ; show that the principal moments at O are*

$$\tfrac{1}{2}[A+B\pm\sqrt{(A-B)^2+4F^2}].$$

Sol. Momental ellipse referred to Ox, Oy is $Ax^2-2Fxy+By=1$

" " " principal axes is $A_1x^2+B_1y^2=1$

where A_1, B_1 are the principal moments.

Hence by invariants

$$\left.\begin{aligned} A_1+B_1&=A+B \\ A_1B_1&=AB-F^2 \end{aligned}\right\} \text{ These give } A_1 \text{ and } B_1.$$

Ex. 8. *A wire in the form of semi-circle of radius a, show that at an end of its diameter the principal axes in its plane are inclined to the diameter at angles* $\dfrac{1}{2}\tan^{-1}\dfrac{4}{\pi}$ *and* $\dfrac{\pi}{2}+\dfrac{1}{2}\tan^{-1}\dfrac{4}{\pi}$.

Sol. O=the centre ; A=moment of inertia about=Ax

$$=M\frac{a^2}{2}$$

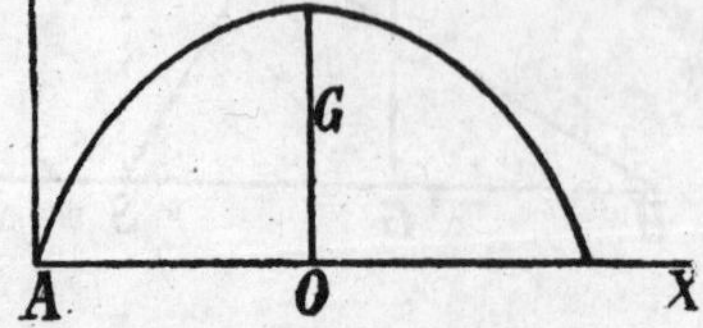

$B=M.I.$ about Ay

$$=M\left(\frac{a^2}{2}+a^2\right)=\frac{3}{2}Ma^2.$$

F=Product of inertia about Ax, $Ay=M\ AO.OG.$

$$=Ma\ \frac{2a}{\pi}=\frac{2\ Ma^2}{\pi},$$

$$\therefore \quad \tan 2\theta = \frac{2F}{B-A} = \frac{\frac{4\,Ma^2}{\pi}}{Ma^2} = \frac{4}{\pi}. \quad \text{Hence the answer.}$$

Ex. 9. *A uniform lamina is bounded by parabolic arc of latus rectum 4a, and a double ordinate at a distance b from the vertex. If $b = \frac{1}{3}a\,(7+4\sqrt{7})$, show that two of the principal axes at the end of a latus rectum are the tangent and normal there.*

Sol. Let L be $(a_1, 2a)$, an end of latus rectum of the parabola $y^2=4ax$. Then tangent at L is $x-y+a=0$ and the normal is $x+y-3a=0$. Tangent and normal at L will be principal axes if

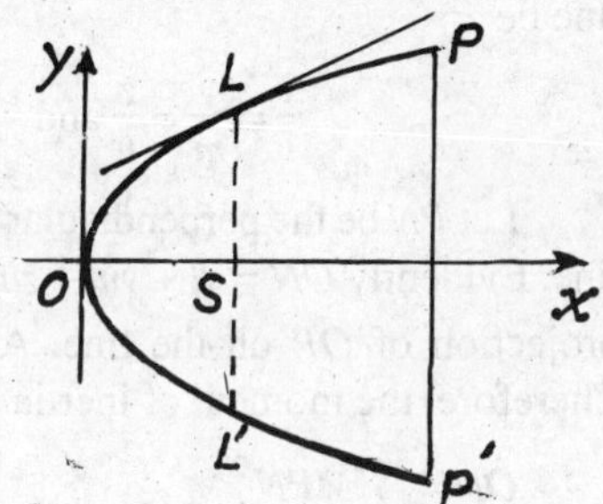

$$\int_{x=0}^{b}\int_{y=-2\sqrt{ax}}^{2\sqrt{ax}} \rho\,dx\,dy \cdot \frac{x-y+a}{\sqrt{2}} \cdot \frac{x+y-3a}{\sqrt{2}} = 0$$

or if, $\iint (x^2-y^2-2ax+4ay-3a^2)\,dx\,dy=0$

or if, $\int_0^b (x^2-2ax-3a^2)\,4\sqrt{ax}\,dx - \int_0^b \frac{1}{3}.2a.8(ax)^{3/2}\,dx = 0$

i.e. if, $b^2 - \frac{14}{3}ab - 7a^2 = 0$ or if $b = \frac{1}{3}a(7+4\sqrt{7})$

EXAMPLES XI (B)

1. Prove that a uniform rod of mass M is equimomental to three particles, $\frac{1}{6}M$ at each end of the rod and $\frac{2}{3}M$ at the middle point.

2. Prove that a triangular lamina of mass M is equimomental to three particles each of mass $\frac{1}{3}M$ placed at the middle points of the sides.

3. Prove that a parallelogram is equimomental with four particles $\frac{M}{6}$ at the middle points of the sides and $\frac{M}{3}$ at the intersection of the diagonals.

4. Lengths AB and AD of the sides of a rectangle $ABCD$ are $2a$ and $2b$; prove that inclination to AB of one of the principal axes A is

$$\frac{1}{2}\tan^{-1}\frac{3\,ab}{2\,(a^2-b^2)}.$$

5. At the vertex C of a triangle ABC, right-angled at C, the principal axes in the plane are inclined to the sides at an angle $\frac{1}{2}\tan^{-1}\frac{ab}{(a^2-b^2)}$.

6. ABC is a triangular area, AD is perp. to BC; E is the middle point of BC and O is the middle point of DE. Show that BC is a principal axis of the triangle at O.

7. Find the principal axes of a right circular cone at a point in the circumference of the base; and show that one of them will pass through its $C.G.$ if the verticle angle of the cone is $2\tan^{-1}\frac{1}{2}$.

11.9. Theorem of the six constants of a body. *If the moments and products of inertia of a body about three mutually prependicular and concurrent axes are given to obtain the moments and products of inertia about any other axis through their meeting point.*

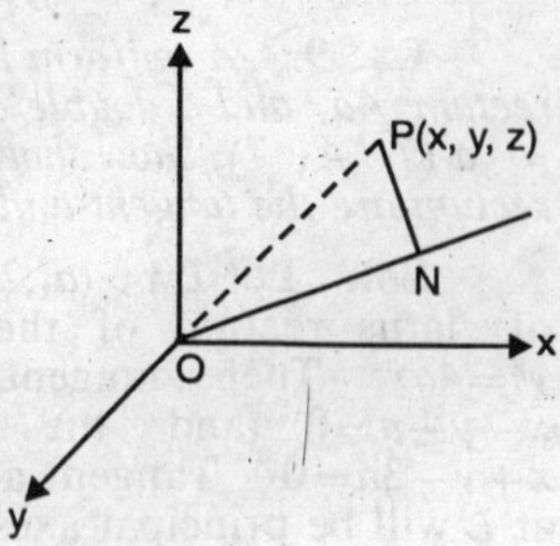

Proof. Take the three straight lines as axes by reference. Let $P(x, y, z)$ be a point of the body where a mass M is situated. Let the given line be

$$\frac{x}{l}=\frac{y}{m}=\frac{z}{n} \text{ and } y$$

Let PN be the perpendicular from P on this line. Evidently $ON = xl + ym + zn$, as ON is the projection of OP on the line. Also $OP^2 = x^2 + y^2 + z^2$ and $l^2 + m^2 + n^2 = 1$. Therefore, the moment of inertia about

$$ON = \sum MPN^2$$

$$= \sum M(OP^2 - ON^2) = \sum M\{(x^2+y^2+z^2) - (lx+my+nz)^2\}$$

$$= \sum M\{(x^2+y^2+z^2)(l^2+m^2+n^2) - (lx+my+nz)^2\}$$

$$= \sum M(y^2+z^2)l^2 + \sum M(z^2+x^2)m^2$$

$$+ \sum M(x^2+y^2)n^2 - 2mn\sum Myx - 2nl\sum Mzx - 2lm\sum Mxy$$

$$= Al^2 + Bm^2 + Cn^2 - 2Dmn - 2Enl - 2Flm$$

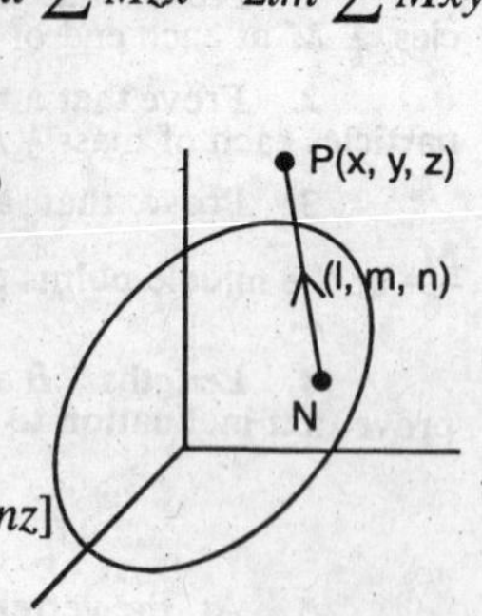

Note. Similarly moment of inertia with respect to the plane through the origin namely

$$lx + my + nz = v \text{ is}$$

$$\sum M PN^2 = \sum M(lx+my+nz)^2$$

$$[\because \quad PN = lx + my + nz]$$

$$= A^1l^2 + B^1m^2 + C^1n^2 + 2Dmn + 2Enl + 2Flm,$$

A^1, B^1, C^1 being the M.Is. with regard to the planes yz, zx, xy respectively.

11.10. Principal Planes. The three planes which has each through the principal axes and called the *Principal Planes* at the given point.

11.11 Momental Ellipsoid. If A, B, C, D, E, F be the moments and products of inertia about the axes, then the moment of inertia about the line OP whose d.c.'s are (l, m, n) is P

$$Al^2 + Bm^2 + Cn^2 - 2Dmn - 2Enl - 2Flm$$

Let the line OP move in such a manner and be of such length that the M.I. about OP is inversely proportional to the square of its length, and let it be Mt^4/r^2, where t is some constant having the dimension of length, $OP = r$ and M is the mass of the body. Hence the polar equation of the locus of P is

$$\frac{Mt^4}{r^2} = Al^2 + Bm^2 + Cn^2 - 2Dmn - 2Enl - 2Flm,$$

or $$Mt^4 = Ar^2l^2 + Br^2m^2 + Cr^2n^2 - 2Drm \,.\, rn - 2Ern \,.\, rl - 2Frl \,.\, rm$$

or $$Mt^4 = Ax^2 + By^2 + Cz^2 - 2Dyz - 2Ezx - 2Fxy$$

(on changing to cartesian coordinates)

This is the equation of a quadric. Thus to every point O of a material body there is a corresponding quadric which possesses the property that the moment of inertia about any radius vector, is represented by the inverse square of that radius vector. Since the M.I. is essentially positive, it is evident that every radius vector r must be real. Hence the quadric is always an ellipsoid. It is said to be the *momental ellipsoid* of the body at the point O.

Thus the momental ellipsoid possesses the geometrical property that any radius vector is equal to some constant divided by the square root of the moment of inertia about the radius vector.

Ex. 10. *Find the momental ellipsoid and the momental ellipse at the centre of an elliptic plate.*

Sol. Here $A = \frac{1}{4} Mb^2,\ B = \frac{1}{4} Ma^2,\ C = \frac{1}{4} M(a^2 + b^2)$

$$D = E = F = 0.$$

$\therefore$ Momental ellipsoid is

$$\frac{1}{4} Mb^2x^2 + \frac{1}{4} Ma^2y^2 + \frac{1}{4} M(a^2 + b^2)\, z^2 = \text{constant}$$

or $$b^2x^2 + a^2y^2 + (a^2 + b^2)\, z^2 = \text{constant}$$

where $z = 0$, this becomes the ellipse similar to the given ellipse. Hence the momental ellipse at the centre of an elliptic lamina is any similar and similarly situated ellipse.

Ex. 11. *Find the momental ellipsoid at any point O of a material straight rod AB of mass M and length 2a.*

Sol. Let O be the origin and the straight line OAB on the axis of x and let $OG = C$. If the rod be indefinitely then

$$A = O\,,\, B = M\left(\frac{1}{3} a^2 + c^2\right) = C.$$

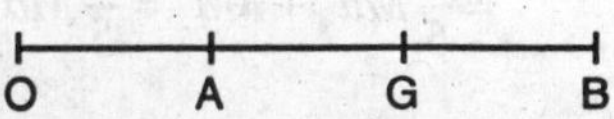

Hence the momental ellipsoid is

$$y^2 + z^2 = \text{const.}$$

The momental ellipsoid is therefore an elongated spheroid which becomes a right cylinder when the rod becomes infinitely thin.

Ex. 12. *The equation of the momental ellipsoid at the corner of a cube of side 2a, referred to its principal axes is*

$$2x^2 + 11(y^2 + z^2) = \text{const.}$$

Sol. M.I. about each of the three perpendicular lines parallel to sides and through G, the centre of gravity of the cube is $\frac{2}{3}Ma^2$.

Hence moment of inertia about any line through

$$G = \frac{2}{3}Ma^2 (l^2 + m^2 + n^2)$$

$$\equiv \frac{2}{3}Ma^2, \; l, m, n \text{ being the d.c.'s of that line.}$$

Let the corner be O. Since M.I. about any time through G is the same, hence the momental ellipsoid at G is a sphere and the product of inertia about any two perpendicular lines through G is zero.

Hence the product of inertia about OG and any line through O perpendicular to OG = that about OG and line parallel to the given line + M.O. = O.

Also the M.I. about any line through O perpendicular to

$$OG = M\left(\frac{2}{3}a^2 + OG^2\right) = M\left(\frac{2}{3}a^2 + 3a^2\right) = \frac{11}{3}Ma^2$$

Hence the momental ellipsoid is $\frac{2}{3}Ma^2x^2 + \frac{11}{3}Ma^2(y^2 + z^2) = \text{const.}$

or $$2x^2 + 11(y^2 + z^2) = \text{const.}$$

Ex. 13. *Prove that the momental ellipsoid at a point on the rim of a hemisphere is* $2x^2 + 7(y^2 + z^2) - \frac{15}{4}zx = C$.

Sol. A = moment of inertia about AOX = $\frac{2}{3}Ma^2$

C = moment of inertia about AZ

$= \text{M.I. about } GOZ + Ma^2$

$= \frac{2}{5}Ma^2 + Ma^2 = \frac{7}{5}Ma^2$

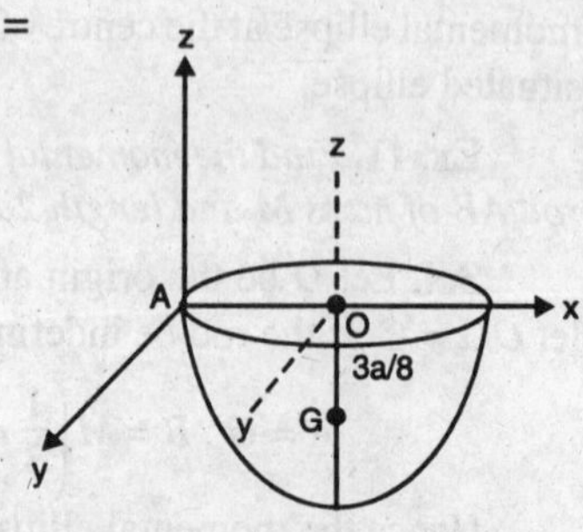

$$B = \text{M.I. about } AY = \text{M.I. about } OY + Ma^2 = \frac{2}{5} Ma^2 + Ma^2$$

$$= \frac{7}{5} Ma^2$$

Ay is symmetrical with respect to body, therefore $D = O = F$.

E = Product of inertia about AX, AZ = about parallel axes through G + $M.a.\frac{3a}{8}$

$$= O + \frac{3}{8} Ma^2$$

∴ Momental ellipsoid is

$$\frac{2}{5}Ma^2x^2 + \frac{7}{5}Ma^2y^2 + \frac{7}{5}Ma^2z^2 - 2 \cdot \frac{3}{8}Ma^2zx = \text{const.}$$

or $$2x^2 + 7\,(y^2 + z^2) - \frac{15}{4}\,zx = \text{const.}$$

EXAMPLES XI(C)

1. Show that the momental ellipsoid at the centre of a solid ellipsoid $x^2/a^2 + y^2/b^2 + z^2/x^2 = 1$ is $(h^2 + c^2)\,x^2 + (c^2 + a^2)\,y^2 + (a^2 + b^2)\,z^2 = \text{const.}$

2. Show that one of the principal axes at any point on the edge of the circular base of a thin-hemispherical shell is inclined at an angle $\frac{1}{8}\pi$ to this circular base. Also show that the equation of the momental ellipsoid is

$$2x^2 + 5\,(y^2 + z^2) - 3zx = \text{const.}$$

3. Four equal particles are placed at the corners of a tetrahedron. If the momental ellipsoid at theoric is a sphere, prove that the tetrahedron is a regular one.

4. Prove that the momental ellipsoid at a point on the circular edge of a solid line is

$$(3a^2 + 2h^2)\,x^2 + (23a^2 + 2h^2)\,y^2 + 26a^2z^2 - 109hxz = \text{const.}$$

where h is the height and a the radius of the base.

12

D'Alembert's Principle and the Equations of Motion

12·1. A rigid body is assumed to be made up of mass particles connected together by a kind of mass-less rigid framework, each pair of particles being always at the same distance apart during the motion. Consider a particle of mass m situated at the point (x, y, z). If f be its acceleration with components $(\ddot{x}, \ddot{y}, \ddot{z})$, then mf is called the *effective* force on the particle m. This effective force is by Newton's Second Law, exactly equal to the resultant forces on the particle. This force is compounded of two kinds of forces, (*a*) an impressed or external force and (*b*) internal force which is the resultant of the reactions of other particles on m. Let the components of force (*a*) be (X, Y, Z) and those of (*b*) be (X', Y', Z'). Thus for the particle m, the forces $(m\ddot{x}, m\ddot{y}, m\ddot{z})$ are equivalent to the resultant of the forces (X, Y, Z) and (X', Y', Z'). Similarly for all the particles. Thus the set of effective forces of which a typical one is $(m\ddot{x}, m\ddot{y}, m\ddot{z})$ is the equivalent of two sets of forces whose typical components are (X, Y, Z) and (X', Y', Z').

Now the *principle enunciated by D'Alembert* may be stated as follows :

Under all circumstances of motion, the internal forces form a system in equilibrium among themselves. Thus the set of forces typified by (X', Y', Z') is in equilibrium. Hence the set of effective forces typified by $(m\ddot{x}, m\ddot{y}, m\ddot{z})$ is the exact equivalent of the set of *external* forces typified by (X, Y, Z) acting at the point x, y, z. Hence the sum of the resolved parts of the effective forces and of the external forces in the directions of ox, oy, oz are equal, also the sum of the moments of these two sets of forces about ox, oy, oz are equal.

These give rise to the six equations :

$$\Sigma m\ddot{x}=\Sigma X, \qquad \Sigma m(y\ddot{z}-z\ddot{y})=\Sigma(yZ-zY),$$
$$\Sigma m\ddot{y}=\Sigma Y, \qquad \Sigma m(z\ddot{x}-x\ddot{z})=\Sigma(zX-xZ),$$
$$\Sigma m\ddot{z}=\Sigma Z, \qquad \Sigma m(x\ddot{y}-y\ddot{x})=\Sigma(xY-yX).$$

These are the general equations of motion of a free rigid body.

In two dimensions *i.e.*, for a plane lamina, the equations of motion are simply

$$\Sigma m\ddot{x}=\Sigma X,\ \Sigma m\ddot{y}=\Sigma Y \text{ and } \Sigma m(x\ddot{y}-y\ddot{x})=\Sigma(xY-yX).$$

Whatever has been discussed earlier can be formally put :

D' Alemberts Principle. *The reversed effective forces acting on each particle of the body and the external forces of the system are in equilibrium.*

Proof. Consider a particle of mass m of the rigid body and let its coordinates be (x, y, z) with reference to three fixed rectangular axes. Let f be the resultant acceleration, so that the effective force on m is mf. If F be the resultant of the external applied forces on m, R the resultant of the mutual action and reactions, then mf is resultant of F and R. This F, R and mf (reversed) are in equilibrium. This is true for all other particles of the body. Hence Σmf (reversed effective forces) acting on the body, the external forces (ΣF) and internal actions and relactions (ΣR) of the body form a system of forces in statical equilibrium.

But internal actions and reactions of a body balance within themselves *i.e.* ΣR will itself be in equilibrium.

Hence External forces will be in equilibrium with reversed effective forces.

Remark. Due to this principle the dynamical problem is reduced to a statical one.

Ex. 1. *A light rod OAB can turn frecly in a vertical plane about a smooth fixed hinge at O ; two heavy particles of masses m and m' are attached to the rod at A and B and oscillate with it. Find the motion.*

Sol. Let the rod make an angle θ with the vertical OZ at any time t.

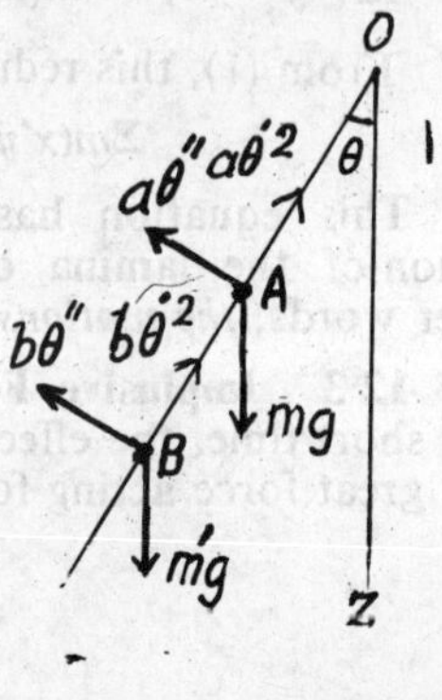

Let $OA=a$, $OB=b$, so that effective forces on m are $ma\dot{\theta}^2$ and $ma\dot{\theta}^3$ and on m' are $m'b\theta^1$ and $m'b\dot{\theta}^2$; external forces being mg and $m'g$ respectively.

Then by D'Alembert's principle, the revesed effective forces should be in equilibrium with the two weights. The internal forces are the reactions at hinge O and at A and B. The reactions at A and B vanish being equal and opposite for the particles and the rod. Now to avoid the reactions at O, taking moments about O, we have the equation of motion as

$$(ma^2+m'b^2)\,\ddot{\theta}+(ma-m'b)\,g\sin\theta=0$$

This equation, when solved, will determine the motion. The constants of integration can be found from initials conditions.

12·2. Independence of translation and rotation.

Taking the case of a plane lamina, let $(\bar{x}, \bar{y})$ be the co-ordinates of its $C.G.$ G and M the mass of the lamina, so that

$$M\bar{x}=\Sigma mx, \quad M\bar{y}=\Sigma my$$

Differentiating twice with respect to t

$$M\ddot{\bar{x}}=\Sigma m\ddot{x}, \quad M\ddot{\bar{y}}=\Sigma m\ddot{y}$$

$\therefore$ the equation of motion takes the form

$$M\ddot{\bar{x}}=\Sigma X, \quad M\ddot{\bar{y}}=\Sigma Y. \qquad \ldots(1)$$

These are the equations of motion of a single particle of mass M placed at the $C.G.$ of the lamina and acted on by forces parallel and equal to the external forces. *Thus the motion of the lamina as regards translation is the same as that of a single particle of mass M placed at its C.G. under the action of the external forces transferred to that point.*

Taking the third equation, namely

$$\Sigma m(x\ddot{y}-y\ddot{x})=\Sigma(xY-Xy),$$

if (x', y') be the co-ordinates of particle m referred to parallel axes through G, so that

$$x=\bar{x}+x', \; y=\bar{y}+y',$$

we get

$$\Sigma m[(\bar{x}+x')(\ddot{\bar{y}}+\ddot{y}')-(\bar{y}+y')(\ddot{\bar{x}}+\ddot{x}')]=\Sigma[(\bar{x}+x')\,Y-(\bar{y}+y')X]$$

Terms containing $\bar{x}$, $\bar{y}$ and their differential co-efficient can be taken outside the Σ and also $\Sigma mx'=0=\Sigma my'$ and $\Sigma m\ddot{x}'=0=\Sigma m\ddot{y}'$ therefore, we get

$$M(\bar{x}\ddot{\bar{y}}-\bar{y}\ddot{\bar{x}})+\Sigma m(x'\ddot{y}'-y'\ddot{x}')=\bar{x}\Sigma Y-\bar{y}\Sigma X+\Sigma(x'Y-y'X).$$

From (1), this reduces to

$$\Sigma m(x'\ddot{y}'-y'\ddot{x}')=(x'Y-y'X). \qquad \ldots(2)$$

This equation has the origin at G. Hence the rotational motion of the lamina does not depend upon the motion of G; in other words, *the motions of translation and rotation are independent.*

12·3. Implusive Forces. When a very great force acts for a very short time, the effect is measured by its impulse. Thus X be a very great force acting for a very short time T, its impulse is

$$\int_0^T X\,dt.$$

Taking the equations for a lamina, namely

$$\Sigma m\ddot{x}=\Sigma X, \quad \Sigma m\ddot{y}=\Sigma Y$$

and integrating, we get from the first,

$$\left[\Sigma m\dot{x} \right]_0^T = \int_0^T \Sigma X \cdot dt = \Sigma \int_0^T X \cdot dt.$$

If u, u' be the velocities of the particle m before and after the impluse, in the x direction, we get from the above

$$\Sigma m(u'-u) = \Sigma X'$$

where X' is the x-component of the impulse on m.

Similarly $\qquad \Sigma m(v'-v) = \Sigma Y'.$

Taking the third equation, namely $\Sigma m(x\ddot{y}-y\ddot{x}) = \Sigma(xY-yX)$ and integrating, we get

$$\left[\Sigma m(x\dot{y}-y\dot{x}) \right]_0^T = \Sigma \left[x \int_0^T Ydt - y \int_0^T Xdt \right]$$

It is to be noted that on the right hand side, x, y have been taken as constants, for during the action of the impulse the particle does not change its position, though its velocity changes. Thus we get

$$\Sigma m[x(v'-v)-y(u'-u)] = \Sigma(xY'-yX').$$

12·4. Motion about a fixed Axis.

Let the plane of the lamina be the plane of the paper and Oz be the axis of rotation perp. to the plane of lamina. Let ZOX be a plane through Oz and fixed in space, G the $C.G.$ of the lamina and P any point of it where a particle of mass m is situated.

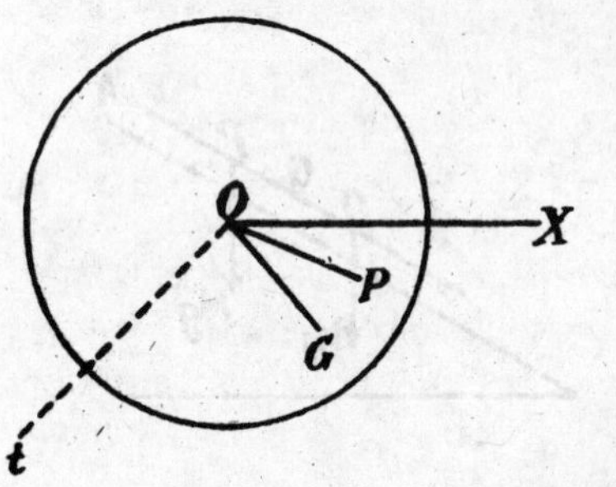

Let $OP = r$, $\angle XOP = \phi$, $\angle XOG = \theta$, $\angle POG = \alpha$.

As the lamina rotates about Oz, angle θ, ϕ change but $\angle \alpha$ remains the same also the distance, r, as P moves in a circle of radius r about O.

Components of acceleration of P are $r\dot{\phi}^2$ along PO and $r\ddot{\phi}$ prep.

But $\qquad \theta = \varphi + \alpha \qquad \therefore \quad \dot{\theta} = \dot{\phi}, \ddot{\theta} = \ddot{\phi}$

Hence the effective forces on m are

$$mr\dot{\phi}^2 = mr\dot{\theta}^2 \text{ along } PO \text{ and } mr\ddot{\phi} = mr\ddot{\theta} \text{ perp. to } PO.$$

Hence the moment of the effective forces about the axis OZ is

$$r \cdot mr\,\ddot{\theta} = mr^2\,\ddot{\theta}$$

Hence the moment of the effective forces of the lamina about OZ is $\qquad \Sigma mr^2\,\ddot{\theta} = \ddot{\theta}\,\Sigma mr^2 = MK^2\ddot{\theta}.$

where MK^2 is the moment of inertia of the lamina about OZ.

Now this moment must be equal to the moment of the external forces about OZ. If L be the moment of the external forces about OZ, we have

$$MK^2\,\ddot{\theta}=L.$$

Also the velocity of m is $r\dot{\phi}=r\dot{\theta}$, therefore, its $K.E.=\frac{1}{2}mr^2\dot{\theta}^2$.

Hence $K.E.$ of the lamina$=\Sigma\frac{1}{2}\,mr^2\dot{\theta}^2=\frac{1}{2}\dot{\theta}^2\,\Sigma mr^2=\frac{1}{2}\,MK^2\dot{\theta}^2$.

Also the momentum of the particle m is $mr\dot{\theta}$ and its moment about OZ is $r\,mr\dot{\theta}=mr^2\dot{\theta}$.

Hence the moment of momentum of the lamina about OZ

$$\Sigma mr^2\dot{\theta}=\dot{\theta}\,\Sigma mr^2=MK^2\dot{\theta}.$$

In the case of implusive forces, if ω', ω be the angular velocities of the lamina about OZ just before and just after the impulse, L the moment of the impulses about OZ, then since change in the moment of momentum is equal to the moment of the impulses, we have

$$MK^2\,(\omega'-\omega)=L.$$

Ex. 1. *A plank of mass M is initially at rest along a line of greatest slope of a smooth plane inclined at an angle α to the horizon and a man of mass M', starting from the upper end walks down the plank so that it does not move. Prove that he gets to the other end in time $\sqrt{\dfrac{2M'a}{(M+M')\,g\sin\alpha}}$, where a is the length of the plank, A the upper end and O the centre of the plank.*

Sol. When the man is at P where $AP=x$, let G be the common C.G., $AG=\bar{x}$

so that $(M+M')\,\bar{x}=M\,\dfrac{a}{2}+M'\,x.$

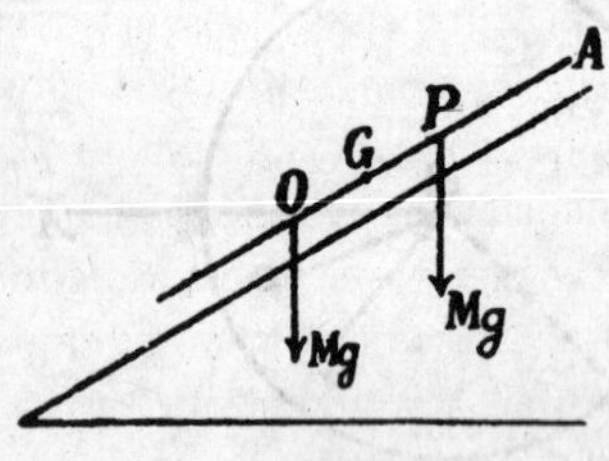

Now the equation of motion is $(M+M')\,\ddot{\bar{x}}=(M+M')\,g\sin\alpha$

or $\quad M'\ddot{x}=(M+M')\,g\sin\alpha.$

Integrating twice, since when $t=0$, $x=0=\dot{x}$, we get

$$x=\frac{M+M'}{2M'}\,g\sin\alpha\,t^2.$$

Hence when $x=a$, we get $t=$answer.

Ex. 2. *A rod, of length $2a$, is suspended by a string, of length l, attached to one end; if the string and rod revolve adout the vertical with uniform angular velocity, and their inclination to the vertical be θ and φ respectively, so that*

$$\frac{3l}{a}=\frac{(4\tan\theta-3\tan\varphi)\sin\varphi}{(\tan\varphi-\tan\theta)\sin\theta}$$

Sol. The elementary mass of an element δx of rod$=\dfrac{M}{2a}\,\delta x.$

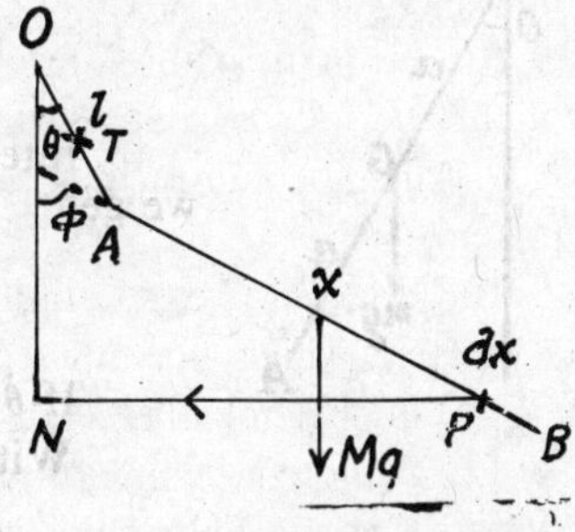

Its Effective force$=\dfrac{M}{2a}\,\delta x\, PN\,\omega^2$ along PN.

$$=\frac{M}{2a}(l \sin\theta+x\sin\varphi)\,\omega^2\,\delta x.$$

The external forces are T and Mg.

Taking moments about A, we have

$$\int_0^{2a}\frac{M}{2a}(l\sin\theta+x\sin\varphi)\,\omega^2\,dx.x\cos\varphi=Mga\sin\varphi \quad ...(1)$$

$$\text{or } \omega^2\left(al\sin\theta+\frac{4}{3}\sin\varphi\right)\cos\varphi \quad ...(2)$$

Now taking moments about O, we have

$$\int_0^{2a}\frac{M}{2a}(l\sin\theta+x\sin\varphi)\,\omega^2\,dx\,(l\cos\theta+x\cos\varphi)$$
$$=Mg\,(a\sin\varphi+l\sin\theta)$$

or from (1), $\displaystyle\int_0^{2a}\frac{M}{2a}(l\sin\theta+x\sin\varphi)\,\omega^2\,l\cos\theta\,dx=Mgl\sin\theta$

$$\text{or } \omega^2\cos\theta\,(l\sin\theta+a\sin\varphi)=g\sin\theta \quad ...(3)$$

Dividing (2) by (3) we have

$$\frac{\cos\varphi\,(al\sin\theta+\frac{4}{3}a^2\sin\varphi)}{\cos\theta(l\sin\theta+a\sin\varphi)}=\frac{a\sin\varphi}{\sin\theta}$$

or $\sin\theta\cos\varphi\,(l\sin\theta+\frac{4}{3}a\sin\varphi)=\sin\varphi\cos\theta\,(l\sin\theta+a\sin\varphi)$

$$\therefore \quad \frac{l}{a}=\frac{(\frac{4}{3}\sin\theta\cos\varphi-\cos\theta\sin\varphi)\sin\varphi}{(\cos\theta\sin\varphi-\sin\theta\cos\varphi)\sin\theta}$$

$$=\frac{(\frac{4}{3}\tan\theta-\tan\varphi)\sin\varphi}{(\tan\varphi-\tan\theta)\sin\theta}$$

$$\text{or} \quad \frac{3l}{a}=\frac{(4\tan\theta-3\tan\varphi)\sin\varphi}{(\tan\varphi-\tan\theta)\sin\theta}.$$

3. *A uniform rod of mass m and length 2a can turn freely about a fixed end. Show the least angular velocity with which it must be started from the lowest position so that it may just make complete revolutions is* $\sqrt{3g/a}$.

Sol. The only external force is its weight mg whose moment about O is $mga\sin\theta$.

Equation of motion is

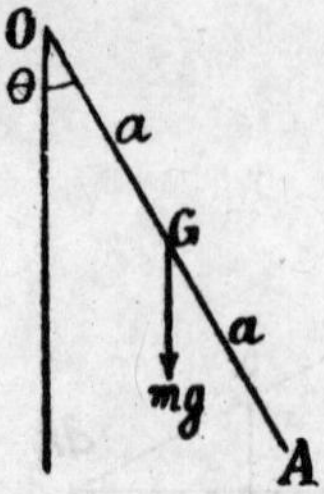

$$mK^2\ddot{\theta}=-mga \sin \theta,\ K^2=\tfrac{4}{3}a^2$$

$$\therefore \quad \ddot{\theta}=-\frac{3g}{4a} \sin \theta.$$

Integrating with condition $\dot{\theta}=\omega$ when $\theta=0$, we get

$$\dot{\theta}^2=\omega^2-\frac{3g}{2a}(1-\cos \theta).$$

If $\dot{\theta}=0$ when $\theta=\pi$, we get $\omega=\sqrt{3g/a}$.
With this ω, we get

$$\dot{\theta}^2=\frac{3g}{2a}(1+\cos \theta)=\frac{3g}{a} \cos^2 \theta/2$$

$$\therefore \quad t\sqrt{\frac{3g}{a}}=\int_0^{\theta} \frac{d\theta}{\cos \theta/2}=2 \log \tan\left(\frac{\pi}{4}+\frac{\theta}{2}\right)$$

A uniform stick of length 2a hangs freely by one end, the other being close to the ground. An angular velocity ω is then given to the stick and when it has turned through a right angle the fixed end is let go. Show that on first touching the ground it will be in an upright position if

$$\omega^2=\frac{g}{2a}\left[3+\frac{p^2}{p+1}\right] \text{ where } p \text{ is an odd multiple of } \frac{\pi}{2}.$$

Sol. When the stick has turned through $\angle\theta$ from the vertical position, the equation of motion is

$$mK^2\ddot{\theta}=-mga \sin \theta,\ K^2=\tfrac{4}{3}a^2$$

$$\therefore \quad \ddot{\theta}=-\frac{3g}{4a} \sin \theta$$

Integrating $\quad \dot{\theta}^2=\omega^2-\dfrac{3g}{2a}(1-\cos \theta)$

$\therefore$ when $\quad \theta=\pi/2.\ \dot{\theta}^2=\omega^2-\dfrac{3g}{2a}=\omega'^2$, *i.e.*, $\omega^2=\omega'^2+\dfrac{3g}{2a}$

If t be the time when the rod touches the ground, then

$$-a=a\omega' t-\tfrac{1}{2}gt^2$$

$$\omega' t=p$$

$$\therefore \quad a=\frac{g}{2}\frac{p^2}{\omega'^2}-pa \quad \therefore \quad \omega^2-\frac{3g}{2a}=\frac{g}{2a}\cdot\frac{p^2}{p+1} \text{ hence.}$$

Ex. 5. *A rod of mass m and length 2a, which is capable of free motion about one end A, falls from a vertical position and when it is horizontal strikes a fixed inelastic obstacle at a distance b from A. Show that the impulse of the blow is* $m.\dfrac{2a^2}{b}\sqrt{\dfrac{2ga}{3}}$ *and the impulsive* *reaction at A is m* $\sqrt{\frac{3ga}{2}}\left(1-\frac{4a}{3b}\right)$ *vertically upwards.*

Sol. When the rod has turned through

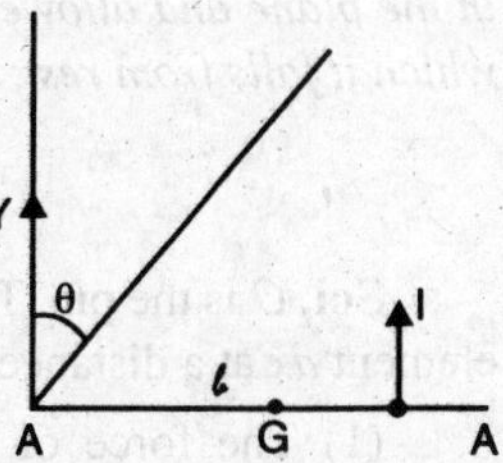

$\angle\theta$ m. $\frac{4}{3}a^2\ddot{\theta} = mga \sin\theta$ or $\ddot{\theta} = \frac{3g}{4a}\sin\theta$

$$\dot{\theta}^2 = C - \frac{3g}{2a}\cos\theta.$$

When $\theta = 0 = \dot{\theta}$ $\quad\therefore \dot{\theta}^2 = \frac{3g}{2a}(1-\cos\theta)$

$\therefore$ when $\theta = \frac{\pi}{2}$, $\dot{\theta}^2 = \frac{3g}{2a} = \omega^2$ $\therefore = \sqrt{\frac{3g}{2a}}$

Let I be the impulse of the blow and Y the impulsive reaction at A. Just before the blow, G has a velocity $a\omega$ and just after zero velocity.

Therefore, the equations of motion are

$$m(O - a\omega) = -I - Y \text{ or } ma\omega = I + Y$$

and

$$m\,.\,\frac{4}{3}a^2(O-\omega) = -Ib \text{ or } m\,.\,\frac{4}{3}a^2\omega = Ib.$$

These give the answers.

Ex. 6. A *uniform beam of mass m and length 2l is horizontal and can turn freely about its centre which is fixed. A particle of mass m′ and moving with vertical velocity u, hits the beam at one end. If the coeff. of restitution be e, show that the angular velocity of the beam just after the impact is*

$$3m'(1+e)u/(m+3m')l$$

and the vertical velocity is $u'(em - 3m')/(m + 3m')$.

Sol. Let v upwards and ω be the velocity and angular velocity just after impact and I be the impulse.

For the rod, $mK^2(\omega - O) = Il$, $K^2 = \frac{l^2}{3}$

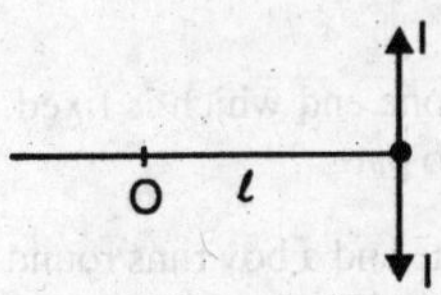

For the particle $m'(v + u) = I$

Newton's law for impact gives

$l\omega + v = -e(O - u) = eu.$

Solving these we get ω and v.

Ex. 7. *A uniform rod OA is freely movable on a smooth inclined plane* *whose inclination to the horizon is i and whose coefficient of friction is μ, about a smooth pin fixed through the end; the bar is held in the horizontal position in the plane and allowed to fall from the position. If θ be the angle through which it falls from rest; show that*

$$\frac{\sin\theta}{\theta} = \mu \cot i$$

Sol. O is the pin. Two forces are acting at an element dx at a distance n from O:

(1) The force of gravity and (2) force of friction $\mu R = \left(\dfrac{mdx}{2a}\right) g \cos i$ perpendicular to the rod. Hence taking moments about O, we have

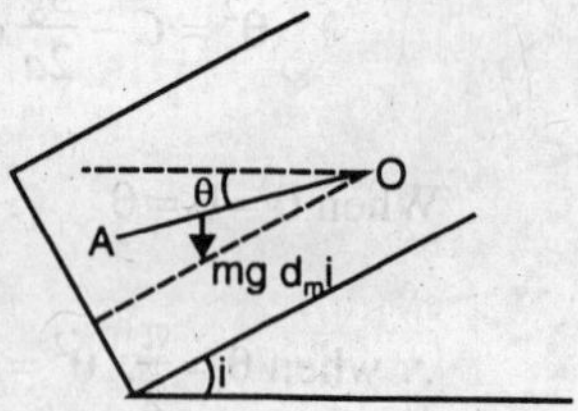

$$mk^2\ddot{\theta} = mg \sin i \, . \, a \cos\theta - \int_0^{2a} \mu \frac{dx}{2a} g \cos i \, . \, x$$

$$= mga \sin i \cos\theta - \mu amg \cos i \text{ (2a is length of the rod)}$$

$$\therefore \frac{1}{2} mk^2\dot{\theta}^2 = mgh (\sin i \sin\theta - \mu \cos\theta i)$$

(on integration and applying the integral conditions)

Rod comes to rest if $\dot{\theta} = 0$.

i.e. when $\sin\theta - \mu\theta \cot i = 0$

or $$\frac{\sin\theta}{\theta} = \mu \cot i$$

EXAMPLES XII (A)

1. A rough uniform board of mass m and length $2a$, rests on a smooth horizontal plane and a man of mass M walks on it from one end to the other. Prove that the distance through which the board moves is $2Ma/(M + m)$.

2. A rod revolving on a smooth horizontal plane about one end which is fixed, breaks into two parts; what is the subsequent motion of the two parts ?

3. A circular board is placed on a smooth horizontal plane and a boy runs round the edge of it at a uniform rate, what is the motion of the centre of the board ?

4. A fine string has two masses m and m', $(m > m')$ tied to its ends and passes over a rough pulley of mass M whose centre is fixed: if the string does not slip over the pulley and the pulley can rotate about the horizontal axis through its centre, show that m will descend with acceleration.

$$\frac{m - m'}{m + m' + M\dfrac{K^2}{a^2}} \cdot g$$

where a is the radius and K the radius of gyration of the pulley about the axis.

5. Show that in Atwood's machine, if account be taken of the inertia of the pulley, the acceleration of the moving masses m, m' is $(m - m')g/(m + m' + \frac{1}{2}M)$, where M is the mass of the pulley.

6. A uniform rod AB of mass m is set in motion by a blow P at A at an angle α to the rod. Find the velocity of A just after the blow and show that energy given to the rod is $P\dfrac{1 + 3\sin^2\alpha}{2m}$.

7. AB, BC are two equal similar rods freely hinged at B and lie in a straight line on a smooth table. The end A is struck by a blow perp. to AB : show that velocity of A is $3\frac{1}{2}$ times that of B.

8. Three equal rods AB, BC, CD are freely jointed and placed in a straight line on a smooth table. AB is struck at A by a blow perp. to AB; show that velocity of centre of AB is 19 times that of CD and its angular velocity is 11 times that of CD.

9. Three equal uniform rods placed in a straight line are freely jointed and move with velocity v perp. to them. If the middle point of the middle rod be suddenly fixed, show that the ends of the other two will meet in time $\dfrac{4\pi a}{9v}$ where a is the length of each.

10. Two equal uniform rods AB, AC are freely joined at A and are placed on a smooth table so as to be at right angles. AC is truck by a blow at C in a direction perp. to AC; show that the velocity of the middle points of AB and AC are as 2 : 7.

11. AB and CD are two equal and similar rods connected by a string BC; AB, BC and CD form three sides of a square. A is struck by a blow perp. to AB: show that the initial velocity of A is seven times that of D.

12.5. Compound Pendulum. *A heavy body moving about a smooth fixed horizontal axis, to discuss the motion.*

Take the vertical plane through the axis as the plane of reference and the plane through the axes and the C.G. G, of the body as the plane fixed in the body. If at any time t, θ be the angle between these two planes, h the distance of G from the axis of rotation, the equation of motion is

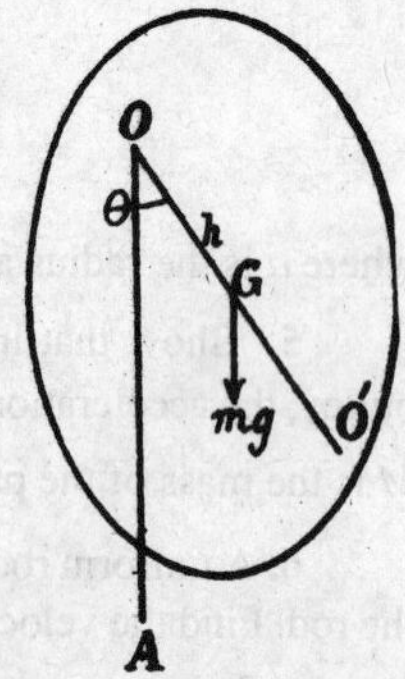

$$Mk^2\ddot{\theta} = -Mgh \sin \theta$$

where Mk^2 is the $M.I.$ of a body about the axis of rotation.

If θ be small, then

$$\ddot{\theta} = -\frac{gh}{k^2}\,\theta.$$

Motion is simple harmonic and the period is

$$2\pi\sqrt{\frac{k^2}{gh}}.$$

The period is the same as that of a simple pendulum of length $l = \frac{k^2}{h}$. This length is that of the *simple equivalent pendulum.*

Even if the oscillation of the compound pendulum is not small, it will oscillate in the same time as the simple pendulum of length k^2/h.

Through G drop a perp. to the axis of rotation cutting it at O. Then O is called the *centre of suspension*. Produce OG to O' and make $OO' = l = k^2/h$, then O' is called the *centre of oscillation.* Thus if the whole mass of the body is collected at O' and suspended from O by a string its angular motion and time of oscillation will be the same as those of the body under the same initial conditions.

If Mk_1^2 be the $M.I.$ about a parallel axis through G, then

$$Mk^2 = Mk_1^2 + Mh^2 \quad \text{or} \quad k^2 = k_1^2 + h^2$$

therefore $$l = \frac{k_1^2 + h^2}{h} \quad \text{or} \quad lh = k_1^2 + h^2$$

Also $$OG.GO' = h(l-h) = hl - h^2 = k_1^2$$

$$OG.OO' = hl = k_1^2 + h^2$$

$$= \text{square of the radius of gyration about } O$$

$$O'G.O'O = (l-h)l = l^2 - hl = (l-h)^2 + lh - h^2$$

$$= (l-h)^2 + k_1^2$$

$$= \text{square of the radius of gyration about } O'$$

The symmetry of these relations shows that if the body were suspended from a parallel axis through O', the point O will be the new centre of oscillation. This is often expressed by saying that *the centres of suspension and of oscillation are convertible, and the times of oscillation in either case are the same.*

The time of oscillation is known when l is known, hence the simple equivalent pendulum is of minimum length and, therefore, the time of oscillation is minimum when

$$\frac{dl}{dh}=0 \quad \text{or} \quad 1-\frac{k_1^2}{h^2}=0 \quad i.e., \; h=k_1$$

and the length of the simple equivalent pendulum is then $2k_1$.

12·6. Torsional oscillations. Suppose a lamina, free to turn horizontally in its own plane about its *C.G.*, which is fixed and let a restoring couple of magnitude $\lambda\theta$, where λ is a constant and θ the angular displacement of the lamina at time t.

The equation of motion is

$$mk^2 \ddot{\theta}=-\lambda\theta$$

where mk^2 is the moment of inertia of the lamina about the vertical axis through its *C.G.*

Thus the lamina will execute torsional oscillations of simple harmonic type and of period $2\pi\sqrt{\frac{mk^2}{\lambda}}$, which is independent of amplitude.

These conditions are realised if a heavy uniform circular disc (or cylinder) is centrally suspended by a fine wire whose other end is rigidly fixed.

Ex. 1. *Three equal particles are attached to a weightless rod at equal distances a apart. The system is suspended and is free to turn about a point of the rod distance x from the middle particle. Find the time of small oscillation and show that it is least when $x=\cdot 82a$ nearly.*

Sol.

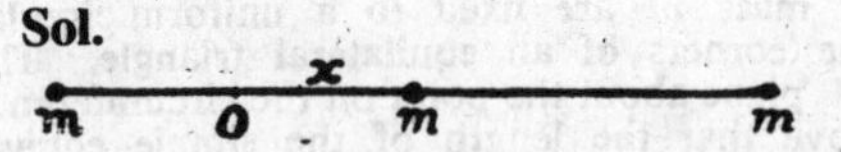

$$l=\frac{K^2}{h}=\frac{MK^2}{Mh}$$

$$=\frac{m(a-x)^2+mx^2+m(a+x)^2}{3mx}=x+\frac{2a^2}{3x}.$$

This is minimum when $1-\frac{2a^2}{3x^2}=0$ or $x=\cdot 82a$ nearly.

Ex. 2. *A simple circular pendulum is formed of a mass M suspended from a fixed point by a weightless wire of length l: if a mass m, very small compared with M, be knotted on to the wire at a point distant a from the point of suspension, show that the time of small oscillation is approximately diminished by $\frac{ma}{2Ml}\left(1-\frac{a}{l}\right)$ of itself.*

Sol. In the first case, $t_0=2\pi\sqrt{\frac{l}{g}}$.

In the second case, $t=2\pi\sqrt{\frac{l'}{g}}$ where $l'=\frac{Ml^2+ma^2}{Ml+ma}$

i.e., $$l'=l\left(1+\frac{ma^2}{Ml^2}\right)\left(1+\frac{ma}{Ml}\right)^{-1}$$

$$\therefore \quad t=2\pi\sqrt{\frac{l}{g}}\left(1+\frac{ma^2}{Ml^2}\right)^{\frac{1}{2}}\left(1+\frac{ma}{Ml}\right)^{-\frac{1}{2}}$$

$$=t_0\left[1+\frac{ma^2}{2Ml^2}-\frac{ma}{2Ml}\right] \text{ approx.}$$

$$=t_0\left[1-\frac{ma}{2Ml}\left(1-\frac{a}{l}\right)\right]; \text{ hence.}$$

EXAMPLES XII (B)

1. ***Find the lengths of the simple equivalent pendulums in the following cases, axis being horizontal :***

(*a*) Circular wire, axis (*i*) tangent $\frac{3}{2}a$.

(*ii*) perp. to the plane at any point of the arc $2a$,

(*b*) Circular disc, axis tangent $\frac{5}{4}a$.

(*c*) Elliptic lamina, axis latus-rectum $\frac{a(1+4e^2)}{4e}$.

(*d*) Triangular lamina, axis (*i*) side BC $\frac{1}{2}b \sin C$.

(*ii*) perp. to lamina through A $\frac{3b^2+3c^2-3a^2}{\sqrt{2b^2+2c^2-2a^2}}$.

(*e*) Cone, axis diameter of base $\frac{2+3\tan^2\alpha}{5}\cdot h$.

2. Three particles of same mass m are fixed to a uniform circular hoop of mass M and radius a, at the corners of an equilateral triangle. The hoop is free to move in a vertical plane about the point on the circumference opposite to one of the masses m. Prove that the length of the simple equivalent pendulum is $2a$.

3. A uniform circular hoop of centre C, radius a and mass m, can move freely about an axis perp. to its plane and horizontal and passing through a point O of the hoop. A particle of m is fixed to the hoop a point A such that angie OCA is a right angle. Prove that the length of S.E.P. is $4a/\sqrt{5}$.

4. A bent lever of arms a and b and angle between them being α, makes small oscillations in its own plane about the fulcrum ; show that the length of S.E.P. is

$$\frac{2}{3}\cdot\frac{a^3+b^3}{\sqrt{a^4+2a^2b^2\cos\alpha+b^4}}.$$

5. A rod of mass of m and length a, carries at one end a sphere of mass M and radius b, the other end being fixed. Prove that the length of S.E.P. is

$$\frac{\frac{a^2}{3}+M[(a+b)^2+\frac{2}{5}b^2]}{m\frac{a}{2}+M(a+b)}.$$

6. An ellipse of semi-axes a, b and a circle of radius b are cut from the same sheet of thin uniform metal and are superposed and fixed together with their centres coincident. The figure is free to move in its own vertical plane about one end of the major axis ; show that the length of S.E.P. is

$$\frac{5a^2-ab+2b^2}{4a}.$$

7. A uniform wire in the form of an arc of a circle is swinging about a horizontal axis through the middle point of the arc perp. to the plane of the arc. Show that time of small oscillation is independent of the length of the arc.

8. A sphere of radius a is suspended by a fine wire from a fixed point at distance l from its centre ; show that the time of small oscillation is

$\pi\sqrt{\dfrac{5l^2+2a^2}{5lg}}\left[1+\dfrac{1}{4}\sin^2\dfrac{\alpha}{2}\right]$ where α is the amplitude.

9. A small weight is attached to a compound pendulum ; show that period of oscillation would be increased or decreased according as the point of attachment is below or above the centre of oscillation.